THE FLOWERING OF A TRADITION: TECHNICAL WRITING IN ENGLAND, 1641–1700

by

Elizabeth Tebeaux

Texas A&M University

Baywood's Technical Communications Series
Series Editor: Charles H. Sides

Routledge
Taylor & Francis Group
LONDON AND NEW YORK

First published 2014 by Baywood Publishing Company, Inc.

Published 2017 by Routledge
2 Park Square, Milton Park, Abingdon, Oxon OX14 4RN
605 Third Avenue, New York, NY 10017

Routledge is an imprint of the Taylor & Francis Group, an informa business

Library of Congress Catalog Number: 2014017425

Library of Congress Cataloging-in-Publication Data

Tebeaux, Elizabeth, author.
The flowering of a tradition : technical writing in England, 1641-1700 / by Elizabeth Tebeaux, Texas A&M University.
pages cm. -- (Baywood's technical communication series)
Includes bibliographical references and index.
ISBN 978-0-89503-843-2 (cloth : alk. paper) -- ISBN 978-0-89503-844-9 (pbk. : alk. paper) -- ISBN 978-0-89503-846-3 (e-pub) -- ISBN 978-0-89503-845-6 (e-pdf) 1. Technical writing--Great Britain--History--17th century. 2. Communication of technical information--Great Britain--History--17th century. 3. England--Intellectual life--17th century. I. Title.
T11.T2954 2014
601'.4--dc23

2014017425

ISBN 13: 978-0-89503-843-2 (hbk)
ISBN 13: 978-0-89503-844-9 (pbk)

Table of Contents

Preface

This monograph marks my second effort to present a sustained description of the history of English technical writing. *The Emergence of a Tradition*, which covered 1475–1640, the heart of the English Renaissance, still stands as the only effort to describe, in broad outline, examples of technical writing in that period. *The Flowering of a Tradition* moves beyond *The Emergence of a Tradition* by focusing on technical writing from 1640 to 1700, even though both works emphasize common themes that underlie development of technical writing. The two books, therefore, form the first attempt to describe technical writing, practical writing, or utilitarian writing as it existed in England during the 1475–1700 era. I chose these demarcation dates, 1475–1640 and 1641–1700, because they reflect the traditional ending date of the English Renaissance, followed by the civil war, the years of Cromwell, and the Protectorate, and then the return of the monarchy and the Stuarts, who held the English monarchy until 1714. Like *Emergence of a Tradition*, *Flowering of a Tradition* seeks to capture and describe technical or practical writing in the milieu that produced it. The visuals I include attempt to show how early printed English technical writing anticipates much modern technical writing in format, organization, and style. In both books, I wish to show that technical writing has a rich and honorable history.

For those of us who have studied English literary history in courses in medieval, Renaissance, and 18th century literature, we can accept the assertion I make in both books that technical writing existed side by side with English literature and shared the same milieu. Unfortunately, practical writing has not received the same recognition as literature has in these periods or in any other. Many senior literature faculty, who remember the heavy literary history requirements, have read Francis Bacon; but fewer have read William Petty, Nehemiah Grew, Joseph Moson, or Robert Hooke and understand the meaning of "plain style" and its sustained presence in English prose. Thus, my two books should fill a background deficiency of many current English and technical communication majors.

Like English Renaissance technical writing, 17th century technical writing served a distinct purpose: to capture and then disseminate practical knowledge. As a result, technical writing of this period, as in any period, shows historians more clearly how people outside the literati lived and worked and viewed their milieu. Through technical writing, we can see the world of work of the period, the new world of science and its associated technologies, and the scientific pursuits that defined the period. Both *Emergence of a Tradition* and *Flowering of a Tradition* portray the issues and the knowledge that defined the 1475–1700 milieu. Technical or practical writing is not inferior to the discourse of literature, religion, history, and philosophy, just different in rhetorical purpose, genres, and style. We can see how techniques characteristic of technical writing developed: plain, unadorned style; paragraphs; page design, format, and visuals. We can follow development of genres that became the hallmark of practical writing—instructions, proposals, and reports. Amid this development, we can see how important practical writers described concepts, conveyed information, and instructed readers about the issues of work and the new science of this period.

LITERATURE REVIEW

My historical research, published from 1991 to 2012 (listed at the end of Chapter 9), provides the majority of the work on English technical writing from 1475 to 1700. Readers interested in this work can find it discussed in three articles: William Rivers' 1994 article, "Studies in the History of Business and Technical Writing: A Bibliographic Essay," *Journal of Business and Technical Communication, 8*, pp. 6–57; Michael Moran and Elizabeth Tebeaux, "A Bibliography of Works Published in the History of Professional Communication from 1994–2009: Part 1," *Journal of Technical Writing and Communication, 41*:2, pp. 193–214, 2011; Michael Moran and Elizabeth Tebeaux, "A Bibliography of Works Published in the History of Professional Communication from 1994–2009: Part 2," *Journal of Technical Writing and Communication, 42*:1, pp. 47–87, 2012. These three essays present and discuss the bulk of work developed on the history of our field and provide a good starting point for those who wish to begin work on the history of practical discourse.

In these reviews, readers can track studies of technical writing by period, in both English and American literature

METHOD

Much of what I have published since 1991 echoes throughout this monograph, which gives me the opportunity to recast and correct points that I presented in 1996. Research, like any writing, is a process—we learn as we study and write.

Since *Emergence of a Tradition*, I have learned more about the development of additional forms of technical and practical writing. I have continued to find and read technical writing of the 17th century and earlier. Studying technical writing in any period requires that historians first find examples of practical or technical writing. Since 1989, I have pursued that goal. In first studying the print versions of both the Pollard and Redgrave's [1] and Wing's [2] *Short-Title Catalogs* (I carefully read each), I located and then read dozens of examples of practical writing, many in facsimile or via microfilm and more recently via *English Books Online* and *Eighteenth-Century Studies*. A fellowship at the Huntington Library enabled me to examine many of the first print editions of these works. My focus the past 23 years has been to determine how these practical or technical books fit into the milieu of the period. Critical to this effort has been my knowledge of 16th, 17th, and 18th century literature and literary history, which formed the bulk of my two master's theses and my doctoral dissertation in English. Without an understanding of the age in which technical writing exists, no historian can find it, understand it, or track its significance. Thus, to those who wish to pursue the history of technical writing, knowledge of the historical period you wish to explore remains essential.

I have chosen to focus mainly on printed texts because of their availability in various facsimile formats. My goal in 1996, and now in 2013, is to encourage other historians interested in the history of practical discourse to go beyond my research, to pursue what I have omitted, and to search for other examples of what could be called technical writing, to suggest other perspectives for "making sense" of these works—their purpose and the context in which they emerged, their intended audiences, and the perspectives of their authors.

Much practical discourse in England during the 1475–1700 period remains to be unearthed. I had difficulty deciding, in many chapters, what works to emphasize. As Malcolm Richardson has shown in his 2011work, *Middle Class Writing in Late Medieval London*, a superb study of late medieval English business writing [3], much incunabula exists in England, in offices such as the English Records Office. These collections are not cataloged, and the cost of that effort has left many collections of letters, reports, instructions, and policies—basic forms of practical writing—unstudied by scholars seeking to understand and describe the beginning of English practical writing. Richardson's newest book offers a convenient starting point for any person interested in the history of practical writing in medieval England and stands as an icon of excellent historical scholarship on the history of medieval practical writing as it embodied business, commerce, and government

JUSTIFICATION FOR MY APPROACH

Please note that I shift from the term *technical* writing to *practical* writing. Scholars such as Douglas Bush [4], Clanchy [5], and Adolph [6] have all reported that practical writing occurs as the dominant form of writing in any era. As Teresa Kynell [7] noted, the term *technical* writing first occurred in the 19th century in the United States to describe non-humanistic writing taught by English departments. The term *technical* writing thus distinguished humanistic writing from engineering writing. We need to recall that the word *technical* suggests writing that describes technique. And the examples I present focus on techniques of early science, medicine, agriculture, mining, the trades, and education reform needed to strengthen these critical disciplines. However, during the 1475–1700 period, *work* or *technique* assumed a different meaning. Work could be medicine, surgery, military science, navigation, glassmaking, mining, pharmacology, agriculture, animal science, production of goods and tools for trades, keeping accounts, and numerous other tasks. The extant documents associated with these kinds of work stand as evidence that practical, how-to writing targeting specific techniques existed in abundance, along with Bibles, almanacs, poetry, literature, and history. Yet outside the histories of established disciplines such as medicine, science, and agriculture, little work has been done to describe the variety and extent of technical writing as a genre embedded in the work-based writing of these disciplines.

As I have discussed in the 20 articles I have published since 1991 and then in *Emergence of a Tradition*, what we now recognize as *technical* writing can be found among the first printed technical books as well as in much English, French, and Latin incunabula. When I first began to unearth many of these examples, I was struck by one finding: Many early forms of practical writing anticipate modern forms of technical writing, particularly the organization and format of these documents. The style of many of these documents exemplifies a plain, concise, and direct presentation of information in both descriptions and instructions, the major forms of practical writing. These forms existed to help readers perform tasks, from surgery to caring for farm animals, to collecting plants needed for experiments, and to building ships. While the plain, or "low" style had been the style early rhetoricians recommended for instruction, the plain, unadorned style can be seen in numerous practical documents. These writers, many who likely had not studied rhetoric, nevertheless knew that in writing to help readers perform tasks, the text needed a style different from that used for humanistic or theological works.

When I selected the title *The Flowering of a Tradition* for this second book, I did not realize the appropriateness of the title. The number of books that appeared after 1640 increased dramatically from the number of books published before 1640. We have only to compare the paper editions of Pollard and Redgrave's [1] *Short-Title Catalog* with the Wing [2] *Short-Title Catalog* to grasp the increase of published material after 1640. While both works have been subsumed by *Early English Books Online*, a major database, the printed bibliographies give readers a grasp of the differences in works printed from 1475–1640 and then 1641–1700. While many works made available via print still focused on religion, government, and politics, an increasing number of works on practical topics continued to appear. Many how-to books covered the same topics as those published before 1640: books about geography, navigation, medicine, and military science; instructions and process analyses on a variety of topics, such as agriculture and husbandry reports on a variety of issues; and the appearance of proposals as a modern genre. As *Flowering* will show, by the end of the 17th century, the major genres of modern technical writing had fully emerged.

The new topics, particularly on English trades and industry, Bacon's legacy to the Royal society, became increasingly important in England. Methods to improve trades and commerce provide a substantial number of new technical writing titles that appeared for the first time after 1640. Many of these, written by the first members of the Royal Society and the precursor group that met at Gresham College, initially focused on developing useful knowledge. Much of it emerged from Bacon's mandate for development of the history of trades, to improve these trades and thus the quality of life for English people. Thus, making and incorporating this knowledge necessarily involved the production of technical writing to show how to perform many tasks associated with various trades. The early Royal Society papers, in topic and content, reveal efforts to provide a history of trades.

While current scholars debate the literacy of 17th century readers, the increasing number of published books irrefutably suggests an increasing reading public and the advancement of knowledge. H. S. Bennett [8], Rudolph Hirsch [9], and Louis Wright [10] comment about the implication of how the numbers of books and their titles apply to the post-1640 period:

> One has only to scan the title of the Short-Title Catalogue to gain some idea of the extent and variety of works produced for the literate public in England between the introduction of printing and 1640. This huge outpouring of books could not have been printed if there had not been an enormous demand from the generality of citizens. The publishers of Elizabethan England could no more live by the custom of learned and aristocratic readers alone than can modern followers of their trade. [9, pp. 82–83]

As Robert Merton noted [10], the primary focus of the Middle Ages and the Renaissance was philosophical and theological, with marked attention to literature, ethics, and art. However, the late 16th and then the 17th centuries marked the transition to the emphasis on science and technology and the utilitarian attitude that supported that shift. This was an age of utilitarian prose as a vehicle for useful information that showed how to apply knowledge. While both these works have been subsumed by Early English Books Online, the paper copies allow historians to view the jump in quantity of printed works after 1641 and should be explored to develop a sense of what was written. natural science, mining, mathematics, agriculture, and trades. This was also an age dominated by Puritan values driven by a spiritual mandate for increasing knowledge to solve economic problems, improve the quality of life, and produce wealth. A survey of technical books of the period (the focus of *The Flowering of a Tradition*) will reveal the major characteristics of technical writing as it continued to evolve. These technical books thus demarcate the spirit of the age and show the relationship between technical writing and scientific writing as both developed during the 1641–1700 period.

PURPOSE AND PLAN

The purpose of *Flowering of a Tradition* is neither to delve into the endless disputes surrounding the development of the Royal Society or the philosophies and religion of its founders, nor is the purpose to determine which specific kind of plain style appeared most dominantly in the technical writing after 1640. The goal is to show that technical writing continued to evolve in the 17th century after the close of the Renaissance; that new forms emerged, such as reports and proposals; that style of technical writing continued to follow the same plain style exhibited in earlier medieval technical writing; and that technical writing, as it had in the English Renaissance and in medieval England, shows us much about life in England, particularly as industry, trades, and science began to permanently reshape the English landscape. With this purpose stated, I offer the following chapters in support of my goal:

Chapter 1 will summarize the major events and changes in England in the 17th century that affected the development of technical writing, with a focus on the 1640–1700 period. This chapter will discuss significant technical writing produced by early members of the Royal Society as their work exemplified Bacon's history of trades and industries.

Chapter 2 will discuss the evolution of English prose style in technical and practical writing and the importance of visual design to the development of English prose style.

Chapter 3 will discuss the contribution that practical, utilitarian discourse made to the development of the modern English paragraph, which emerged much earlier than historians of composition have noted.

Chapter 4 will focus on page design, format strategies, visuals, and technical description exemplified in technical writing that anticipates modern documents designed to enhance usability and readability.

Chapter 5 will discuss development of instructions, beginning with some of the first texts, and focus on the development of agrarian and estate management works, how-to writing that had critical importance to the economy of England.

Chapter 6 will explore the development of shipbuilding instructions as these moved from oral to textual renditions. Technical discourse, like literary discourse, evolved from a rich oral history.

Chapter 7 will discuss the evolution of proposals as these emerged from petitions launched during the reign of Edward II after 1290.

Chapter 8 will discuss the emergence of printed reports as these developed in the Renaissance and then 17th century reports, many written to Parliament or to the Royal Society.

Chapter 9 will offer reflections on my research and recommendations for further research, a question often asked by technical communication faculty.

REFERENCES

1. A. W. Pollard and G. R. Redgrave, *Short-title catalogue of books printed in England, Scotland, & Ireland and of English books printed abroad, 1475-1640* (3rd ed.), Bibliographical Society, London, 1976.
2. Donald G. Wing (comp.), *Short-Title Catalogue of Books Printed in England, Scotland, Ireland, Wales, and British America, and of English Books Printed in Other Countries, 1641-1700* (2nd ed.), Modern Language Association, New York, 1994.
3. M. Richardson , *Middle-Class Writing in Late Medieval London*, Pickering & Chatto, London, 2011.
4. D. Bush, *English Literature in the Earlier Seventeenth Century* (2nd ed.), Cambridge University Press, New York, 1962.
5. M. T. Clanchy, From *Memory to Written Record 1066-1307*, Basil Blackwell, London, 1993.
6. R. Adolph, *The Rise of Modern Prose Style*, MIT Press, Cambridge, Massachusetts, 1968.
7. T. Kynell, *Writing in a Milieu of Utility: The Move to Technical Communication in American Engineering Programs, 1850-1950* (2nd ed.), Ablex Publishing Corporation, Stamford, Connecticut.
8. H. S. Bennett, *English Books & Readers 1475-1557*, Cambridge University Press, London, 1952.
9. R. Hirsch, *Selling and Reading, 1459-1550*, Harrasowitzz, Wiesgaden, Germany, 1975.
10. L. B. Wright, *Middle-Class Culture in Elizabethan England*, Cornell University Press, Ithaca, New York, 1935.
11. R. Merton, *Science, Technology, and Society in Seventeenth Century England*, Howard Fertig, New York, 1970.

Acknowledgments

The existence of this monograph results from the influence of a number of individuals who have shaped my intellectual life over the past 50 years. Many have passed on. Yet I find myself hoping that they would be pleased by the historical research to which I have devoted over 20 years of my academic career. Their influence gives me hope, after a bad or exasperating day or week of teaching, that perhaps I have made a difference in the lives of a few students over the past four decades in the classroom. I have shared some of my research (usually at the end of the term) with honors students whom I have had the privilege of teaching, and many have shown amazement at some of the works I have shown them. In short, even standard undergraduate technical writing courses can benefit from faculty discussing the history of standard course topics: plain style, proposals, reports, letters, and visuals. Without an understanding of history, few students and faculty can appreciate the struggle early technical writers and printers faced in developing charts, tables, and drawings needed to aid artisans in making instruments used in fields such as navigation, metallurgy, building construction, and mining.

To Dr. Christine Fall, Professor of English at Baylor, I owe my passion for great prose and my introduction to Victorian prose, which gave me my first exposure to the history of ideas as an undergraduate. Without her sharing her voluminous knowledge of 19th century English writing during afternoon tea at her apartment, I doubt that graduate school in English would have beckoned me. To Dr. Amy Turner, Professor of English at University of Houston-University Park, perhaps the most demanding professor under whom I ever studied, I owe my knowledge of Milton's prose and its connection to the 17th century milieu. Knowing how to read great prose is a technique that must be taught, and to her I am deeply indebted.

Dr. Laurie Zwicky, a Milton scholar and Professor of English, also at the University of Houston, showed a small but devoted class of graduate students the role of the 17th century in England in the development of modernity through the prose of the century. She directed my first thesis and spurred my interest in Renaissance English style and English pulpit oratory from Andrewes through Tillotson. Her influence led to my second thesis and then my dissertation on Donne's sermons.

To Malcolm Richardson, a friend for more than three decades, I owe inspiration. Malcolm is one of the most careful, analytical thinkers I have known. He embodies his thinking in humorous but memorable prose that should be required reading for anyone exploring the history of practical writing. He has repeatedly challenged my theories and provided me exquisite examples of medieval writing, such as instructions for making lace, a short document that supports the literacy of medieval women, a point which he has developed in his research and I discuss in two of my articles on Renaissance women writers.

Finally, to my editor, Charles Sides, I owe gratitude for friendship and the pleasure of working with someone who understands that technical writing is not an exercise in arcane theory but an important form of communication, one that has persevered despite the efforts of humanities departments to crush it. To both Malcolm and Charles, I offer thanks for helping me keep my sanity as I proposed new theories for the development of practical writing. My hope is that a few technical communication graduate students will see the many opportunities open to them in helping build the history of technical/business/practical writing and establish its importance in conveying the history of ideas. Knowing the history of any field we teach makes us better thinkers, researchers, and teachers.

CHAPTER 1

Industry, the History of Trades, and the Development of Technical Writing

As I discussed and illustrated in *Emergence of a Tradition*, much technical or practical writing existed in the English Renaissance, and the quantity continued to increase throughout the 17th century. Demand for practical writing grew from 1475 to 1700 because of (1) the expansion of knowledge and reading and (2) the improvement in print technology that allowed more books to be printed and distributed to a population that became increasingly literate. Comparing the print versions of *The Short-Title Catalogs* [1, 2] makes clear the increase of the printed word and the variety of documents, and the reading audience.

General descriptions and examples of some of the first Royal Society works exemplify the Society's contribution to knowledge through its writing, much of it exemplifying early attempts at process analysis, instructions, descriptions, proposals, and then analytical reports needed to develop English industries and government, one of the original goals of the Society. These genres appeared in medieval and Renaissance documents, but this chapter will describe the changes in the 17th century English milieu, the rise of the new science, and the resulting works of many Royal Society members. To present their research, these thinkers developed documents that anticipated modern technical writing. These major genres initially existed to assist industrialists, artisans, craftsmen, seamen, physicians, and agriculturalists in improving their work.

Thus, *Flowering of a Tradition* will begin by focusing on the milieu of the new science with an overview of the technical writing and the virtuosi/scientists who developed it.

Chapters 2, 3, and 4 will discuss what are often called the "elements of technical writing": evolution of English plain style, content development via the paragraph, and a variety of visual design features (drawings and tabular efforts) that helped readers "see" the text and the concepts it presented. The discussion of technical writing's contribution to the English paragraph is the first of its kind.

Chapters 5–8 illustrate the development of technical writing "applications": instructions, proposals, and then reports. Chapters 5 and 6 focus on the development of instructions through two topics of critical importance to England: agriculture and shipbuilding. While agricultural instructions emerged in written form as early as the 12th century in England, shipbuilding instructions remained orally transferred until the end of the 17th century to protect design secrets. Both topics illustrate methods that would continue to appear in instructional material but show how instructions, as a genre, develop differently, depending on the context and readers surrounding the content for which how-to writing was needed.

Chapters 7 and 8 will suggest a theory for the development of proposals and then reports, major genres of modern technical writing, as these emerged from the petitions of medieval England. By the second half of the 17th century, both proposals and reports existed as documents that anticipated their modern forms.

THE GROWTH OF WEALTH AND PRACTICAL KNOWLEDGE IN THE RENAISSANCE AND THE 17TH CENTURY

The reluctance of Oxford and Cambridge to adjust their curriculum to embrace change—to move away from scholasticism—led to the founding of Gresham College and a large number of continuing education schools in and around London in the latter decades of the English Renaissance (1600 to 1640) [3]. These schools led to a number of cheaply printed books on practical topics, many written by men whose expertise was part of the early technological conversation that developed at Gresham College. These men and their practical works formed the incubator for the Oxford group destined to become the Royal Society in 1663. The number of practical and "new science" works can be seen as icons for the changes driving England from a medieval culture in 1500 to the beginnings of a modern culture by 1700 [4]. In many of these practical books, we see the needs of a seafaring nation like England where mathematically based navigational aids and technical advances could improve emerging industry, seafaring, commerce, military power, and agriculture.

If we examine the history of the Royal Society and the Royal Society papers from 1665 onward, we can see examples of writing from a host of practical and technical topics. Members of the Royal Society, many renowned in English history, prepared these papers in addition to separately printed reports and proposals they distributed outside the Society to share with Parliament and to encourage widespread change. The writings of many of the first Royal Society members anticipated modern technical documents. This chapter attempts to show the contributions of many Fellows through their technical reports, technical letters, instructions, and process descriptions. Thus, a number of "Greshamites" and then Royal Society members produced works that characterize important genres in technical writing before 1645 and then after the founding of the Society in 1663.

Gresham college, founded by the estate of London merchant Sir Thomas Gresham in the 1580s, stands as the first English college devoted to practical studies. The college housed laboratories and employed instrument makers [5]. According to Gresham's will, courses in astronomy, navigation, and mathematics were to be taught; and courses in anatomy, physics, and pathology were soon added to the curriculum. Students, adult London citizens, wanted useful education that they could apply to their careers, a point discussed by both Houghton [5] and Parry [6, pp. 135–165]. Many wanted their children prepared for a trade. The early faculty—Gwynne, Gunter, Briggs, and Gellibrand—focused on teaching and experimentation that continued through the 1630s. Gresham, and then Oxford, when it came under Puritan control in the later 1640s, became directed by thinkers and early scientists who sought to reform society.

The continued meetings and work of the Greshamites, as early as 1642 [7], provided a forum for nurturing ideas that would culminate in the establishment of the Royal Society in 1663. Their work exemplifies the climate for new concepts developed and presented via print to enhance England's intellectual and economic development. This precursor group included Henry Briggs, whose work in logarithms is well known; mathematicians Seth Ward, John Wallis, and William Oughtred; Thomas Blundeville, a virtuoso who wrote extensively about horsemanship and astronomy as needed by young gentlemen; William Barlow, who developed the concept for the compass and described its use in *The Navigator's Supply* (1597); Edmund Gunter, who published *The Description and Use of the Crosstaff* (1623); and Henry Gellibrand, who focused on developing navigational dials. Works of many of these men provide examples of writing that would evolve into genres important to modern technical writing, that is, technical description needed to help individuals and artisans design and construct these instruments, and then instructions to show users how to operate them.

By the 1650s, the "Invisible College" at Gresham had become an effective scientific community characterized by synergy among an increasingly diverse group of members. All supported strong utilitarian interests. By 1660, the link between the Gresham mathematicians and the mercantile and naval community had cemented the attempts of Henry Gellibrand and Lawrence Rook to improve navigation methods. The first members focused on the manual arts. Other members, reformers rather than scientists, included Samuel Hartlib, William Petty, Gabriel Platts, and John Evelyn. They thought less about experiments than improvements for the practical needs of middle-class society [6, pp. 135–140]. Hartlib and Petty proposed major changes in education that would include education in the trades. These documents exemplify full-fledged early proposals, discussed in Chapter 7. Ultimately, this group linked directly to the work of Nehemiah Grew (botany), Robert Hooke (microscopy), and the early Newtonians. These virtuosi and an increasing number of mathematicians and scientists produced at Oxford in the 1650s were supported by the instrument makers interested in ingenious mechanical devices and those committed to agricultural improvement.

Bacon's Influence on the Development of Trades

Sir Francis Bacon, the intellectual leader of this group and a friend of John Evelyn, desired a program to improve English industry. Bacon believed that the study of industry, what he called the history of trades, would increase knowledge of the trades and nurture improvements based on experiment. These improvements would benefit practitioners and ultimately society. Bacon's argument appeared in *Novum Organum* as well as *The Advancement of Learning.*[1] In a broad sense, we can see Bacon's influential works as the first of the major proposals for improving life for English people by improving industry. He understood the implications of the discoveries of Kepler, Tycho Brahe, and Galileo. While their discoveries rapidly altered current knowledge of the cosmos, their knowledge gave sailors, particularly English sailors, the ability to explore far from land and increase commerce. Applications of astronomy emerged in the work of Gunter, Barlow, and John Dee, all committed to improvements in navigation [5] and who wrote instructions for making nautical instruments and for sailing by using these instruments. The findings of Vesalius and Harvey changed knowledge of human anatomy based on dissection, not on centuries of error, a point emphasized in Chapter 4. Velasius' work provided critical impetus to the development of medical writing throughout the 17th century as well as to technical description. Vesalius' visual display of anatomy was so powerful it became art, a topic also discussed in Chapter 4.

Bacon died in 1626, but his prowess emerged from his prose, which convinced the precursive thinkers who fostered the Royal society that intellectuals and virtuosi of the age could use their writing to produce a vast edifice of information needed to understand the natural world. These discoveries would lead to applied knowledge, better quality of life, and increasing commerce with Europe.

Bacon formed his prose into an instrument of precision and utility. It lacked the eloquence of baroque style but achieved power through austere phrasing. As Basil Wiley noted in 1934, Bacon was neither a mathematician nor metaphysician but a master of language. As a result, his works have been more influential than works of many philosophers and early scientists. Bacon's prose of utility generated among his followers intellectual commitment to the ideas he advocated [8, p. 32]. His powerful style provided a rationale by which the new science could proceed. Scientists such as Galileo, Harvey, and Gilbert were

[1] *The Works of Francis Bacon*, Eds. James Spedding, Robert Leslie Ellis, and Douglas Denon Heath. 14 Volumes. New York, Garrett Press, 1968.

steadily advancing science-based knowledge, and Bacon used his great literary power to establish the scaffolding for accepting their work.

Bacon believed that education should lead to the improvement of society; and Bacon's initial followers, such as John Evelyn, Samuel Hartlib, and William Petty, took Bacon's views and shaped them into the launching ground of what would emerge as the Royal Society. Bacon's earliest followers—Evelyn, Hartlib, and Petty—developed the following plan: (1) write the history of the trades; (2) use the histories to understand the trades; and (3) begin experiments, based on extant observations captured in the histories, to improve basic processes used by the trades. Bacon never advocated "how-to" projects divorced from knowledge of their processes,[2] as seen in many of the *Philosophical Transactions*. From the themes in both *The Adavancement of Learning* and *Novum Organum*, we can see how Bacon, and then his disciples, rejected the scholastic curricula of the universities, to focus on change that would profit England [6, pp. 135–155].

The initial group that adopted Bacon's theory included eight men: Robert Boyle, Henry Oldenburg, Sir William Petty, Sir Robert Moray, John Beale, John Evelyn, Robert Hooke, and Christopher Merrett [9, p. 131]. All would become central to the foundation of the Royal Society and produce work on many concepts important to Bacon: agriculture, dyeing, glass manufacture, enameling, gunpowder, construction, and fireworks, for example [5, pp. 36–37]. The works of these individuals illustrate methods for writing the history of trades. Discussion of the works of several Greshamites who became early Royal Society members will illustrate their contribution to knowledge and technical writing.

John Evelyn—Principal Virtuoso of the Royal Society

Evelyn, often called a virtuoso, exemplified the well-educated Renaissance English gentleman who sought to advance knowledge to serve the practical needs of middle-class society. Evelyn was not a scientist but a visionary, who shared Bacon's vision for methods of civil and commercial improvement needed to move England forward as an economic power. As a landowner, Evelyn's interests lay in land improvement. His major work, *Sylva, or a Discourse of Forest Trees*, [10] argued for the need to replenish England's forests, ravaged by wars and the country's insatiable need for timber. Evelyn then wrote *Terra, or a Philosophical Discourse of Earth* [11], a study of soil, which surveyed crops and suggested plants best suited to different soils. The work also included suggestions to improve soil fertility. He presented both *Sylva* and *Pomona* (see Figure 1-1), the latter, a proposal for developing the cider industry in England, to the Royal Society in 1664 [6, pp. 127–132]. Both works provide 17th century examples of

[2] Robert Moray, An Account, How Adits & Mines are Wrought at Liege without Air-Shafts, *Philosophical Transactions* (1665–1678), Vol. 1, pp. 79–82. This paper stands as an icon to the type of research Bacon advocated.

process description and elegant examples of early how-to manuals. Evelyn's mannered, ornamental style, possibly derived from his scholastic education, lacked the power of Bacon's, a difference that may explain why Evelyn's place in the emergence of practical writing has been minimal. Nevertheless, Evelyn stands as an icon for the intellectual direction the Royal Society would take and the genres it would generate. In describing "Stocks," in Sylva, his major work on agriculture, he discusses, through stocks for propagation purposes. He combines the plain style that would be advocated by the Royal Society, but he retains his Latinate ornamental style:

> CHAP. II
> *Of* Stocks.
>
> The former thus establish'd, after all *humours* and *varieties* have been sufficiently wearied, we shall find the *Wilding* to be the hardiest and most proper stock for the most delicate *Fruit*: This confirm's by *Varro,* lib. 1. Cap. 40. *In quameung; arborem inservas, &c,* and 'tis with reason: However, they do in *Herefordshire,* both in practice, and opinion, limit this Rule; and to preserve the gust of any delicate *Apple* (as of the *Pear-main, Quince-Apple, stockin,* &c.) rather grass upon a *Gennet-moyle* or *cydoddin-stock,* (as tyhere call'd) than a *crab-stock;* but then indeed they conclude the Tree last not so long; and 'tis observ'd, That Apples are better tasted from a clean, light land, &c. than from stiffer clay, or the more pinquid and luxurious *soil,* whence we may expect some assistance from the civility of the stock, which is a kind of prepared soil, or foundation to the Graft, even as our very *Transplantations* into better ground is likewise a kind of *Grafting.*
>
> Thus in the manner of our *Master Varro,* loco citato *concerning Pears; si in pyrum sylvaticam, &c.* The *Wild-stock* does enliven the dull and phlegmatic *Apple,* and the Stock of a *Gennet-Moyle* sweeten and improve an *Apple* that seems *over-tart,* as the *Pome-roy,* or some *Greening* &c. or may rather seem to abate at least some *Apple* over-tart and severe.
>
> Your *crab-stock* would be planted about *October,* at thirty two Foot distance, and not *grafted* till the third *Spring* after, Or at least not before the *second.* [6, p. 27]

THE FIRST TRADE HISTORIES WRITTEN BY THE ROYAL SOCIETY FELLOWS

Other Royal Society members assumed responsibility to write a number of trade histories. A small number of the monographs were published, but numerous articles were written and published in the *Philosophical Transactions* and Sprat's *History of the Royal Society* (1667). These papers exemplify partial histories of trades. While the founding group planned a number of full-length histories, only two members produced book-length histories: (1) Christopher Merrett translated

Antonio Neri's *The Art of Glass* (1662), to which he added extensive information about his study of English glass manufacturing. (2) Walter Charleton wrote *The Mystery of the Vinters* (1667). However, (3) Samuel Colepresse provided the best study of mining based on his description of the Devon and Cornish tin mines.

The history of trades did not achieve its goal of improving important trades and industrial processes. Much of the content proved too complex, a problem clearly visible in Merrett and Colepresse's histories. In addition, craftsmen did not communicate well with members of the Royal Society, and interviews and observations of processes in action provided the bulk of the information that the writers had to organize and present. Nevertheless, documents that resulted from efforts to study the trades yielded a number of instructions and process descriptions that, as technical historian Oakes asserts, helped develop new attitudes toward the need for industry. These histories helped revolutionize industries and connect science with industry [9, p. 130]. As Oakes also states, the focus of these early Society members was chemistry and biology [9, p. 131] and their applications to manufacturing processes. Because Merrett's history of glassmaking and then Colepresse's study of mining have been deemed highly influential, a discussion of the histories of both is warranted, along with Charleton's.

Christopher Merrett—Director of the History of Trades for the Royal Society

Christopher Merrett held an MD from Oxford. A member of William Harvey's circle, he was active in both the Royal Society and the College of Physicians. However, Merrett's most important contribution was his English translation of Antonio Neri's 1612 *L'arte vetaria, The Art of Glass, How to Colour Glass* (1662), to which he added his own *An Account of the Glass-drops* [12]. Merrett had his work on cold published as an appendix to *Robert Boyle's New Experiment Touching Cold.* He headed the Royal Society's committee on the history of trades [5, p. 52].

For physicians like Merrett, active in both the Royal Society and the College of Physicians, the new philosophy and medicine complemented each other. Merrett never wrote a book on prescriptions and their preparation, and his reputation as a scientist resides in his work on glassmaking, which provides extensive instructions for making numerous types of glass.

Merrett's Translation of Neri's *The Art of Glass*

This work illustrates early efforts by Society members to begin writing the History of Trades. Merrett's effort lacks polish but does offer early written explanations of how to make glass. After the dedicatory front matter, Merrett begins *The Art of Glass* by presenting a numbered list of five "quality controls" to aid users in avoiding errors he has made that would produce repetitions of the process. Merrett's quality control statements anticipate quality and safety warning sections found in modern technical instructions in the opening pages. He explains the errors he has made that he considers important to quality. He appears to be thinking about the needs of his readers:

Merrett's Translation of Neri's *The Art of Glass*

This work illustrates early efforts by Society members to begin writing the History of Trades. Merrett's effort lacks polish but does offer early written explanations of how to make glass. After the dedicatory front matter, Merrett begins *The Art of Glass* by presenting a numbered list of five "quality controls" to aid users in avoiding errors he has made that would produce repetitions of the process. Merrett's quality control statements anticipate quality and safety warning sections found in modern technical instructions in the opening pages. He explains the errors he has made that he considers important to quality. He appears to be thinking about the needs of his readers:

To avoid our Authors Repetitions, Observe

1. All the fires must be made with dry and hard Wood.
2. When the Glass is coloured before you work it, mix
The colours well (which otherwise sink to the bottom
Of the pot) with the metal that the Glass may be colored
Throughout. This must be observed all the time you
Work the Glass into any vessels.
3. The sign that Brass or Copper are well calcin'd it, that
They being put into the metal, make it swell and suddenly rise,
If they be calcin'd too much or too little, those signs are
Wanting, and Glass made therof will be Black and foul.
4. Manganese consumes the natural greenness of Glass.
5. Copper, Brass, Lead, Iron, and all compositions of them
As also Manganese, must be put into the metal, but a little at
a time, and at convenient distances, and the pot must be large,
and not filled too full, Bbause they all swell and rise much, and
so are apt to run over, into the fire to your loss. [12, p. 1]

Each chapter then operates as a process description explaining how to make different sorts of glass. Many read like recipes. As the list of "quality controls" shows, Merrett did not blindly follow Neri's instructions. He added his own procedures to his translation. My sense is that Merrett was attempting to adapt Neri's work to an audience of apprentices rather than master craftsmen. Note that Merrett's style reflects the direct, unadorned, precise prose advocated by the

Royal Society, as illustrated by Wilkins in 1676, a point that will be discussed in Chapter 2.

> *Another way to extract the Salt of Polverine, which makes a Crystall as fair and clear as naturall Crystall: This was my invention.*
> CHAP. III
>
> Take Polverine of the Levant well ferced, and put into great glass bodies, luted at the bottom, with ashes, or sand, into the furnaces, filling themat first with common water, give them a temperate fire for some hours in the furnace, and let them stand till half the water be evaporated, the furnace being cold, gently decant off the water into earthen pans glased, putting new water upon the remainder of the Polverine, and let it boil (as before)(this is to be repeated till the water hath extracted all the salt. . . . [12, p. 10]

Merrett's translation does not include a table of contents, and he does not begin to list specific steps until the end of the translation, 196 pages. He uses linear text, a characteristic of the majority of 17th century instructions. However, at the end of Chapter 74, instructions for "The way to prepare natural Crystall," he adds this interesting list:

These rules often repeated together

> 1. That the whole be done cleanly, to this end lute all the pots wherein the Crystall is calcin'd, and wherein the past are baked with lute well dryed, before they be set to calcine or bake.
>
> 2. Take pots that will bear the fire.
>
> 3. Grind all on Porphyrie-stone, and not in metal, to a most implapable powder, first singly, and then together.
>
> 4. Keep a just proportion in the dose of the Ingredients.
>
> 5. Mix the materials well before you bake them, and if the past be not sufficiently baked the first time, bake it again a second time in the potters furnace, and never break the pot till you see 'tis baked; for if you do so the pasts will be foul, and full of blisters.
>
> 6. Leave the vacuity of a fingers thickness in the top of the bot, especially where 'tis said it swells much, or that you must put it in leasurely, lest the materials run our into the fire, or stick to the cover, and so make a foul colour. [12, p. 127]

Merrett apparently thought that Neri's instructions needed clarification for English readers. Merrett's translation suggests that he was attempting to prepare instructions for readers who needed them in the glass houses. In addition, many instructions in the glass houses had been oral instructions, from master to apprentice, and perhaps never preserved in written form. Merrett seems to be trying to decide what information apprentices might need. He includes this feature again in Chapter 93, "The Material wherewith all Enamels are made" [12, p. 147].

Following the instructions for glassmaking, Merrett then includes an analysis of many of Neri's instructions, a brief (unhelpful) index, a 10-page discussion of the furnace required for making glass (no drawing or description), a lengthy commentary of items in chapters on glassmaking, and his Royal Society paper for making glass drops, which could be used for medicinal preparations. He concludes this work of 365 pages with a clarifying commentary on topics discussed in the sections on the furnace. The last half of his translation and commentary appears disorganized and lacks the clarity of the initial instructions for making glass and colors. The entire work, except for the recipes, lacks careful structure and organization. Reading this work, we can sense Merrett's struggle to explain glassmaking—what topics to include and how much description would be needed. This work was likely Merrett's first effort to write instructions for making glass. Nevertheless, the work exemplifies one of the early printed industrial manuals.

As I will discuss in Chapter 5, instructions were perhaps the oldest form of writing, but industrial manuals did not appear until the last half of the 17th century.

According to Charleston, Merrett's 1662 translation gave the English glass industry a boost by providing a step-by-step guide to the manufacture of glass. Although, as Charleston points out, theoretical recipes are one thing and use inside the glasshouse is another, the availability of this book could have encouraged at least some potential manufacturers to enter the glass industry [13, p. 132). In discussing John Baker's late 17th century glass house at Vauxhall, Tyler and Wilmott note that Baker's glass house "was close to a brewery, a distillery, a gunhouse, a marble processing plant, pothouses and a soap works" [14, p. 7]. "It is no surprise that glasshouses should be spied amongst those various manufactures" [14, p. 8]. As the authors also note, England had 88 glasshouses in 1696. Glassmaking had become a buoyant industry that employed 709 apprentices between 1665 and 1739. Oral instructions likely would not have sufficed, as Merrett's translation suggests. The problem he faced reflects the challenge faced by any writer attempting to textualize instructions that have existed predominantly in oral form.

Samuel Colepresse—The First History of Mining

From an economic and historical perspective, the most important histories, often either process descriptions combined with instructions or instructions only, were Royal Society papers that discussed mining and metallurgy. As Oakes states, Samuel Colepresse wrote the best study of mining, a description of Devon and Cornish tin mines. His history can be categorized as a process description: how miners found and then dug an exploration shaft; how they removed the ore, smelted, and refined it. Adits (horizontal passages that allowed entrance to mines) drained the shallow hillside mines. The milling operations, more developed than the mines, could be no deeper than a depth to which a man could throw a shovelful of earth. Next, the ore was milled in water-driven stamping mills. After milling, the unwanted mineral gangue was separated from the ore by washing the lighter weight gangue over the side of a pan. The ore was then roasted in a furnace fired with charcoal and finally in a furnace, probably a blast furnace, which used bellows. Colepresse diagrammed the mines as they were situated in the hills, but he illustrated neither the miners' tools nor the furnaces. As Oakes states, Colepresse's description still has value to historians of technology [9, p. 137] because of the precision of the process descriptions.

The technical communication quality of many of the histories, as recorded in the early *Philosophical Transactions*, stems from their organization, their clarity of description. Most of these exemplify process analysis as we teach it today. In several, procedures can be inferred. Oakes [9] argues that Samuel Colepresse's description of the tin mines at Devon and Cornish exemplifies one of the best of the histories, possibly because of the clarity of the presentation. Colepresse lays out his description much like a modern technical writer would do. The title announces the plan of the process description: "An Accompt of some mineral observations touching the Mines of Cornwal and Devon; where is described the Art of Trayning a Load; the Art and Manner of Digging the Ore; and the Way of Dressing and of Blowing Tin" [15].

Colepresse first discusses the Art of Trayning a load, a process that he divides into 13 steps. He begins with three suppositions about finding the ore, which he lists as First, Secondly, and Thirdly.

He begins Step 1:

> 1. Where we suspect any Mine to be, we diligently search that Hill and Countrey, its situation, the earth, or grewt, its colour, and nature, and what sort of stones it yields; the reason hereof being only this, that we may the better know the grewt, and stones, when we meet with them at a distance in the neighbouring valley; for mineral stones may be rfound 2, 3, 4, 5 miles distant from the Hills or Loads, they belong unto. [15, p. 2097]

His Second heading, The Art and manner of Digging up the Ore, he divides into 16 specific steps. For example,

> 1. The difficulty of this is not considerable to that of *Training*. When we have found our *Load*, that last Essay-hatch looses or Rather exchanges its name for that of a *Tin-shaft*, or *Tin-Hatch*, Which we sink down about a fathom, and then leave a little long Square place, termed a *Shamble*, and so continue sinking from the Cast to cast, (*i.e.* as high as a man can conviently throw up the Ore With a shovel,)till we find either the Load to grow small, or Degenerate into some sort of weed, which are diverse; as *Mundick*, Or Maxy (corrupted from *Marchasite*) of 3 sorts; white, yellow, and Green; *Daze*, white, black, and yellow: *Iremould*, black, and rusty: *Caul*, red: *Glister*, bloud-red, and black. [*See these hard names Explained below*. No. 3]. [15, p. 2102]

As a good technical writer, Colpresse defines terms he believes his reader will not know, as he does in this first step. Then, in Step 3, he tells readers to "be pleased to cast an eye on the annexed Scheme, in Tab. II, as here, by the Alphabet, explicated; which may give some information to those that have not been conversant in Mines" [15, p. 2103]. See Figure 1-1, a drawing of the mine, which correlates with his verbal description.

After the steps for locating the mine, he presents "the Art and manner of Digging up the Ore." This segment contains 16 steps, with a description of the drawing in Figure 1-1 that shows the location of the mines. The locations on the map are labeled as follows, and the foldout drawings appear at the end of this paper, a common placement in the early *Philosophical Transactions* and in many modern technical reports (see Figure 1-2).

The next section, "The Manner and Way of Dressing Tinn," includes six steps, followed by "The Manner and Way of Blowing Tinn." This final section, composed of two lengthy paragraphs, is less easy to follow and the process less verbally organized. As in Merrett's instructions for glassmaking, we find Colepresse struggling to find the best method to structure his explanation.

Walter Charleton—The History of Winemaking

Charleton's Work, *Two Discourses: I. Concerning the Different Wits of Men; II. Of the Mystery of the Vinters* (1675) [16] stands as one of the best works on analysis of winemaking in the 17th century. While earlier works relied on classical works on winemaking, Charleton, in the spirit of the Royal Society, provided an analytical study of wine. As agricultural historian Michael Best remarks in his discussion of the history of winemaking,

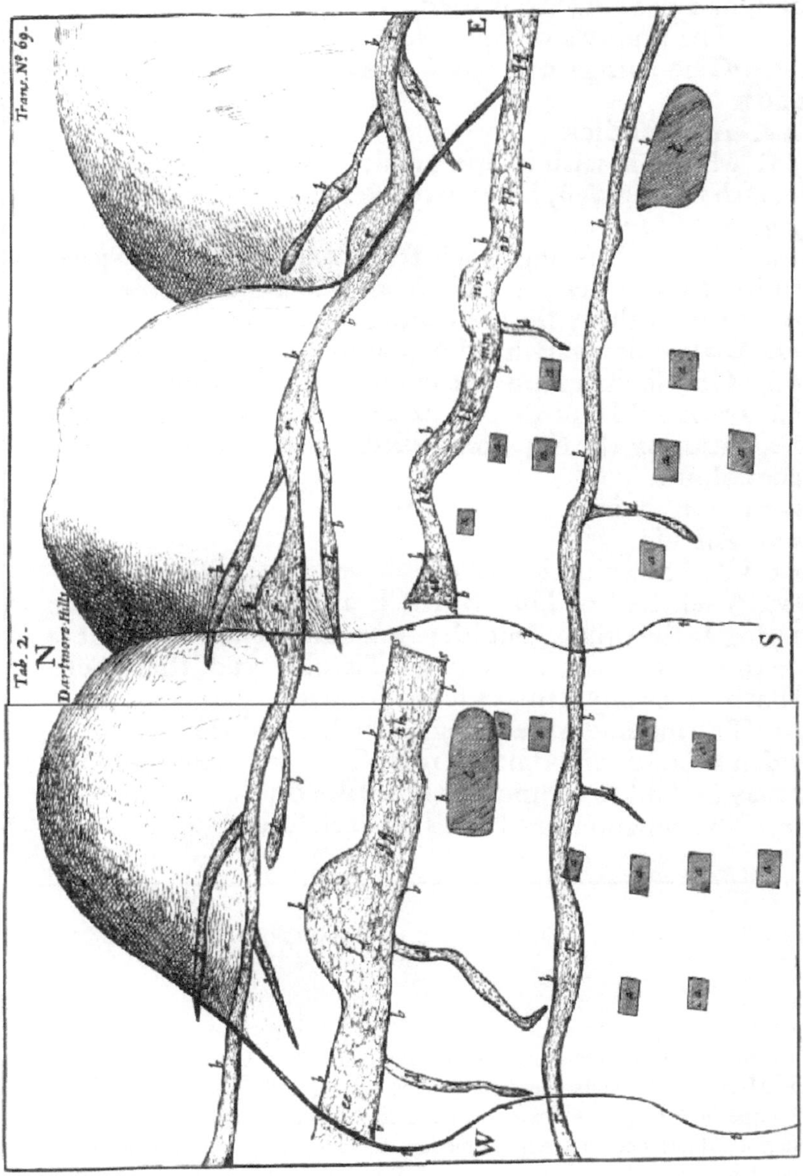

Figure 1-1. Colpresse mine drawing. Permission from the Royal Society.

a a. The Eſſay-Hatches.
b b. The wall which the Shelf makes on both ſides the Load, and the Load ſo walled.
c c. The Bonny's or Squatts.
d d. The ſtrings or little Veins of the greater and leſſer Loads.
e e. All Mundick.
f f. Moſt Tin with its Spar, which places prove all good Tin, if the ſtring *d, b,* happen to be Tin.
g g. All Clay.
h h. Caul; differing both from *Marcaſite* and *Sparr*: it endureth the fire, which *Marcaſite* or *Mundick* doth not. *Sparr* is a flinty ſtone of different colours.
i i. Clay which may hold a quantity of Tin.
k k. Tin; ſuch flexures are commonly well tinned.
l l. Ire-mould and Daze. *Daze* is a kind of glittering ſtone, enduring the fire, ſome ſofter, ſome harder, of different colours.
m m. Tin again.
n n. All Tin.
o o. Clay carrying Mundick.
p p. A Sell-bed of Tin, which is all Tin, and needs no ſtamping as the other, but dry knacking (*i. e.* without a grate or Cock-water, as anon.) 'Tis obſerved, that a Sell-bed hath never any ſtrings iſſuing from it.
q q. The innumerable ſtrings, like little Capillary veins, in which a Load ſometimes in reſpect of its uppermoſt ſurface may be ſaid to terminate and ſtrike out.
r r. The concomitant Loads on each ſide.

Figure 1-2. Textual description of Figure 1-1 drawing of a mine [15, p. 2103].

Vinter, in 1669, published two works on wine: Two Discourses and Of the Mysterie of Vinters. He delivered the second to the Royal Society to describe how to care for wines. Charleton synthesized information from ancient (Pliny and Plutarch) to modern sources (Bacon and Boyle). He added his own observations, as many Royal Society members did. He emphasized specific remedies, which he found in the manuscripts of quality vinters, he found. He advocated wines as remedies, which echo Observations on wines by Christopher Merret. [17, pp. 367–368]

These two works—Merrett's and Charleton's—proved highly influential. However, as a trade history, *The Mystery of the Vinters* [16] fulfills the requirements of a history, as directed by the Society, and uses pristine language, clear organization—division of each topic into specific subsections—and a sustained focus on how to ensure purity of wine. While the topic is less complex than either glassmaking or mining, the work exemplifies a quality trade history, from the perspective of its writing.

Charleton divides his treatise into four sections: natural purification of wine, sicknesses that result from impure wine, the decay of wine into vinegar, and recipes to address the first three problems. The work focuses on what he deems the causes of spoilage and recipes to correct the problem. For example,

> To help *Stinking* wines, the general Remedy is *Racking* them from their old and corrupt Lee. Besides which, some give them a fragrant smell or *Flavor*, by hanging in them little bags of spices, such as *Ginger*, *Zedoary*, *Cloves*, *Cinnamon*, *Orras* roots, *Cubebs*, *Grains of Paradise*, *Spinkard, &c. Aromatics*, Others boyl some of these Spices in a portal of good found wine of the Same sort, and tunn up the decoction hot., Others correct the ill favour of rank-leed *French* wine With only a *few Cinamon* canes hung in them. Others again for the same end use *Elder Flowers*, and Topps of *Lavender*. ¶, [17, pp. 196–197]

As Best notes, the work contributed to the knowledge of wine quality, but its value did not last. Pasteur's work on pasteurization, which appeared 75 years later, provided understanding of the fermentation process and the types of spoilage to which wine is susceptible [17, p. 376].

Samuel Hartlib, William Petty, and John Dury—Proposals to Reform Education

A number of the Greshamites and then the Society founders/members contributed to change for improved education by proposals (a genre that will be discussed in Chapter 7) as well as in trades studies. The principle reformers, which included Samuel Harlib, William Petty, John Dury, and John Evelyn, thought less about experiments than ways to improve trades. Their goal, as I have already noted, was to apply knowledge to the immediate and practical needs of middle-class society [5, pp. 39–43]. Hartlib sought to develop an Office of Address and Communications to address and share ideas that would benefit society [9, p. 131], an office similar to outreach and extension services in modern universities. Hartlib and his circle sought to move from words to deeds, from philosophy to practice, from verbal logic to experiments that would yield practical knowledge. One main thrust of this perspective was to reform education based on the study of reality and vocational training.

Hartlib focused on creating a commonwealth in which government and people collaborated in creating prosperity generated by practical applications of knowledge. He is best remembered for his work in imparting scientific ideas and information. While not successful in his efforts to reform education and government, he did successfully infuse new ideas into husbandry—planting new leguminous crops, experimenting with fertilizers and manures, and using seed drills and new plows. He also advocated the possibilities of apiculture (rabbit farming), fruit-tree propagation, and silk cultivation. His pamphlets should be read as models of how the processes should be implemented. Thus, his process descriptions, analysis, and procedures provide extended examples of 17th century technical writing.

Hartlib recorded his vision in *Marcaria* (*A Description of the Famous Kingdome of Macaria*) in 1641 [9, p. 131]. This work, a broad educational proposal, influenced two specific educational proposals, discussed in Chapter 7, by William Petty and John Dury. In Petty's educational proposal, the History of Trades would have been the main text used to prepare students for an increasing range of trades as more were written by Society members.

While Bacon's idea for a college devoted to experimentation to produce improvements in industry, trades, and agriculture never came to fruition, the idea for such an institution continued to be discussed and proposed by John Dury, Abraham Cowley, as well as John Evelyn and William Petty. Their ideas for improving England by improving education by adding experimental science provide modern readers with some of the first examples of project proposals, which will be discussed in Chapter 7.

William Petty—Developer of Educational Proposals and Statistical Reports

In the first effort to launch the histories, members were assigned preparation of a trade history that suited their interests. William Petty, another virtuoso, projector, and Hartlib's protégé, distilled Bacon's ideas into a full proposal for utilitarian education. As Houghton concludes, "Is it too much to say that in Bacon's conception of the History of Trades, Petty has condensed and focused the ideal of the scientific middle-class society which was born in the Interregnum and grew up in the Restoration?" [6, p. 45]. As we shall also see in Chapter 8, Petty's utilitarianism continued to develop, and to him and John Graunt, another RS member, can be attributed the development of statistics and the first printed English statistical reports.

The commercial and social motives combined with science as the foundation of educational reform emerged as full-scale proposals to be discussed in Chapter 8, such as Dury's *The Reformed School* and Petty's *The Advice of W. P. to Mr. S. Hartlib for the Advancement of Some particular Parts of Learning*. Petty, a shrewd businessman, developed his educational proposal to embody the new science, the history of trades as reading material, and a new approach to education that would prepare the young for a new society based on commerce, not liberal education. In collaboration with William Petty, a businessman who possessed extensive knowledge of clothesmaking, dyeing, smithing, watch-making, carpentry, and joining, Hartlib pressed forward with a plan for a scientific college (actually a college of tradesmen) that benefited both scientists and practitioners. The main written product would be the history of trades, which should include "the whole process of manual operations and applications of one natural thing. . . with the necessary instruments and machines, . . . all instruments and tools must be pictured, and colours added, when the descriptions cannot be made intelligible without them." In short, these full-fledged proposals exemplify the spirit of utilitarianism that defined the early work of the Royal Society.

TECHNICAL WRITING AND THE GROWTH OF WEALTH IN THE 1660–1700 ERA

Commerce, defined as the need for more trade, enhanced England's development of navigation and ship design, a topic that will be discussed in Chapter 6. Impetus for technical writing also emerged from technical developments encountered by industry and navigation. At first, scientific activity in England centered on new navigational problems, traditional problems of accounting, calendar making, surveying, and military engineering. These problems, largely ignored by the universities, were studied at Gresham. When Oxford came under Puritan control in the later 1640s, the first generation of "natural philosophers" came from Gresham.

While many of these first Gresham scientists were Puritans or Latitudinarians, the needs of society to aid "gentlemen as employ their estates in those chargeable adventures of Drayning, Mines, Cole-pits, etc." continued. However, the importance of the Interregnum and the rise of Puritanism on the development of both science and technical writing to present this new knowledge cannot be overestimated. As Parry notes, "A new intellectual phase began after the capture of Oxford in 1645, when the parliamentarians weeded out the Laudians and their associates and replanted the university with fresh stock. In this changed atmosphere, men of a scientific cast of mind began to prosper" [5, p. 144]. The Oxford group began to join forces with experimental scientists, such as Robert Boyle, who recruited Robert Hooke as his research assistant in the late 1650s. An impressive group of mathematicians continued the work of William Oughtred, who valued practical applications and worked closely with instrument makers, surveyers, and navigators.

John Wilkins, another Baconian, envisioned an interlinking of the sciences across a broad experimental range. He too sought developments to benefit England. However, Wilkins stood for reform of language to achieve precision, objectivity, and mutual understanding. His goal was elimination of affected phrases that obscured meaning. His influence is discussed in Chapter 2. Thomas Spratt, in his History of the Royal Society, published in 1667, reemphasized Wilkins argument for language reform: A rational discovery of nature cannot proceed unless writers use a style both clean and sharp. Sprat restates Bacon's arguments from the *Advancement of Learning* with refinements developed by Wilkins, his mentor at Wadham. Spratt and Wilkins rejected use of metaphor, imagery, and verbal embellishment of all kinds [6, pp. 145–146].

While the major figures of the Royal Society group have received the most attention—Evelyn, Petty, Boyle, and Hooke, for example—other lesser figures contributed to the foundation projects developing technical writing. Excellent examples of technical writing, even by today's standards, can be found in works by Joseph Moxon and John Woodard. The works of both show how rapidly technical writing developed by Society members and produced highly useful technical writing.

Joseph Moxon—Hydrographer and Industrial Technical Writer

Another Royal Society member, Joseph Moxon, contributed numerous printed instructions in his support of the history of trades. Moxon, best remembered for his instruction manuals in printing, was the first tradesman elected to the Royal Society in 1678. An astute student of mathematics, between 1654 and 1684, he published more than 30 popular scientific expositions and technical handbooks. His instructional books on architecture, astronomy, and globes ran to several editions in his lifetime. He was well known for his ability to devise and present tabulated data. Because of his skill and understanding of mathematics, the members of the Royal Society supported his petition to Charles I for appointment as hydrographer, a title that reflects his skill in his work with maps.

Moxon's instructions appeared in six different versions from 1677 to 1700. Each began with instructions on blacksmithing, *Mechanick Exercises, or, The Doctrine of Handy Works . . . of the Smiths Trade* (illustrated accounts of the trades of the smith, joiner, carpenter, and turner) and illustrate his support for the History of Trades [18, 19].

One of Moxon's instruction books, *Exercises: Or, the Doctrine of Handy-Works*, first printed in 1693 and in several other edition until 1700 [20]. This work covers four distinct topics specific to the trades. Each begins with a separate title page, followed by a page of drawings of tools that will be needed in the tasks for which he provides instructions. In his preface, he explains his rationale for providing instructions for processes essential to trades. His books provide excellent examples of modern how-to books in terms of consistent organization from version to version, clarity in style and definition, and effective use of visuals.

In his instruction for Smithing, he begins with the definition:

Definition

Smithing is an Art-Manual, by which an irregular Lump (or Several Lumps) of Iron, is wrought into an Intended Shape.

This Definition, needs no Explanation; therefore I shall proceed to give you an Account of the Tools a Smith uses; not but that (they being so common) I Suppose you do already know them; but partly because They may require some pre-caution in setting them up Fitted to your use, and partly because it behoves you to Know the Names, Smith call the several parts of them by , that when I name them in Smith's Language (as I shall oft have occasion to do in these Exercises) you may the easier understand them, as you read them. [20, p. 1]

He then partitions his discussion into 14 segments: "Of Setting up a Smith's Forge"; "Of the Anvil"; "Of the Tongs"; "Of the Hammer, and the Sledge"; "Of the Vice"; "Of the Hand-Vice"; "Of the Plyers"; "Of the Drill, and Drill-Bow"; "Of the Screw-late, and its Taps"; "Of Forging in general"; "Of the Several Heats Smiths take of their Iron"; "Of Brazing and Soldering"; "Of Several Sorts of Iron, and their proper Uses"; "Of Filing in general." Moxon's approach and his style exemplify his commitment to providing sufficient information for beginners. Throughout each section, he uses first person and writes as if he is addressing his reader, as shown in the above excerpt from page 1. He uses italics for smithing nomenclature and refers readers to the visual located after the title page. See Figure 1-3 which portrays the tools for smithing:

After his definition of smithing, he describes each item required for the smithing process in "Of setting up a Smith's Forge." He next describes each of the implements required for smithing. For example, he describes tongs and classifies them into two categories.

Of the Tongs.

There are two sorts of Tongs used by Smiths; the one The Straight nosed Tongs, used when the work is Short, and somewhat flat, and generally for all Plate Iron. The others Crooked nos'd Tongs, to Be used for the forging small Bars, of such thicker Work, as will be held within the Returns of their Chaps. The Chaps are placed near the Joint, Because that considering the length of the Handles, they hold the Iron faster than they would do, were they placed farther from the Joint, as in the Fig. 3, 4. A the Chaps, B the Joint, CC the Handles. [20, p. 3]

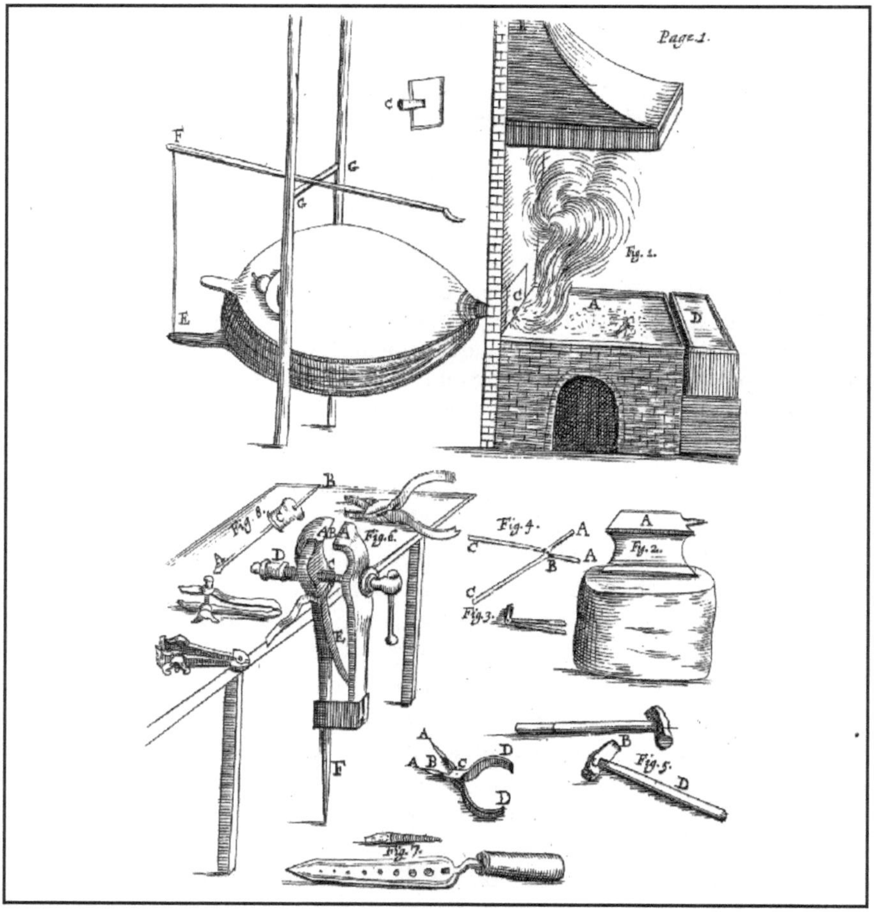

Figure 1-3. Moxon's drawings of smithing tools.

After describing each tool, he describes the different processes: "Of the several Heats Smiths take of their Iron," "Of Brazing and Soldering," "Of several Sorts of Iron, and their proper Uses," and finally "Of Filing, in general." Moxon uses headings throughout. His style is direct and concise and reflects modern English, the legacy of the Royal Society. Moxon develops each segment following this pattern, which provides an excellent example of the use of definition.

Moxon's second segment, "Mechanick Exercises: Or, the Doctrine of Handy-Works," develops in the same way: an introductory page with the tools he will describe, as these are needed for various processes; Hinges, develops his segment on "joinery" with the same method; and a page with drawings of various hinges, locks, keys, and screws, a description of hinges, which includes how to make hinges. Remaining sections, which include technical descriptions and instructions for making each one, are "Of Hinges," "Of Locks and Keys," "The manner of Rivetting," "The making of Screws and Nuts," and "The Rules and manner of Cutting Worms upon great Screws." From a format perspective, the work uses italics consistently: main terms appear in italics in both the text and in the descriptive headings. Moxon's style reflects the unadorned style supported by the Royal Society.

Moxon concludes his final segment of this 1700 version of Mechanick Exercises with instructions on Bricklayers Work. His instructions for cement are particularly interesting, as they fully anticipate instructions that can be found today in procedures published by the American Society of Civil Engineering.

I shall conclude this Exercise with the Art of making two sorts of Cements, for the Cementing [of] Bricks.

There are two sorts of Cement, which some Brick-layers use in Cementing of Bricks for some kind of Mouldings, or in Cementing a block of Bricks, as They call it, for the Carving of Scroles or Capitals or such like, &. One is called Cold Cement, the other is called hot Cement, because the former is made and used Without Fire, but the latter is both made and used with Fire; the cold Cement being accounted a Secret, is known but to few Bricklayers, but the hot Cement is common.

To make the cold Cement

Take ½ a Pound of Old Cheshire-Cheese, pair of the Rine, and throw it away, cut or grate the Cheese very small, and put it into a Pot, put to it about a Pint of Cows-Milk, let it stand all Night, the next Morning get the Whites of 12 or 14 Eggs, then take ½ a Pound of the best Unslackt or Quick Lime that you can get, and beat it to Powder in a Morter, then sift it through a fine Hair Sieve into a Tray or Bole of Wood, or into an Earthen Dish, to which put the Cheese and Milk, and stir them Well together with a Trowel, or such like thing, breaking the Knots of Cheese, if there be any, then add the Whites of the Eggs, and Temper all well together, and so use it; this Cement will be of a white Colour, but if you would have it of the Colour of the Brick, put into it either some very fine Brick-Dust, or Almegram, not too much, But only just to colour it.

To make the hot Cement.

Take one Pount of Rozin, one quarter of a Pound of Bees-Wax, half an Ounce of Chalk-Dust, or powder of Chalk, fist both the Brick-Dust and Chalk-Dust through a fine Hair Sieve, (you may beat the Brick and the Chalk in a Morter, Before you fist it) boil altogher in a Pipkin, or other Vessel, about a quarter of an hour, stirring it all the while with an Iron or a piece of Lath or such like, then take it of, and let it stand 4 or 5 Minutes, and 'tis fit for use.

Note, That the Bricks that are to be Cemented with this kindnd of Cement, must be made hot by the Fire b efore you spread the Cement on them, and then rub them to and fro on one another, as Joiners do, when they Glew two Boards together. [20, p. 45]

John Woodward—First Royal Society Laboratory Manual for Researchers at Sea

John Woodward, a member of the Royal Society, prepared a decidedly modern handbook for those who would collect samples for study by Royal Society members: *Brief Instructions for Making Observations in All Parts of the World; as Also for Collecting, Preserving, and Sending Over Collections of Natural Things* (1696) [21]. This little laboratory manual of 20 pages opens immediately with instructions for keeping a journal at sea and then provides similarly direct instructions for collecting items from seashores. He partitions the items into 10 specific collection types. He discusses items that may be found in countries such as New Guinea, the East and West Indies, and Greenland; then provides instructions for preserving these items. Woodward's little book stands as one of the first purely instructional or laboratory manuals of the 17th century. It avoids theory and focuses on methods. However, Woodward does explain the rationale for instructions. For example,

> *Coralls, and Shells to be Sent.*
>
> 11. As to *Sponges, Brain-stones, Sea-fanns, Sea-roses, Corals* of all sorts, Crabs, Lobsters, Sea, River, and Land Shells, whether common or uncommon, great or small, send five or six of each, wherein there is any Difference in Figure, Colour or Bignesse. Of the Shells, where they are easie to be got, choose those that have the Creatures still living in them (which yet ought to be pluckt out, or they will putrifie and stink) such being by much the freshest and fairest: but where such are not to be got, take the empty and dead Shells as you Find them; only of the Bivalves or double Shells, endeavour to send both the upper and under Shell together. [21, p. 14]

His final section lists tools necessary for collection that should be taken by researchers who will do research beyond English shores. He describes each tool and its purpose (see Figure 1-4).

He also provides detailed instructions for how to conduct experiments in the order of the tasks. For example, he explains how to keep research records, what to record (see Figure 1-5).

CONCLUSION—TECHNICAL WRITING AND THE GROWTH OF SCIENCE IN INDUSTRY, 1640–1700

Development of commerce within England increased through the last two decades of the 17th century with the Royal Society providing an important influence. Increases of consumer goods like pottery, home building,

A Liſt of ſuch Inſtruments, *and other things, as may be ſerviceable to thoſe Perſons who make Obſervations, and Collections, of Natural Things.*

1. THE *Weather-glaſs* now lately contrived by *Robert Hook*, M. D. Profeſſor of Geometry in *Greſham* College, and S. R. S. of uſe at *Sea* as well as at *Land*.

2. The common *Barometer*.

3. The common *Thermometer*.

4. The *Hygrobaroſcope*, or ſmall Glaſs Inſtrument, with the Neck or Stemm graduated into ſmall Diviſions, ſerving to try and compare the *Gravity* of *Liquids*, *Waters*, *&c*.

5. A *Dipping-needle*; in order to obſerve the ſeveral Degrees of its *Inclination* in all Parts, both at Sea and Land.

6. A large *Quadrant*, for *taking the Height of Mountains*, *&c*. alſo for the making *Aſtronomical Obſervations*, *&c*.

7. A *Level*, whereby to judge of the *Situation* (in reſpect of the Horizon) of the *Beds of Earth*, *Stone*, *Marble*, *Cole*, *&c*. in *Mines*, *Quarries*, *Cole-pits*, *&c*.

8. A *two-foot-rule*, or other like *Meaſure*, exactly graduated, as well to take the Depth of the Water in the *Experiment* related above *page* 3. as for other Uſes.

9. *Hammers*, bigger and ſmaller, to *break*, and *examine* the interiour Conſtitution of *Ores*, *Native Minerals*, *Stones*, *&c*. with a *Chiſſel* wherewith to diſſever or ſtrike off *ſhells* from the Maſs of Stone, Marble, *&c*. conf. pag. 6. and pag. 11. above.

10. *Crucibles*, *Fluxing-powders*, &c. for *melting* and *tryal of Ores*.

11. An *Eradicator*, or ſmall Iron Inſtrument to take up the *Roots* of Herbs out of the Earth.

12. Several *Quires of brown Paper* to dry and keep *Plants* in.

13. Several Neſts of *Dutch*, or *Pill-Boxes*: alſo *Deal Boxes* of ſeveral Sizes, in which to put up, and *ſend over Natural Bodies*.

14. *Bottles*, *Jars*, *Gally-Pots*, or other like *Veſſels*, wherein to put the more *tender Creatures*, with Spirits *&c*.

15. *Spirit of Wine*, *Rum* or *Brandy*, to preſerve the ſaid Creatures, *e. g.* Inſects, Lizards, Serpents, *&c*.

Figure 1-4. Woodward instructions on what items researchers should take with them when they go to sea to do research [21, pp. 18–19]. Permission from The Royal Society.

I. *At Sea.*

KEEP a *Journal* of the *Ship's Courſe* : Of the *Latitude*, as often as taken: Of the *Variation of the Compaſs* : of the *Soundings*, obſerving what ſorts of Shells, Sand, or other Matter is brought up with the Plummet. In *Calmes*, or with any other Opportunity, both at Main Sea, or elſewhere, *ſound* to the Bottom, if all the Line or Tackle you have will reach it; but if not, only note what Length of Line you uſed In the ſaid Journal alſo keep an Account of the *Currents* ; of the *Brizes*, and other *Winds*, as well thoſe which are *ſettled* and conſtant, as thoſe which are *accidental* : of *Storms* and *Hurricanes* : of the Riſe and Fall of the *Weather-glaſs* : Of the *Weather*, *Heat* and *Cold*, *Fogs*, *Miſts*, *Snow*, *Hail*, *Rain*, *Spouts* or *Trombs*, vaſt *Diſcharges of Water* from the *Clouds*, *Thunder*, *Lightning*, *Meteors*, *&c.* Obſerve whether ſome Seas be not *ſalter* than others, or diſtant Parts of the ſame Sea differ not in *Saltneſs* : Whether the Water of the Sea be not *warmer* than ordinary, or ſuffer not ſome unuſual *Bubblings* or *Commotions* before *Storms*, or there do not ſome other like uncommon Accidents forerun them, whereby they may be foreſeen : What Voragines or *Whirlpools* appear at Sea, to what *Diſtance* the Force of their *Flux* or Stream extends, whether they only receive and *ſwallow in* the Water, or *ſpue it forth*, or, if *both*, with what Periods, *i. e.* at what time doth *each* begin and end : What ſorts of *Fowls* occurr at *Main Sea* : What *Fiſhes*, what *Weeds*, *Shrubs*, or other things. In brief, take notice of every obſervable natural *Occurrence* throughout the whole *Voyage*, and this too in as *full* and circumſtantial a *manner* as may be.

Obſervations to be made at Sea.

Figure 1-5. Woodward's instructions for keeping research records [21, p. 1].

and new inventions that had commercial value expanded. New patents for inventions grew, along with the British navy. Government military contracts and naval construction stimulated the economy [22, p. 266] and produced large-scale enterprises such as sailcloth making, saltmakers, and silk manufacturing. And as Claggett remarked, "In a more backward technological setting the scientific revolution could not have occurred" [23, p. 12].

While Hooke's *Micrographia*, discussed in Chapter 4, was not particularly utilitarian, it did announce the "pure science" direction that the Royal Society would take by the end of the 17th century. And the research-based utilitarian focus of the early RS members promoted the kind of scientific investigations that did take place.

In short, science was not an autonomous pursuit or a rarefied endeavor. The "science" pursued targeted practical problems for navigating ships needed to broaden commercial interests and to measure land, as pursued by William Petty in his survey of Ireland in the 1650s. The establishment of Greshman College exemplifies the failure of Oxford and Cambridge to respond to the needs of a new society that focused on commerce, industry, and development of technologies to aid that development. As a result, industrial and technical advances facilitated scientific progress by making available more reliable instruments and consequently more accurate measurements. Without earlier improvement in glassmaking and metallurgy, Robert Hooke, after the Restoration, would have had difficulty conducting his fundamental research into microscopy [2, pp. 93–94].

For historians of technical writing, these advancements preserved in the technical writing of the Greshamites and then the Fellows allow us to track the development of genres now firmly established in our field. This overview leads next to the rationale and development of the sentence style that became the hallmark of the Royal Society and most technical writing.

REFERENCES

1. D. Wing, *A Short-Title Catalogue of Books Printed in England, Scotland, Ireland, Wales, and British America, and of English Books Printed in Other Countries 1641-1700* (2nd ed.), Index Committee of the Modern Language Association, New York, 1972.
2. A. W. Pollard and G. R. Redgrave, *A Short-Title Catalogue of Books Printed in England, Scotland, & Ireland, and of English Books Printed Abroad, 1475-1640* (2nd ed.), W. J. Johnson, F. S. Ferguson, and W. F. Panzer (comps.), Two Volumes, Bibliographical Society, London, 1976 and 1986.
3. E. Howes, The Third Vniversity of England, Or a Treatise of the Fovndations of All the Colledges, Ancient Schools of Priviledge and Houses of Learning, and Liberall Arts, Within and About the Most Famovs Cittie of London, in *The Annales or General Chronicle of England,* I. Stow, London, 1615.
4. R. Merton, *Science, Technology: Society in Seventeenth Century England*, Howard Fertig, New York, 1970.
5. W. E. Houghton, The History of Trades: Its Relation to Seventeenth Century Thought, *Journal of the History of Ideas, 2,* pp. 33-60, 1994.

6. G. Parry, *The Seventeenth Century: The Intellectual and Cultural Context of English Literature, 1607-1700*, Longman, London and New York, 1989.
7. F. R. Johnson, Gresham College: Precursor of the Royal Society, *Journal of the History of Ideas, 1*:4, pp. 413-438, 1940.
8. B. Wiley, *The Seventeenth Century Background,* Doubleday Anchor, Garden City, New York, 1935.
9. K. H. Ochs, The Royal Society of London's History of Trades Programme: An Early Episode in Applied Science, *Notes and Records of the Royal Society of London, 39:2,* pp. 129-158, 1985.
10. J. Evelyn, *Sylva, Or A Discourse of Forest-Trees, And the Propagation of Timber in His Majesty's Dominions,* London, Printers to the Royal Society, 1670.
11. *J. Evelyn, Terra, Sylva, or, A discourse of forest-trees, and the propagation of timber in His Majesties dominions as it was deliver'd in the Royal Society the XVth of October MDCLXII, upon occasion of certain queries propounded to that illustrious assembly, by the honourable the principal officers, and commissioners of the Navy: Terra, a philosophical essay of earth, being a lecture in course: to which is annexed Pomona, or, An appendix concerning fruit-trees in relation to cider, the making, and several ways of ordering it, published by express order of the Royal Society*, London, 1679.
12. C. Merritt (trans.), *The Arte of Glass, Wherein Are Shown the Ways to make and Colour Glass, Pastes, Enamels, Lakes and Other Curiosities,* London, 1662.
13. R. J. Charleston, Our Forefathers in Glass, *Glass Technology, 21:1,* pp. 27-36, 1980.
14. K. Tyler and H. Wilmott, John Baker's Late 17th Century Glasshouse at Vuxhall, Museum of London Archaeology Society Service, London, 2005.
15. S. Colepresse, Devon: Wherein is Described the Art of Trayning a Load: The Art and Manner of Digging the Ore: And the Way of Dressing and of Blowing Tin, *Philosophical Transactions (1665-1678), 6,* pp. 2096-2113, 1671.
16. W. Charleton, *Two Discourses: I. Concerning the Different Wits of Men; II. Of the Mystery of the Vinters,* London, 1675.
17. M. R. Best, The Mystery of Vintners, *Agricultural History, 50*:2, pp. 362-376, 1076.
18. J. Moxon, *Oxford Dictionary of National Biography*: in Association with the British Academy: from the Earliest Times to the Year 2000, H. C. G. Matthew and B. Harrison (eds.), Oxford, Oxford University Press, v. 29, pp. 602-603, 2004.
19. G. Jagger, Joseph Moxon, F.R.S., and the Royal Society, *Notes and Records of the Royal Society of London, 49*:2, pp. 193-208, 1995.
20. J. Moxon, *Mechanick exercises, or, The doctrine of handy-works applied to the art of smithing, joinery, carpentry, and turning / by Joseph Moxon . . .* , printed for, and sold by J. Moxon, London, 1700.
21. J. Woodward, Directions for Observations and Experiments to Be Made by Masters of Ships, Pilots, and Other Fit Persons in Their Sea-Voyages, *Philosophical Transactions (1665-1678), 2*, pp. 443-448, 1666-1667.

22. C. Hill, *Century of Revolution, 1603-1714,* Thomas Helson and Sons, Ltd, Edinburgh, Scotland, 1962.
23. A. R. Hall and E. M. Claggett, *Critical Problems in the History of Science,* Institute for the History of Science, Proceedings, Madison, Wisconsin, 1959.

CHAPTER 2

The Evolution of English Plain Style in the 17th Century: A Confluence of Sound, Format, Diction, Syntax, and Purpose*

Any discussion of the history of technical writing must deal with the matter of style used in the technical and other practical documents of the period(s) studied. In discussing *style*, we can define the term as the various devices an author employs in his/her writing: word arrangement, period and sentence structure, word choice, sentence patterns, clause length, use of schemes and tropes. In Chapter 1, I noted examples of clear or plain style, a quality of modern technical writing, in the works of several of the first members of the Royal Society. Writing that we now call "plain style" existed in much technical writing of the English Renaissance and then the 17th century.

In *Emergence of a Tradition*, I argued that what emerged as plain style—unadorned sentences emphasizing tight subject-verb-object structure—was fundamental to Old English, Middle English and finally the 15th century plain style. Three styles were advocated by Greek Rhetoric: the "high" style, also called "oratorical," elaborate, or Ciceronian; the second, usually clear and direct, often called anti-Ciceronian or low style; and the third style, the "middle style," blended characteristics of Ciceronian and anti-Ciceronian. Practical writing, such as technical writing, occurred in "low" or plain style throughout the Renaissance and the 17th century. As Jones points out, this shift may have occurred in deference to the middle-class English reader who lacked instruction in Latin. However, rhetoric's use of the "low style" for instruction or teaching may also have had

*This chapter began as an article published in *Journal of Business and Technical Communication, 18*:2, 2004: 165–197, doi: 10.1177/1050651903260738. See http://online.sagepub.com. The article is used with permission from SAGE Publications.

influence. Literary historians have, in general, talked about plain style as a shift from the "luxuriant prose of the Commonwealth into that of a diametrically opposite nature in the Restoration" [1, p. 977]. But, as I have shown elsewhere, this "diametrically opposite" kind of prose had existed in practical (or technical) writing for well over 300 years. Thus, technical or practical writing remains fundamental to the development of modern English prose style, which matured in the last half of the 17th century. Examples of this mature style occur in Merrett, Colepresse, Moray, Moxon, and Woodward, whose works I discussed in Chapter 1.

Since completing *Emergence of a Tradition*, I have found additional examples of plain, unadorned writing in medieval estate management books, such as those by Walter of Henley, who targeted a range of readers, a point discussed in Chapter 5. Writers of many how-to books before and during the 17th century seemed to understand that readers who used written text to perform tasks needed a concise, direct style and that verbose, decorous language would not achieve that purpose. Even the early technical writers recognized the importance of an appropriate rhetorical stance in presenting information.

In this chapter, I discuss the contribution of practical writing to the development of modern English prose style. This chapter continues my study of plain style in *Emergence of a Tradition*. In Chapter 3, I will show that technical writing in the first printed English books illustrates paragraphs and organizational principles of paragraphs that exist today. These paragraphs echo the plain English style I discuss and emphasize in this chapter and anticipate requirements for modern English paragraphs that appeared in the late 19th century. Understanding the emergence of English style requires both a study of sentences and then paragraphs, the focus of Chapter 3.

Existing studies of the evolution of modern English prose have misrepresented the history of English style by focusing on literary, humanistic, and philosophical texts. The focus on literary and humanistic works, to the exclusion of nonliterary works, reveals the conflict within English studies over the legitimacy of noncanonical writing and its value in the history of discourse. For example, scholarship by Jones [1] and Croll [2] from the 1940s to the 1960s focused specifically on texts that emerged from classical humanism. Croll flatly stated that Montaigne and Bacon were

> the first writers in the vernacular languages who employ a style which renders the process of thought and portrays the picturesque actuality of life with equal effect . . . Bacon, Hall, Johnson, and Wotton in England . . . are the actual founders of modern prose style. [2, p. 184]

This focus distorts the story of our language because it eliminates business, administrative, and other forms of how-to writing entrenched in the English culture of work that existed before the Royal Society. And, as I will discuss in

Chapter 3, the lack of awareness of technical writing led Alexander Bain and other early compositionists to state that the English paragraph, as we know it today, did not exist before the 18th century. I will also show in Chapter 3 that clearly developed paragraphs existed in many printed English medical texts of the early 16th century.

The argument for the importance of examining practical and humanist writing in tracking the development of English prose style has had other supporters. As Douglas Bush commented,

> We have only to think of the vast bulk of plain writing in books of travel, history, biography, politics, economics, science, education, religion, and most popular literature. Plain prose was the natural medium for most kinds of utilitarian writing, and most writing was utilitarian. [3, p. 192]

Elton, in assessing the value of English historical sources from 1200 to 1640, argued that nonliterary texts remain the best source of history because of their writers' intent to portray rather than embellish the world as they saw it. "The history of the people of England, high and low though more the relatively high, is deposited in the materials arising from the efforts of her kings to finance their governments" [4, p. 53]. The style of these documents is clearly significant because we can assume that writers of administrative and legal documents recorded reality, or facts, as accurately as they could. Unaware of the world of deconstruction, they believed that the world exists and that it could be described in ways that could be understood; thus, writing that attempts to record fact is as important in the study of style as humanistic works. Fisher [5], who, along with Richardson [6, 7], gave us the seminal work on the influence of Chancery English, went beyond both Clanchy and Bush by stating bluntly that nonliterary texts are critical to the development of English prose style:

> Historians of the English language are agreed that the genesis of the standard language is not literary, even though our predilections as literary scholars lead us to study most closely and to take examples largely from belletristic materials. . . . Except for the small quantity of writing related to literature and learning, virtually all writing before the seventeenth century related to business. [5, pp. 60–61]

Albert C. Baugh summarized the situation in this way:

> Ciceronism substituted slavish imitation for what had been a natural and spontaneous form of expression. Not only was the vocabulary of Cicero inadequate for the conveyance of modern ideas, but when Wycliff refused to carry on his argument with the Church in any language but English, the power of Latin, as emanating from the church, was also lost. [8, p. 247]

In short, nonliterary writing from the year 900 onward existed in greater quantity and was more pervasive and influential in English society than were classical and literary writing, a point recently emphasized by Richardson [7, pp. 25–26]. But in considering nonliterary texts, we have to broaden the canons of style to include features ignored by studies of literary writing, namely, the confluence of sound, format, syntax, and diction, as these are dictated by purpose. I suggest that five factors shaped modern English: (1) brevity induced from accounting/administrative format; (2) aural/oral based text, written to be heard as well as seen, that produced conversational style; (3) persistence of the indigenous subject-verb-object syntax found in the earliest English documents, religious and secular; (4) the growing Renaissance book market of literate middle-class readers who responded to speech-based prose; and (5) English scriptural renditions surfacing during the late Renaissance that associated colloquial speech with Protestantism. These factors suggest that studies of the shaping of English style should include attention to visual and aural elements that underpin diction, syntax, and discourse purpose. These factors therefore engendered five canons of style through which English evolved and must be studied: format, sound, diction, syntax, and purpose. These factors, discussed here, will show the evolution of modern English.

THE EVOLUTION OF ENGLISH PLAIN STYLE: A CONFLUENCE OF SOUND, FORMAT, DICTION, AND SYNTAX

Nonliterary texts also provide additional insights into the relationship between speech and writing and the shift from text produced from sound (text heard rather than seen) to text prepared to be read silently rather than heard, and the importance of the visual aspects of text as these direct meaning. The conversational quality of modern English style likely evolved from the need for sound-based text that could be understood by those unable to read and who relied on hearing the text read aloud. But as text supplanted speech as a means of communication and documentation, text had to be retained in visual memory, as opposed to sound retained in acoustic memory [9, pp. 114–115]. Because the form of language followed its function, the design of text has been critical since the earliest English texts that attempted to use space, format, and words to capture the sound and vision of what was said. This perspective means that format also persevered in written English and evolved from orally based design to visually based design. Today we know that text that ignores auditory quality remains difficult to understand and remember. An examination of several early nonliterary texts illustrates the importance of the visual to the development of English, a characteristic that would continue in works by the Royal Society and other late 17th century technical writers.

THE PRACTICAL AND VISUAL ROOTS OF EARLY ENGLISH TEXT AND PROSE STYLE

The earliest English prose was financial, not religious. Noncanonical texts, such as financial and estate-management records, were among the first English texts. As Clanchy argued, "Lay literacy grew out of bureaucracy and the need to document financial transactions" [10, p. 1]. E. M. Miller directly attributed the eradication of illiteracy to capitalism: Most people learned to read for utilitarian reasons, and they read what they believed was beneficial to their success [11, pp. 32, 41].

By the 12th century, writing as documentation for financial and administrative transactions, occurred regularly in England [10, Ch. 2]. Beadles, stewards, and bailiffs prepared lists of property and property owners. Figure 2-1, from a mid-13th century estate book of Richard Hotot in Northamptonshire and Beffordshire, shows "rents and farms from Turvey at Michaelmas" [9, Plate XV]. Beneath the title appear two columns of tenants' names and their annual rents. The symbols preceding many of the names correlate with the footnotes, appearing in small cursive script, which specify the size of each tenant's holding. This document may exemplify one of the first uses of footnotes in a business text.

Increases in commerce, government, and estate administration meant that in the period between the years 1000 and 1300, England became remarkably document-minded [10, p. 53]. As Chapter 5 will note, the first texts were administrative lists. Thus, we should not be surprised to find that common English documents of the 1000–1300 period included identification letters; charters (a public letter issued by a donor recording a title to property); chirographs (recorded agreements between two parties) [10, p. 87]; certificates, private letters, writs, reports of various types, legal memorandums; depositions, warrants, financial accounts of receipts (of money, crops, livestock) and expenditures; and surveys (of property, land, possessions) and lists of who owned what. The documents, along with chronicles and yearbooks, were much more prevalent than were learned works and literature because of the need to track people, property, obligations, and possessions.

Although clerks did much of the writing, they prepared documents to be read aloud. The illiterate could thus hear and participate in the business process [10, p. 53]. These resulting texts, which incorporated writing that could be heard as well as read, demonstrated the conversational nature of much early practical writing. Text, divorced from its oral/aural form, did not shift until the late 16th century. By the close of the Renaissance, much practical English had lost its oral (conversational) residue and was replaced by an unadorned functional style designed to be read, understood, and used in private reading separate from spoken intervention. This shift testifies to the increase in literacy during the 17th century.

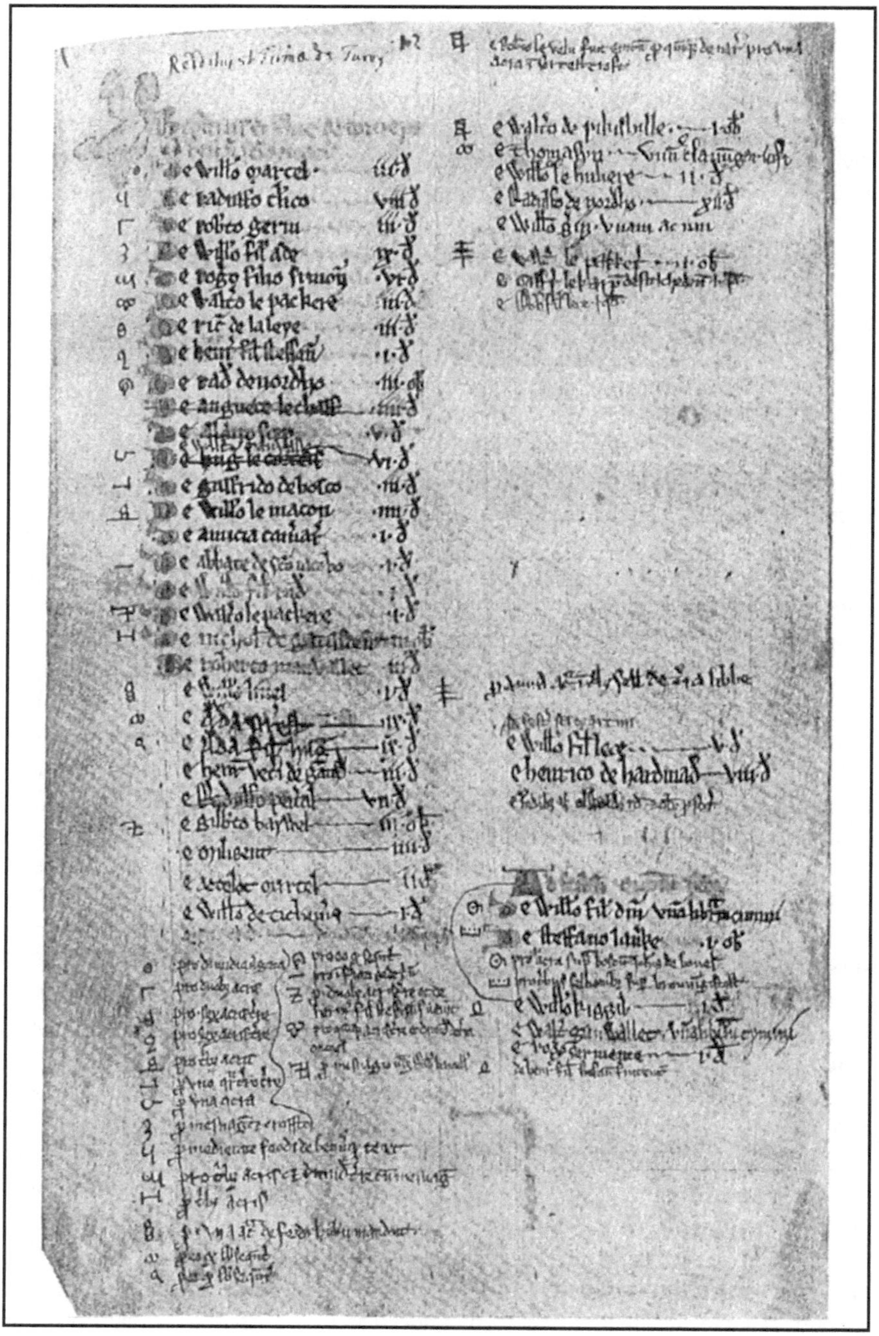

Figure 2-1. Estate Book of Richard Hotot (ca. 1240).

Format and Visual Display in Early Business Documents

Because writing in early business documents incorporated and attempted to capture activities seen and heard, such as the checkered cloth, used to track taxes paid[1] (shown in Figure 2-2), format became a means of preserving the visual and the oral in text. Perhaps the most important documents in the history of text and style were the Exchequer accounts. Founded at the beginning of the 12th century, the Exchequer (the treasury), England's earliest government department, existed until 1833. Its name came from the great checkered cloth, an abacus on which the sums of money owed in the king from the various sources were computed using counting blocks. The accounts were audited (spoken) in the Upper Exchequer by the chancellor and barons of the Exchequer. The sheriffs' accounts were then written on the Pipe Rolls according to a specific format, illustrated in Figure 2-2, that continued until 1833 [12, p. 9].

This visual aspect of accounting—documentation through arrangement of objects—was indigenous to the Exchequer system. The specific, sustained format lent credibility to the documents, as these attempted to capture the arrangement. The Dialogue of the Exchequer (Dialogus) written by FitzNeal, the treasurer of England, noted that format was an important as content. Fitzneal's instructions showed that specific, concise language was critical to format:

> It is the duty of the scribe who sits nearest the treasurer to prepare the rolls For Writing . . . the Rolls are then ruled from the top almost to the bottom and on both sides, the lines being drawn at a suitable distance from each other, and at the top of the roll are marked the countries bailiwicks whose accounts are listed below. A medium-sized space about the width of three or four fingers is left, and in the middle of the line is inscribed the name of the county which is to be dealt with first. Then, at the head of the following line is set down the name of the sheriff, followed by this form of words, "such-and-such a sheriff renders on in the same line" is written, "in the treasury," and nothing more is added until the account has been completed. . . . Next, at the head of the following line, is set down how much of the "farm" of the county has been expended in alms and fixed tithes and also how much in salaries. After these, at the head of the lower line, are noted under the demesne-lands those the munificence of kings has bestowed on churches or on those who performed military service for them. [13, pp. 543–544]

The combination of text and format, first used in account and administrative texts, was destined to produce a visual relationship that would continue in the development of modern English prose for the next 500 years in other kinds of

[1] Final settlement . . .

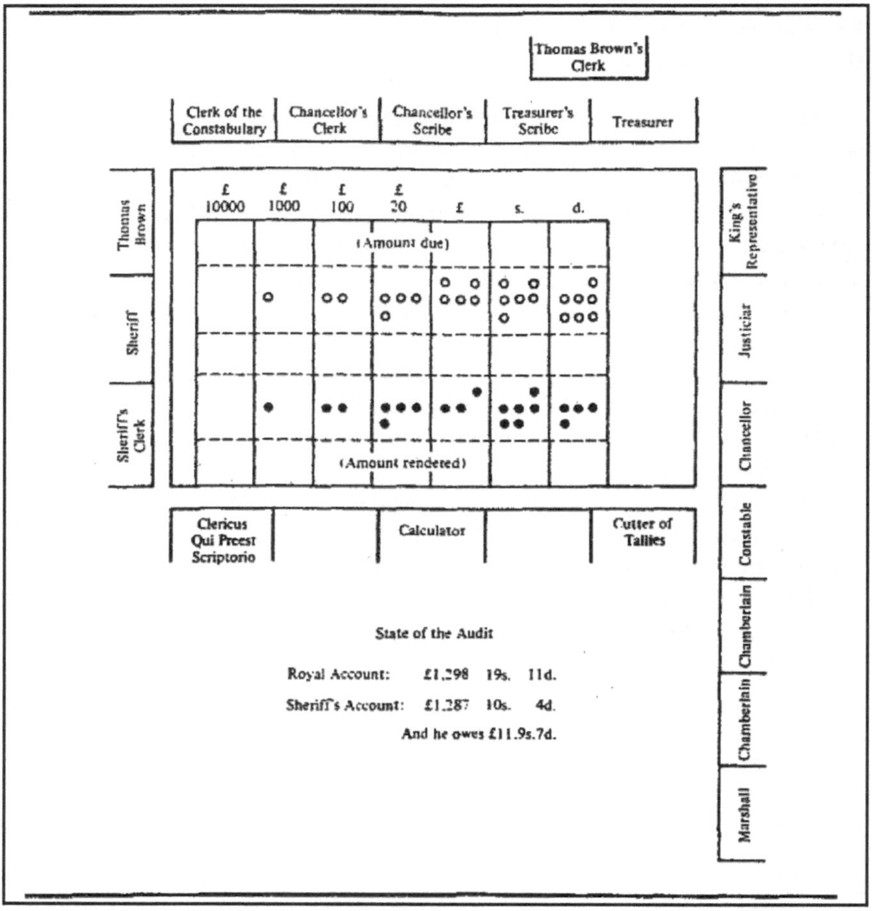

Figure 2-2. The Exchequer Table.
Michael Chatfield, *A History of Accounting Thought,* Krieger, Huntington, New York, 1977.

writing. An early English account (1500–1501) illustrates the persistence of the listing arrangement and the brevity of account descriptions. Note the parallel presentation of *items*:

> The account of John Paty, Thomas Shabetor, John Cliff and Thomas Ascot, Keepers Of the goods of the store of the parishioners of the parish church of Ashburton, for one Complete year from the Feastday of John at the Latin Gate in the year of our Lord 1500 to the same Feastday in the year of our Lord 1501.

Item Willielmo Erle for making the account	ij s iiij d
Item for mending the boat-shaped incense holder	xiij d
Item for Rushes	iij d
Item for tightening Bells on their stocks	vj d
Item for molding wax at various times of the year	iij s j d
Item in payment for various men that caste the Steeple and for their expenses when they made the Bargain	iij li viij d
Item in Payment for rock work	ij s iij d
Item in Payment for lime Sacks	
Item for mending of a bell wheele & for carriage of Tymber and For wood	ij s ij d
Item in expences at Exeter and London against the people of Bickington this year	iij li viij s
Item for lime to the steeple when it was rough-cast	xvj s viij d
Item for Buckets for a lime sieve for roofing stones & for plastering of the church house and for wood strips, nails, Y wooden pegs	v s iiij d ob
Item for Tin for the organ	xvj d
Item for wax	xx viij s iij d
Item I paide to the plummer	v s j d
Item for timber nails making bell ropes and for grease to the bells	iij s

[14, p. 46]

While format became intrinsic to accounting and administrative texts, monastic writers also used format—two-or-three-column layout and glosses—to organize text. Monks often combined glosses and Scripture by using different sized fonts to separate Scripture from commentary. They also used ornate, factotum initials to announce major divisions within Scripture, a technique captured in printed works on all subjects. However, use of multiple columns in financial writing was from its beginning more visually oriented.

Accounting's use of visual presentation was a natural extension of the shift from graphic communication—visual techniques as vestiges of preliterate pictorialism [10, pp. 226–229]. Drawings used as communication evolved to words used as icons, then to spatial arrangement of words and tables used to capture meaning. Organized textual and numerical presentation became critical to the acceptance of text as a substitute for the "truth" of objects and auditory statements. Seals attached to official letters, agreements, and charters enforced their "official" status. Acceptance did not come without difficulty, as words (for well over a half century) were deemed less reliable (accurate) than sticks, objects, seals, and the checkered cloth, all of which could be seen and verified while representing status and numerical reality. Growth of commerce fueled this shift from using objects and auditory statements to communicate financial and administrative information. As Prestwich remarked, from the 13th century forward,

> There is an ever increasing amount of material [written documents] from the royal Household, from the chancery and the law courts, as well as further series of Records from the exchequer, in an astonishing profusion that cannot be matched by any other European country. [15, p. 95]

Visual text, neatly penned and well organized, became critical to the acceptance of writing and documentation. Scribes recorded many of these documents on large parchment. Wallingford's roll of tradesmen, dated 1220–1230, lists people by trades, recording how much each person owed for the year to practice his trade in the town. Figure 2-3 shows an excerpt from one of these rolls that lists, under headings in the left-hand margin, the glovers (Wa[n]ti[ere]s) and mercers (Mercat[iari]i), or general traders. Other portions of this roll (not shown here) list butchers, carpenters, hosiers, smiths, weavers, and other trades. The sum due, followed by the details, is entered after each person's name. The rapid shift toward formality of the roll over several years suggests that the Wallingford clerks considered rolls as permanent records rather than casual memoranda supplementing receipts on wooden tallies. A tradesman who saw his name engrossed, as if in a liturgical book, and his payment accumulating from left to right on the parchment in a scrupulously ordered format, was likely impressed by the dignity of the written record [10, Plate X]. Financial scribes quickly learned that well-organized and formatted texts improved the acceptance of written records.

Textualized, rather than auditory, accounts thus nurtured literacy, which became increasingly necessary with expansion and complexity of financial operations too complex for hearing only. But the shift demanded that text produce clarity and a single meaning for writers, auditors, and readers. The creation of text that could be trusted—whether it was heard or read—formed the basis of credibility, because text carried one meaning.

The Perseverance of Plain Words

Because English was first a tool of business, government, and daily existence for recordkeeping, Old English kept words for fundamental concepts—*man, wife, child, house, bench, meat, grass, leaf, bird, good, high, strong, drink, sleep*, and *fight* [8, p. 63]. Words like *tax, subsidy, revenue, tally, exchequer*, along with the names of office titles (*chancellor, treasurer, warden*) and words relating to the economic organization of society (*manor, demesne, bailiff*, and *bondman*) were adopted by the emerging English language [8, pp. 202–203]. Borrowings from the first influence of Christianity (e.g., *abbot, angel, candle, hymn, mass, pall*) were as concrete as terms absorbed from the Scandinavian incursions (e.g., *cap, sock, beet, lentil, pear, cook*) [8, p. 99]. Baugh explained that

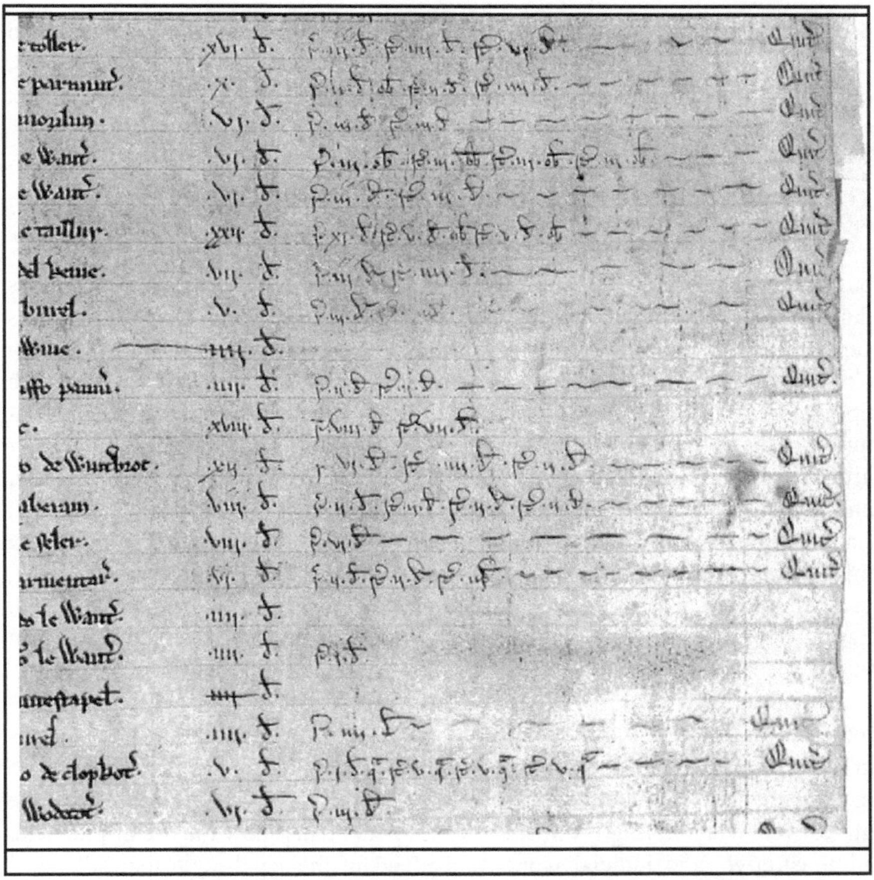

Figure 2-3. Wallingsford's Second roll of tradesman (1129–1230), Plate X, Clanchy.

> words are generally taken over by one language from another in answer to a definite need. They are adopted because they express ideas that are new or because they are so intimately associated with an object or a concept that acceptance of the thing involves acceptance also of the word. [8, pp. 98–99]

Danish words adopted by the English language include similarly specific words (e.g., *bull*, *dirt*, *egg*, *leg*, *root*, *scab*) in addition to a host of adjectives and verbs (e.g., *bait*, *glitter*, *nag*, *ransack*, *screech*, *sprint*) [8, p. 117]. The majority of words absorbed into English from other languages were concrete, operational words needed to describe common activities and objects used by people in their daily lives. For example, English absorbed many common French words for clothing, food, combat, law, and church [8, pp. 202–215]. The English that became the language of the common people, the "mother tongue," developed from an assortment of functional words drawn from many languages [8, pp. 123–124]. These words were then parsed in the basic subject-verb-object pattern and its variants used in many practical documents.

The Perseverance of English Over French and Latin

Following the Norman Conquest in 1066, French made a number of contributions to shaping English, but English still remained the language of the masses during the 12th century. What occurred during the 11th through the 15th centuries can be summarized as follows: During the century and a half after the Norman Conquest, French had been not only natural but rather necessary to the English upper class. While French and English coexisted during the 11th and 12th centuries, in the 13th and 14th centuries, maintenance of French became increasingly artificial. In the late 14th century, English won its way back into universal use; and in the 15th century, French all but disappeared [6, pp. 150–151]. Although Latin remained the province of the learned, its isolation from the emerging middle classes became its doom as a language of influence. For example, in 1362, law courts were ordered to use English for their proceedings; in 1363 and 1365, Parliament was opened in English; and in 1399, Henry IV accepted the crown with a speech in English.

A southwestern deed of 1376 remains the oldest surviving private legal instrument in English. The oldest English petition to the Crown is that of the mercers guild of London in 1313, the oldest English wills in the London court of probate date from 1387; and returns of the ordinances, usages, and holdings of the guilds were made in English in 1389. These documents show the use of practical, administrative, and judicial words embedded in the legal style that still used subject-verb-object sentence patterns. As Richardson noted [6], plain style can be seen in a range of texts, such as in the Plumpton, Paston, Celey, and Stoner letters and in Chancery warrants. For example, a Warrant for the Chancery, dated 1454, reads as follows:

> Memorandum quod die et anno subswcriptis hec billa liberate fuit domino liberate fuit domino Cancellario Anglie apud Westmonasterium exequeda
>
> The vij day of Novembre the yere etcetera xxxiij. At Westminster in the Sterret Chambre. The Reuerent Fader in god the Bisshope of Ely be- soughte my lordes to hauc him discharged / of thoffice of the kinges procuratour in the Court of Rome insomoche as he may not from henceforth entende therto. And forthwithe my lm des had communicacion to // ordeyne a newe and considering the honour wisdome and experience that Maister Vincent Clement nowe the Popes Collectour in this Royaumee is ordained aduised // and appointed that the sade Maister Vincent should be entreated to take upon him the saide occupacion. He to haue suche

> power of the King as other the Kinges / procuratours have had in tyme passed with wages and fees therto accustomed. [16, Vol. 1, p. 230]

The following business letter, written by Wylliam Cely in 1488, illustrates the tight workaday character of English:

> Ryght worschyppfulle Syr
> and myne esspccyalle good mastyr
>
> With all dew / recommendacion had I vmbly Recommend mee vnto yowre good mastyrschypp / Fyrdyr please hytt yowre good mas- tyrschypp to vnderstond that I haue / Receyued by Thomas Graungers seruante yowre letter datyed at londone the 15th day of merche the whyche I hawe well vnderstoud. And . . . haue not / as yet spookene with Johne Hubard theroff but withyn this 15 days I wylle / goo to guynesse and speke with hymme therffor . . . I hawe nott yet seync hytt but I spake with stryke / hys wyffe here at merkett. And sche says her husband hathe / a quytaunce of sybsone the whyche I schallc *see* whanne I Come theder. [16, Vol. 1, pp. 244–245]

The market for books accessible to middle-class readers provides a critical piece of the character of English as it developed [17]. Booksellers knew that sales of English books far surpassed sales of Latin books. The flood of printed works, an increasing number in English after 1600 and then again after 1640, fueled the rise of literacy among the middle classes [18]. Literacy, in turn, fueled the demand for books. Thus, the basic character of English—its plainness, concreteness, and utility—and its growing demand by an emerging literate class ran deep. The continued rejection of the aureate that was also obscure can be seen as underpinning the rejection of "Inkhorn Terms" (i.e., arcane, opaque Latinate words deplored by English people such as Ascham and Mulcaster, champions of an English free of what today would be called academese). Diocese documents, which recorded events such as childbirths, marriages, divorces, rapes, and ratifications similarly illustrate plain, unadorned English [19].

The Perseverance of the Subject-Verb-Object Pattern

Separating discussion of English diction from English syntax is difficult because the two are tightly related. Modern prose style, to use Mueller's definition, on one hand, must be "a serviceable instrument of its writer's thought and, on the other, must show a vital connection with spoken English, to the extent of sustaining an idiomatic if not always colloquial or conversational mode of discourse" [20, p. 1]. Thus, in examining text, we must look at both words and sentence structure. The structure of the English sentence is critical, and English prose is built on "a single, intact clause that can be isolated within a larger sentence unit" [20, p. 34]. "English syntax, then, works as follows: noun phrase

+ verb + noun phrase (+ modifier), the basic subject-verb-object pattern. Text that veers away from this form becomes difficult to understand because syntax serves—and perhaps also projects—the functional capacities and limits of our minds" [20, p. 34]. As linguists have also discovered, even Old English clauses show movement toward subject-verb-object, verb-object, or verb-subject. Plain style and plain diction seem to enforce each other. This pattern emerged by the 9th century. For example, as Chambers (1950) noted, Alfredic English set the style of English [21, p. lix], as is illustrated in *The Complete Works of Alfred* (ca. 871–899) by Giles. The following excerpt comes from the 32nd chapter of King Alfred's translation of Boethius:

> I. When Wisdom had sung this lay, then began he again to speak, and thus said: Therefore there is no doubt that this present wealth obstructs and hinders those men who are intent upon tru felicities: and it can bring no one where it promised him, that is, to the highest good. But I can in a few words declare to these with how many evils these riches are filled. What meanest thou, then, by covetousness of money? When thou no how else canst acquite it, unless thou steal it, or take it by force, or find it hid; and wheresoever it increases to thee, it decreases to others. Thou wouldest, then, be illustrious in dignity? But if thou wilt have this, then must thou very meanly and very humbly flatter him who is able to help thee thereto. If thou wilt make thyself greater and more honourable than many, then must thou suffer thyself to be inferior to one. Is not this, then, somewhat of misery, that a man so anxiously cringe to him who has the power of giving to him? Of power art thou desirous? [22, p. 476]

This style-subject-verb-object syntax and substantially mono and dysyllabic diction prevailed in Exchequer and estate documents and in early Biblical translations, such as this excerpt from the parable of the prodigal son taken from the Late West Saxon Old English of 1050:

> 11 He quoth [spoke] smoothly [truly] some man had two sons. 12 then quoth the younger to his father. father sell [give] me my deal [part] of my property. That me to belongs, then dealed [gave] he him his property. 13 then after few days all his things gathered the younger son & fared abroad in far-off country. & there spilled [wasted] his property living in his luxury. [23, p. 3]

The earliest English prose Psalter, compiled from manuscripts dated approximately 1226, shows the same sustained directness:

PSALM 1.

1. Blesced be the man, that hath not go in the counseil of wicked, ne stode nought in the waid of sinners, ne sit not in false judgement.

2. And his wylie was in the wylle of our Lord, and he schal btand in hiw lawe both daye and nyght.
3. And he schal be as the tre, that he sett by the ernynges of aters; that schal give his fruit in his tyme. [24, p. 1]

Little difference exists between the prose of this psalter and that used in Renaissance cookbooks published in the late 1500s. The style suggests that the cook dictated the recipe to the writer. The tone exemplifies oral instruction carefully recorded as text:

> Gyngerbrede.-Take a quart of honey, & seethe it, & skemi it clene; Take Safroun, pouder Pepit& throw ther-on; take gratyked Brede, & make it so chargeaunt [stiff] that it wol be y-lechyd; then take poudcr Canclle, & straw ther-on y-now; then make it square, lyke as thou wolt leche it; take when you lechyst hyt, an caste Box !eve a-bouyn, y-stykyd ther-on, on cloys. And if thou solt haut it Red, colour it with Saunderys ye-n ow. [25, p. 35]

Tyndale, in his Scriptural rendition from the early 16th century, was continuing this homely and colloquial sense of the Word as speech: "Seke the kingdome of heaven therefore and the righteousnesse of the same, and be sure thou shalt ever have sufficient, and these thynges shalbe ministered unto thee" [as quoted in 16, p. 184]. In contrast, also in the early 16th century, Erasmus, whose writing remains an icon of grandiloquence, wrote this same verse as follows:

> Let your chefe care be aboute that good thing, in comparison of the whiche, these worldly thynges be of no value nor reputacion. The kingdom of god must be set up, that is to say, the doctrine of the gospel, by the which we attayne unto the heavenly inheritaunce. [as quoted in 20, p. 184]

Wycliff, in the early 15th century, exemplifies less conciseness than Tyndale in his version of Scripture: he used nearly a thousand Latin words in his Biblical translation. But he so effectively selected and integrated them into plain English syntax that they were easily absorbed into the English language. Perhaps the persistence of subject-verb-object syntax led to borrowings that worked with this entrenched grammar, a theory that may explain why "golden prose," with its aureate diction, made little impact on the character of modern English as it existed by the latter decades of the 17th century. As Baugh stated, "While it was an English changed in many important particulars from the language of King Alfred, its predominant features were those inherited from the Teutonic tribes. . . . Great and basic elements of the vocabulary were still English" [8, pp. 221–223]. In many ways, the utilitarian style of the 11th and 12th centuries and then Protestant Scripture provided the foundation of clarity and precision required by the Royal Society, intent on meshing words with objects.

English continued to grow with the infusion of words from European cultures, but by the late 14th century, the principal changes were not in syntax but in diction. These same patterns continued in Chancery, Exchequer, and routine business documents. For example, the following excerpts from a guild ordinance [26] contains the sustained subject-verb-object pattern with variants and the use of thought division (paragraphing) to separate topics:

> There be the poyntes and the articles ordeilgned of the brotherheed of Saint ffabian and Sebastian in the Cite of Londone, the which is founden in the chirche of seint Botulf with-out Aldrichesgate.
>
> Johannes Dancastre, lymenor
> Magistri ffraternitatis
> Ricard us Spaigne, pelliparius
>
> The furst point is this; that whan a brother or a suster schal be received, they schul be swore vpon a book to the brotherhede, for to holde vp and meyntene the poyntes and the articles that be write after folwynge, eche to his power, sauynge his estat; [and] that euerich brother and suster, in tokenyng of loue and charite and pees, ate resceyuynge schul kiss other of them that ben there.
>
> . . . .
>
> Also, what type that a brother is ded, or a suster, that they come and offer with hem, when they be warned to come, vpon the foreseid peyne, but if he haue a verrey request to be excused.
>
> Also, if any brother dye that hath nought of his owene to be beried with, if it mowe be so a ta ken, thane that he be buried of the commune box. [26, pp. 9–10]

In an early chancery proceeding (1436–1450), English syntax surfaced despite remnants of Latin legal style:

> Vn to the ryghte worschupfulle and reuerent fadyr / in godde the Bysshoppe of Bathe Chaunceller of Englonde /
>
> Besekethe meekly William Symthe of Hanam where that William Tournoure and Alice his wife in the morowe of allesowlynne in the yere of the regne of oure soueraygne lorde the kynge that now is xiiij *//* areryde a fynd at Westminster by four William Babyngtone at that tyme being Chefe Justice of the Comynne bynche of oure Soueraynge lorde and other of his felowys Justices vn to the sayde beseker Johan *//* Tournoure ande Robert Tournoure of the maner of Brokenboroughe wythe the appurtenauncez in the Shire of Gloucestre. [16, Vol. 2, pp. 224–225]

The entrenched plain character of English, having dropped its French syntax, can be seen in estate management works such as Walter of Henley's *Husbandry* and its related versions *Sencschaucie* and *Grossectesec's Rule,* originally written in the 12th and 13th centuries. These works will be discussed in Chapter 5.

The works tell us much about the evolution of style because the authors wrote for one purpose: to describe how the ordinary business of an estate should be conducted. The writers were not following rules of rhetoric. In their effort at clear description, they represent a genuine statement of English as it emerged through translation from Norman French, the language in which the husbandries were initially written. In *Husbandry,* Walter of Henley surveyed each of the departments of a rural economy and discussed plowing, harrowing, and other operations. He suggested ways the lord of the estate could avoid the leakage that occurs so easily without careful supervision. The work was rightly titled *Husbandry,* not because it has to do with tillage but because it shows how the lord may husband his resources and manage them effectively [27, p. xviii], as the following excerpt illustrates:

View of Account.

> Have an inspection of account, or cause it to be made by some one in whom you trust, once a year, and final account at the end of the year. View of account was made to know the state of things as well as the issues, receipts, sales, purchases, and other expenses, and for raising money. If there is any let it be raised and taken from the hands of the servants. For often it happens that servants and provosts by themselves or by others make merchandise with their lord's money to their own profit and not to the profit of their lord, and that is not lawful. And if arrears appear in the final account let them be speedily raised, and if they name certain persons who owe arrears, take the names, for often it happens that servants and provosts are debtors themselves, and make others debtors whom they can and ought not, and this they do to conceal their disloyalty. [27, pp. 33–35]

Caxton also became a benchmark in the development of English. He preferred aureate prose in many of his introductions to attempt to infuse English with the prestige of French, but he lapsed into his own voice when giving his own exposition [28, pp. 170–175]. Thus, writers often shifted styles—from aureate to plain—because plain English was the *prose* of common usage while *aureate* prose was consciously used for situations in which stylistic gilding was deemed appropriate [28, p. 224]. But even gilded writing became less prevalent than plain style. Words such as *abusion*, *equipotent*, and *tenebrous* were incompatible with the concrete nature of French borrowings such as *bar*, *beak*, *fool*, *fury*, *horror*, *river*, *rock*, *cruel*, *fierce*, *rude*, *safe*, and *tender* [28, p. 225]. Again, despite numerous French influences, English adhered to its tradition of concrete diction and tight subject-verb-object syntax. The Inkhorn Terms likely foundered when writers attempted to use them in the basic English syntactical arrangements. As I have discussed, changes in style showed that writers were sensitive to their readers' educational and social levels [29]. Because English was used by members of both high and low classes, the level of plainness often indicates the educational level of the intended readers.

Ramism and the Rise of Utilitarian Prose

Since 1991, I have argued that the Ramist method, with its emphasis on spatial arrangement to achieve clarity, was critical to the evolution of modern technical writing. Formatting strategies from incunabula as well as accounting documents developed further with the rise of printing [30]. The visual character of English prose, entrenched by way of accounts and administrative documents, may have contributed to the rapid acceptance of Ramist logic and spatial display of content evolving from Agricola. For example, Exchequer records from 1556 show the influence of Ramist brackets as a means of recording and tracking pensions before ruled columns took the place of brackets, but the first English translation of Ramus's *Diaeccticac* (1574) was not published until 1584 in English. The use of brackets for organization is not surprising given the Exchequer's reputation, its commitment to format and organization, and Ramus's influence on the Continent and in learned circles in England.

Also, in *The Castle of Health,* Elyot (1534/1535) [31], used brackets for organizing material, which may suggest his awareness of Ramus' work well before Ramus' influence in England became significant, after 1580. *The Castle of Health* is one of the first printed English works showing the use of Ramist spatial display. Content was partitioned with brackets. Each partition was then further partitioned with brackets indicating the grouping of each partition. Partitioning continued until the parts became indivisible. Medical books used the Ramist method extensively to help students see and understand the relationships between types of wounds, for example, a strategy that can be seen in Figure 2-4.

Because of the confluence of text and visual display in Ramist-inspired works, texts that emphasized stringent organization and spatial arrangement of content contributed to the shift from oral-based text to efficiently displayed reader-based text that appealed to the eye (visual memory) rather than the ear (acoustic memory). As I have also shown, in utilitarian text, undifferentiated prose was the exception rather than the norm, particularly by the closing years of the 16th century [30]. Writers seem to have perceived that organized, visually appealing content helped readers—often middle-class readers who were the target of many technical (utilitarian) books—to comprehend and remember what they had read. As "reading to learn to do" grew in popularity, books became organized so that the relationship between the parts could be opened, or displayed. Title pages, tables of contents, and indexes became common in works of substantial length. Graphics improved in quality and value in supporting the meaning of content. Format and graphics were, therefore, inextricable from English style and the development of typographical text.

Text itself also continued to change in other ways. In oral societies, poetic embodiment of knowledge enabled the nonliterate to hear, to internalize, and to

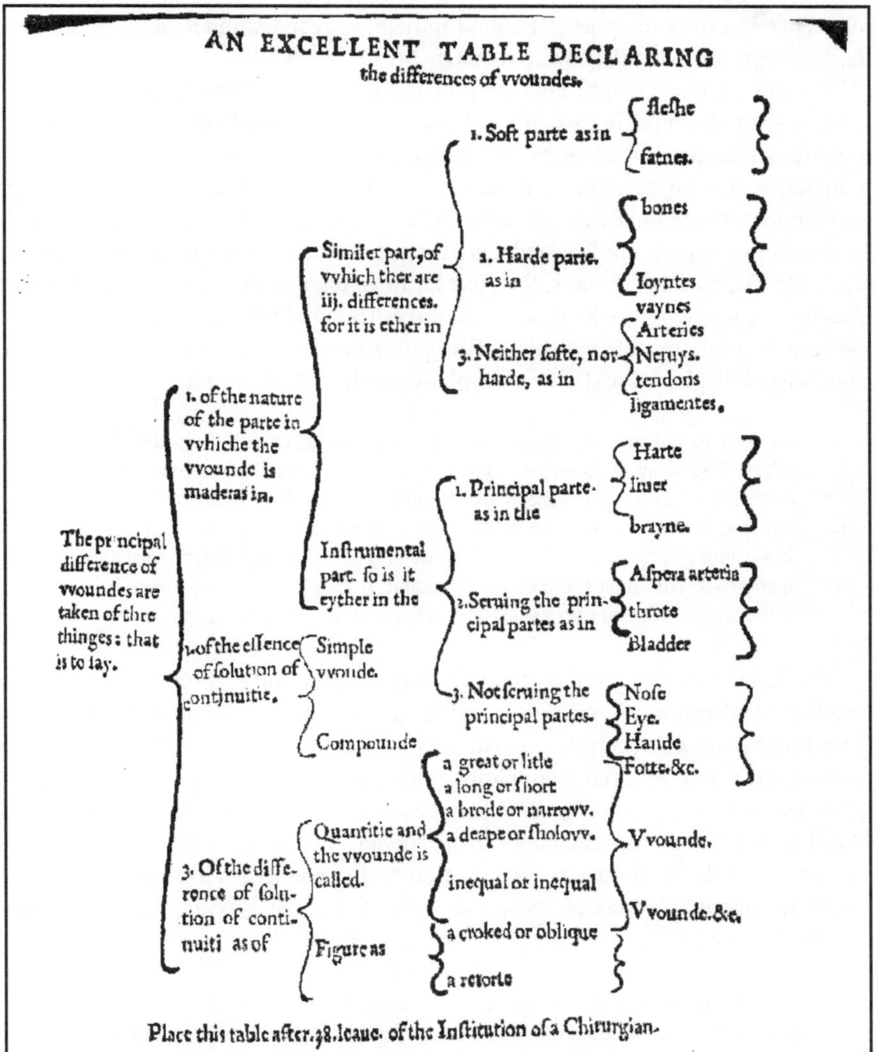

Figure 2-4. Ramist table for summarizing kinds of wounds.
Gale, Certaine Workes of Chirurgeria (1563).

remember based on sound. In a text-based society, writing that could be retained in visual memory took the place of oral-based writing. As technology and knowledge advanced, poetry and other oral mnemonic devices became less able to serve as tools for transmitting information. In the emerging textual society that was predominantly utilitarian, visual display supported understanding of concepts and took the place of sound in helping readers understand, remember, and perhaps use what they were reading.

But did Ramist thought also support the evolution of the English sentence? The answer, I think, is yes. Ramist logic, with its emphasis on division and classification, developed texts by laying things out in a descending order of generality, moving from the general to the concrete and using extended headings as signposts to announce the division and classification process. Disciplines such as medicine, which emphasized classification, were drawn to Ramist display. Gale (1563), an English physician, used Ramist display extensively in presenting medical concepts in books that targeted medical students, as Figure 2-4 shows. Accompanying text also displayed organization of thought that mirrored the tight visual display of text. For example, Gale described "wounds in the face":

> Wounds in the face are eyther in the fleshy partes or not, if in the fleshie partes, The wounde is either drye or moiste. If in the fleshye parte and the wounde be moist, you must of force stitch it with a fine small round needle and fine thread; but if it be in the fleshy part and by you shal in the place of a round needle use a square. If the wound be not in the fleshy partes then in any case you must leave of stitching and in the place of it Joyne the lippes of the wound together you shall doe in thys maner. [32, Bk. 2, p. 21]

The Ramist method was particularly welcome in 16th century English when academe had run amok with ornate disputation that had no practical outcome. The foundering of degenerate Aristotelianism opened the door for a new form of text, supported by print technology, which provided efficient ways to transfer knowledge between experts and from teachers to students. Ramism provided the theory that text should move from "given" to "new," a theory that would emerge not only in the macro-organization of documents but also in the development of coherent sentences and paragraphs. The great Harvard historian Perry Miller observed that

> the essence of the Ramist system was exactly this belief that logic is no more than the distinguishing of entities and the joining of them together, that the distinguishing of entities and the joining function of thinking is primarily discerning and disposing, not investigating or deducing. [33, p. 134]

Invention does not create or devise. It discovers, or lays open to view, that dichotomous or logical partitioned arguments such as these exist. Logic derives from nature and the natural mind. As Miller put it, "God has created the world by creating entities and then placing them in sequences, relations, and patterns that can be visually or spatially displayed" [33, pp. 149–150].

In short, Ramism saw writing as a source of discovery, a view that anticipated the modern view that writing is thinking. For an evolving, newly literate Christian society, Ramism provided tools for univocal presentation and anticipated the objectivist view of reality: The world exists; by using our minds, we can know it, and by language, we can code reality correctly because our minds are part of God's mind. The result was text and ideas that could be seen. The Ramist insistence on the value of clarity and organization meshed well with the concrete nature of English diction and the prevailing subject-verb-object syntax.

Although Ramist brackets lost favor in less than a century in England, complex prose did not lend itself to bracketed display. The idea of the visual display of information emerged in other, more practicable ways, particularly in writing designed to be learned, remembered, and used to perform tasks. What emerged were pages that emphasized a pronounced use of white space, better word and page division, headings, subheadings, fonts, drawings, and tables, characteristics that emerged in the writing of many of the early Royal Society Fellows.

Many texts that incorporate some Ramist display further illustrate careful use of page design to reveal content. For example, Figure 2-5, from a book on surveying by Leybourn (1656) [34], shows a decidedly modern page layout designed to make the information accessible. And Nehemiah Grew's seminal work *The Anatomy of Plants* reveals this careful organization, direct sentences, and a Ramist bracket, in Lecture VI, A Discourse of the Diversities and Causes of Tast[e]s chiefly in Plants [35, p. 294–295], which tracks the differences among tastes in plants (Figure 2-6).

BACON, PLAIN STYLE, AND THE CHARACTER OF ENGLISH CULTURE

By considering the larger range of printed texts, and not just classical or literary texts, we can see that plain style appeared in the earliest English texts and was sustained in medieval and Renaissance English documents that ran a gamut of genres. In addition to private estate accounts and the established tradition of the Exchequer documents previously described, another office supported and extended the use of plain style—the Chancery—the third powerful influence on the evolution of English because of the quantity of documents introduced. As Fisher stated,

> Chancery English grew up outside the orbit of either the church or the schools, nearly all of whose reading and writing continued to be in Latin until well into the sixteenth century. In the absence of any other national model for writing in the vernacular, and the view of the enormous prestige and ubiquitous presence of Chancery writing, it is not surprising that chancery set the fashion for business and private correspondence. [36, p. 891]

In crediting Bacon for the entrenchment of plain style based on his attacks on scholasticism, historians have ignored an important point in Bacon's background. As Fisher (1996) [36] observed, Bacon was well acquainted with the Chancery and its English, having studied law at Gray's Inn and served between 1603 and

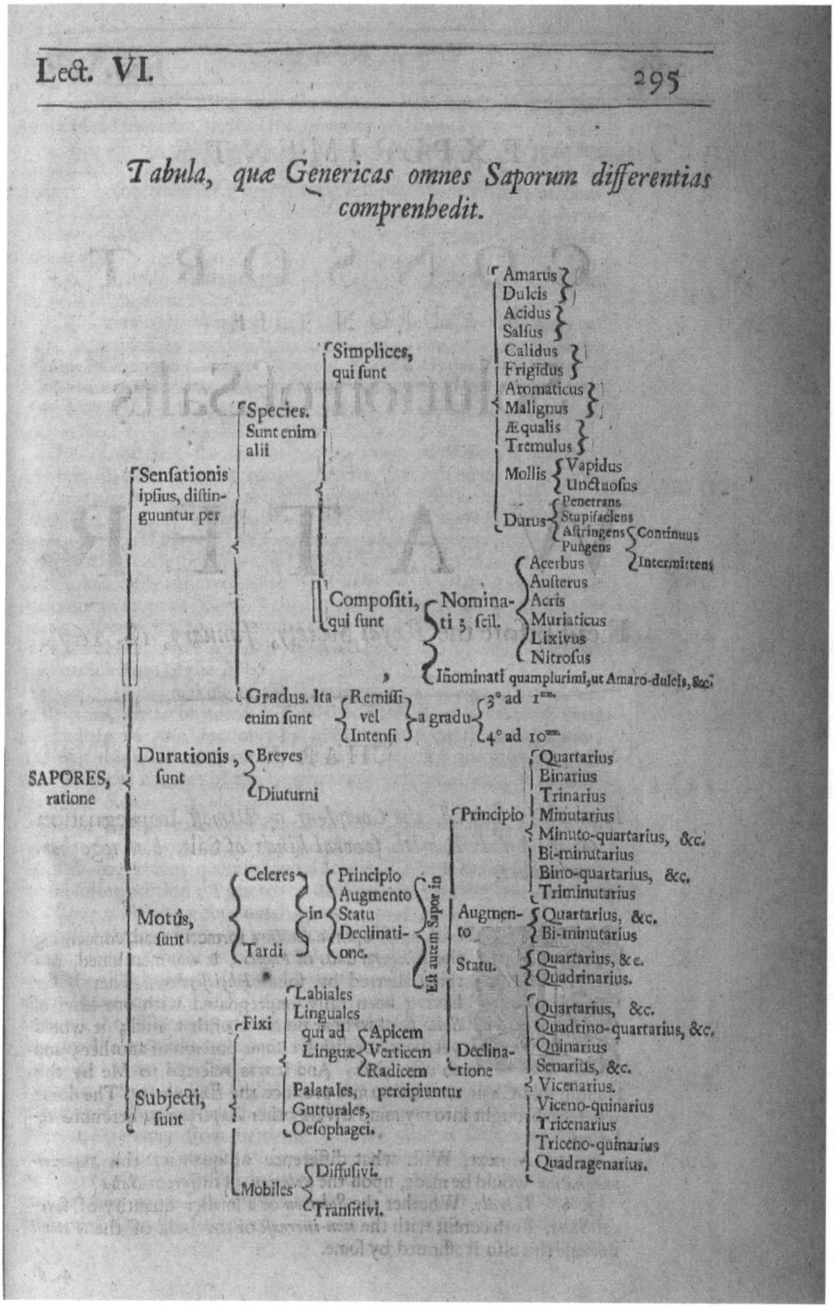

Lect. VI. 295

Tabula, quæ Genericas omnes Saporum differentias comprenhedit.

SAPORES, ratione
- Sensationis ipsius, distinguuntur per
  - Species. Sunt enim alii
    - Simplices, qui sunt
      - Amarus / Dulcis
      - Acidus / Salsus
      - Calidus / Frigidus
      - Aromaticus / Malignus
      - Æqualis / Tremulus
      - Mollis { Vapidus / Unctuosus
      - Durus { Penetrans / Stupifaciens / Astringens / Pungens { Continuus / Intermittens
    - Compositi, qui sunt
      - Nominati ; scil. Acerbus / Austerus / Acris / Muriaticus / Lixivus / Nitrosus
      - Inominati quamplurimi, ut Amaro-dulcis, &c.
  - Gradus. Ita enim sunt { Remissi vel Intensi } a gradu { 3° ad 1um / 4° ad 10um
- Durationis, sunt { Breves / Diuturni
- Motûs, sunt { Celeres / Tardi } in { Principio / Augmento / Statu / Declinatione }
  - Est autem Sapor in
    - Principio { Quartarius / Binarius / Trinarius / Minutarius / Minuto-quartarius, &c. / Bi-minutarius / Bino-quartarius, &c. / Triminutarius
    - Augmento { Quartarius, &c. / Bi-minutarius
    - Statu. { Quartarius, &c. / Quadrinarius.
    - Declinatione { Quartarius, &c. / Quadrino-quartarius, &c. / Quinarius / Senarius, &c. / Vicenarius. / Viceno-quinarius / Tricenarius / Triceno-quinarius / Quadragenarius.
- Subjecti, sunt
  - Fixi
    - Labiales
    - Linguales qui ad Linguæ { Apicem / Verticem / Radicem
    - Palatales, percipiuntur
    - Gutturales,
    - Oesophagei.
  - Mobiles { Diffusivi. / Transitivi.

Figure 2-5. Page from Leybourn's *Art of Surveying* that shows accessible and readable page design.

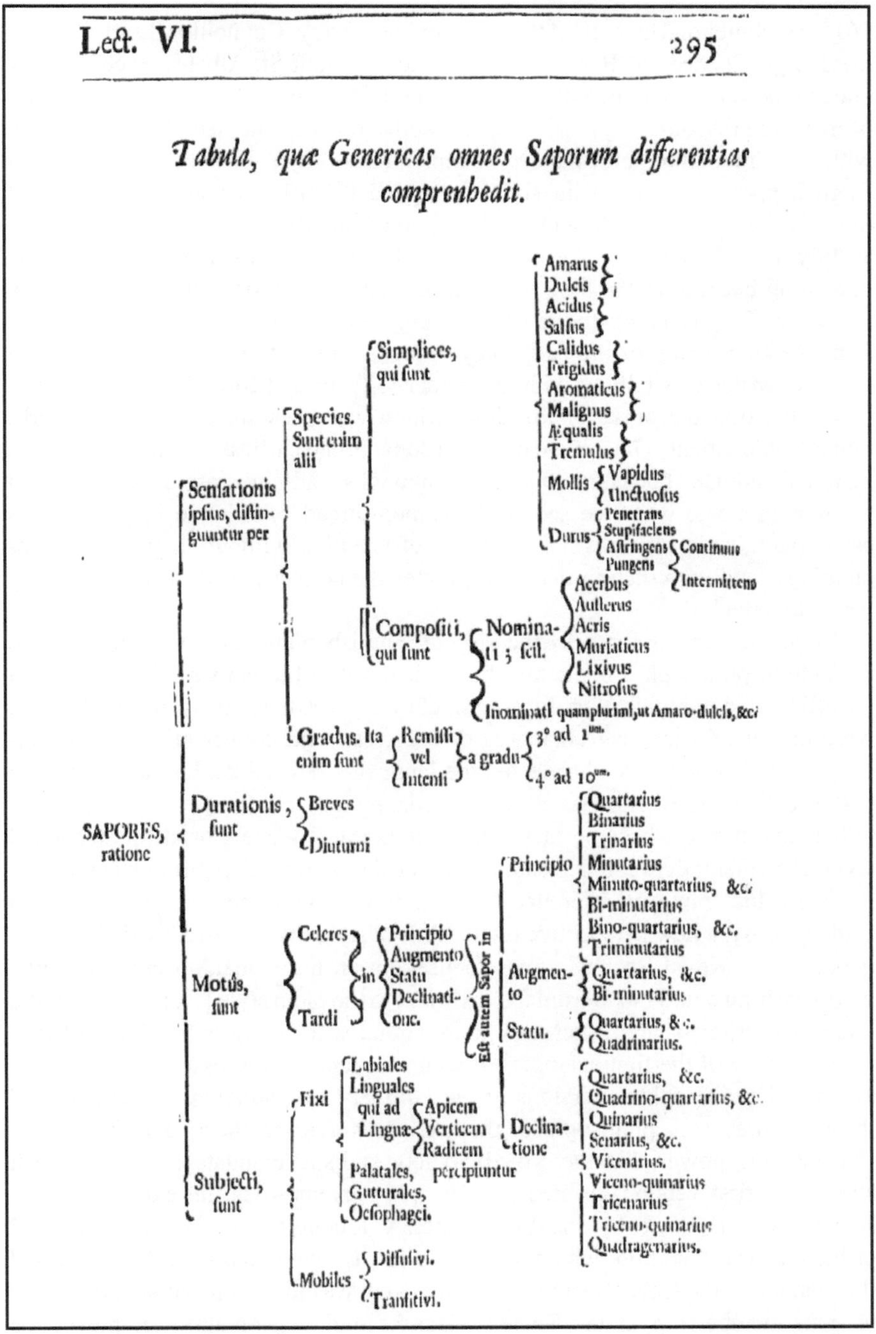

Figure 2-6. Grew's ramist classification of plants.

1616 as Solicitor-General, Attorney-General, Privy Councillor Lord Keeper, Lord High Chancellor, Baron Verulam, and Viscount St. Albans. Bacon clearly understood the importance of appropriateness in style and was himself a master of matching discourse style with context and audience. The rise of science, along with the embedding of Chancery administrative style in a large quantity of English prose, allowed plain style to expand its influence in the mainstream of utilitarian writing. As MacDonald [37] noted, the Royal Society provided only a "tidying influence" on the emergence of plain style as a preferred method of writing because of the long tradition of plain style prior to the Royal Society debates on language, which focused on things, not words.

In addition, many of the early prayer books and maternal advice books read as if the writer was talking with the reader [38]. In that form, they consistently reveal the oral discourse of the time, which was predominantly subject-verb-object/complement. Thus, if we compare some of the earliest textualized prayers, maternal advice books, published cookbooks, administrative reports, and accounting texts, what we see is a common-thread plain style in prose. This early plain style delivers its message efficiently, without augmentation or tautology, and with diction that incorporates the conversational terms shared by writer and reader.

While the central force in the rise of English plain style was not science (or natural philosophy) or literary humanism but utilitarianism, which required an efficient communication form. Middle-class readers, many newly literate, were drawn to a language that was direct, clear, and conversational: The language they learned to read reflected the language they spoke. Utilitarian writers believed that the world existed and that language could code reality. Clarity, rather than artifice, became increasingly important. Writers used this technical style whenever they wanted to capture univocal representation through language.

Given this frame of reference, we can find, from Jones' research on 17th century prose, a new perspective on the shift in style that occurred with the emergence of the Royal Society. Jones focuses on an individual, Joseph Glanville, whose writing could move from the luxuriant to the plain style [1], a characteristic that other writers shared, such a William Grant, who, along with William Petty, provide some of the first examples of technical reports (discussed in Chapter 8). Joseph Glanville's three versions of the *Vanity of Dogmatizing*, which gathered the new threads of philosophical thought that traversed the mid-17th century, illustrate the power that the Royal Society stylistic mandate had on English style. The first version, printed in 1661 [39], exemplifies the exuberant style so profusely attacked by the Royal Society founders. But in 1665, Glanvill printed a second edition, which he called *Scepsis Scientifica* [40]. He addressed this version to the Royal Society. Glanvill's goal was to use this version of *Vanity* to gain membership in the Society. This second version used simple words rather than Latinisms. Then in 1676, Glanvill published a third abbreviated version of the *Vanity of Dogmatizing*, which appeared as the first of seven essays combined to form a volume entitled *Essays on Several Important Subjects in Philosophy and Religion* [41].[2] This third version shows what the language requirements of the Royal Society had done to English prose. Several excerpts will illustrate Glanvill's ability to shift style:

First version, 1661:

> Nor is the composition of our bodies the only wonder; we are as much nonplus by the most contemptible Worm and Plant, we tread on. How is a drop of Dew organiz'd into an Insect, or a lump of Clay into animal Perfections? How is a drop of Dew organiz'd into an Insect, or a lump of Clay into animal Perfections? How are the Glories of the Field spun, and by what Pencil are they limn'd in their unaffected bravery? By whos direction is the nutriment so regularly distributed into the respective parts, and how are they kept to their specific uniformities? If we attempt Mechanical solutions, we shall never give an account, why the J Wood-cock doth not sometimes borrow colours of the May-pye, why the Lilly doth not exchange with the Daysie, or why it is not something painted with a blush of the Rose? Can unguided matter keep itself to such exact conformities, as not in the least spot to vary from the species. That divers Limners at a distance without either copy, or designe, should draw the same Picture to an undistinguishable exactness, both in form, colour, and features; this is more conceivable, than than matter, which is so diversified both in quantity, quality, motion, site, and infinite other circumstances, should frame it self so absolutely according to the Idea of it and its kind. And though they fury Appelles, who threw his Pencil in a desperate rage upon the Picture he had essayed to draw, once casually effected those lively representations, which his Art could not describe; yet 'tis not likely, that one of a thousand such praecipitancies should be crowned with so an unexpected an issue. For though blind matter might reach some elegancies in individual effects; yet specific conformities can be no unadvised productions, but in greatest likely hood, are regulated by the immediate efficiency of some knowing agent. [39, pp. 44 ff.]

Third version, 1676:

> Blind Matter may produce an elegant effect for once, by a great Chance; as the Painter accidentally gave the Grace to his Picture, by throwing his Pencil in rage, and disorder upon it; But then constant Uniformities, and Determinations to a kind, can be no Results of unguided Motions. [41, p. 11]

As Jones commented, "Here, indeed, is merciless pruning" [1, p. 996]. In several places in the third edition, as we compare it to the first, we find Glanvill seemingly asking himself, "What am I trying to say? The result produces an

[2] A collection of all three versions can be found in the following work: *The Vanity of Dogmatizing: The Three 'Versions,'* by Joseph Glanvill, The Harvester Press, Hove, Sussex, England, 1970.

extremely terse summary that "gets to the heart of the matter" with no flourishes or circumlocutions. Gone are specious topes, excessive phrases and sentences: native English; conversational, direct statement emerges. Sentence length diminishes. Amplified style is pruned to the bone. No verbal superfluity; lack of metaphor and other rhetorical ornaments; words have a single meaning:

Other changes emerge as Jones compares phrases from the first edition to the same material in the third edition:

> 1. "Which to us is utterly occult, and without the ken of our Intellects"
> 3. "to which we are stranger."
>
> 1. "those abstrusities, that lie more deep, and are of a more mysterious allon'
> 3. "the Difficulties that lie more deep."
>
> 1. "those principiate foundations of knowledge"
> 3. "The Instruments of knowledge."
>
> 1. "Plato credits this position with his suffrage; affirming"
> 3. "Plato affirms."
>
> 1. "is but the Birth of the laboring Mountains, Wind and Emptiness"
> 3. "stands yet unresolved"
>
> 1. "preponderate much greater magnitures"
> 3. "outweigh much heavier bodies." [1, pp. 996–997]

The above excerpts anticipate many before/after style examples seen in modern technical writing and technical editing texts. When we read the first version of *Vanity of Dogmatizing* [39], we would describe the essay as "wordy." Glanvill's revisions [40, 41] show his focus on achieving a style that conveys precision, not an emphasis on words that represent affectation. What the scientists pursued specifically was what early technical writers had done intuitively: realization that truth depended upon accuracy of language. And the advancement of science depended in precise words. Linguistic reformation required strict definition, objective rather than subjective meaning. As Sprat states, the Society "have therefore been most rigorous in putting in execution . . . to reject all the amplification, digressions, and swellings of style."

The goal,

> A close, naked, natural way of speaking; positive expressions; clear sense; a native easiness; bringing all things as near the Mathematical plainness, as they can: and preferring the language of Artizans, Countrymen, and Merchants, before that, of Wits, or Scholars. [42, p. 113]

As Jones notes, the luxurious style was characterized by use of various rhetorical devices such as figures, tropes, metaphors, and similes; the sentences, long, often obscurely involved. This style, associated with the wordiness of peripateticism, was seen by the first members of what evolved into the Royal Society to be antagonistic to the aim of science. Thus, in the statutes of the Royal Society, published in 1652, the following statement appears in Chapter 5: Of Experiments, And the Reports thereof: "In all Reports of Experiments to be brought into the Society, the matter of fact shall be barely stated, without any prefaces, apologies, or rhetorical flourishes; and entered so in the Register-book, by order of the Society" [43, p. 290]. What happened, in the rejection of luxurious style as an appropriate vehicle for conveying the "new knowledge," was that the style of much practical discourse, as easily found in many published works of the 16th and early 17th centuries, began to surface in the papers of many members of the Royal Society.

FORMAT, SOUND, DICTION, SYNTAX, AND PURPOSE: THE CANONS OF STYLE

Examining the evolution of English style from a broad array of texts shows that we cannot ignore the role of the oral and the visual in the elements of diction and syntax. Because form follows function, the form in which text emerges results from the purpose of the discourse. Writers of early technical books apparently realized that writing to enable the reader to perform a task differs from writing to be read slowly, meditatively, or leisurely, such as fiction and religious works. Many of the early technical writers, like Elyot, were capable of a range of styles appropriate to audience and purpose. And, as Elyot's work demonstrates, the emergence, development, and persistence of plain style reveals the development of a style appropriate for conveying information, a style that preserved the spoken quality of instructions while being direct and concise.

Prose, used in many early books, replaced poetry as a medium for conveying knowledge after reading skills made memorization of instruction less necessary. Prose replaced poetry as the best communication method when increasing knowledge exceeded the capabilities of poetic mnemonics. Oral residue, with its additive clauses and phrases, decreased and was replaced by interlocking prose clauses and phrases designed to be read rather than heard [30]. English style, as it emerged from the 10th to the late 17th century, echoes in our contemporary works on style. Books, such as those by Gunning (1952–1968) [44], Flesch (1946–1980) [45], Lanham (1987–2000) [46], and Williams (1989–2003) [47]—to name a few of the notable and extremely popular contributors to the elements of plain English—show that *readable text* is still defined as writing that can be easily heard, easily read, and easily spoken. When people want information, they still want plain, direct, concrete style. The trends in these books echo 1,000 years of English style. (Alfred's psalter would have earned a high readability score.)

The canons of style must, then, include format, sound, diction, syntax, and purpose. The sustained need for utilitarian instructions transmitted textually rather than orally led to the emergence of a plain style that foreshadowed modern technical writing style—concrete, visual, descriptive nouns and action verbs; subject-verb-object sentence order; and active-voice clauses. Examined chronologically in multiple forms, many texts show the transition from art/graphic representation to written language to capture meaning, which could then be retained in visual memory. Style necessarily embodies syntactical arrangement that portrays organization. In another sense, style is a cultural icon that captures the visual and the aural in word arrangements that are pleasing and useful to the intended context. Ramistic rhetoric was the first theory of English style to use graphics combined with text and page design to convey meaning in non-aural-dependent text. Perhaps Ramus's theory appealed to English people's love for the concrete and the organized. Written English owes its commitment to the importance of logic embodied in syntax and page design to Ramus. Sentence structure that moves from given to new reflects Ramist rhetoric and its goal of tightly interlocked clauses.

CONCLUSION

Reflecting culture, language becomes a tool for sharpening the portrait of the historical terrain that has shaped that culture. Modern English prose style thus developed its basic character outside the university and the church. Despite the existence of classical humanism in England, humanist writing did not lead to modern prose style. By 1250, the use of English by craftsmen and the merchant class was augmented by the spread of English among the upper classes and the rise of the English middle class, which flourished in self-governing communities. These groups, engaged for the most part in trade or in the manufacturing crafts, banded together into commercial fraternities or guilds for their mutual advantage [26, p. 171]. Because these groups spoke and then wrote English, the use of English spread. By the beginning of the 14th century, English had become the dominant language among the middle class. French and Latin were still used in the courts, church, and universities, but the overall dominance of French and Latin is questionable when the total number of documents produced is considered.

Ultimately, growth in size, activity, and power of a manufacturing, commercial, and bureaucratic section of the population and the organization of guilds, trading companies, municipal corporations, and a civil service during the 14th and 15th centuries led to the entrenchment of an English-speaking political and economic power base formerly controlled by the nobility and the ecclesiastical hierarchy. Modern trends embracing urbanization and secularization and the increasing popular participation in the national economy and culture became motives for various kinds of English prose texts in this period. By the 16th century, the inability of the common English person to speak or read either Latin or French, the improving literacy of the middle class, the increasing availability of cheaply printed books for a newly literate population eager for works in the vernacular, and the growth of knowledge in nearly every field all contributed to the rise of vernacular English as the prose of utility.

From a discussion of the plain English sentence of the 17th century, we then need to examine the existence of the English paragraph as it contributed to the clarity, coherence, and unity of English technical writing.

REFERENCES

1. R. F. Jones, Science and English Prose Style in the Third Quarter of the Seventeenth Century, *PMLA, 45,* pp. 977-1009, 1940.
2. M. W. Croll, *Style, Rhetoric, and Rhythm: Essays by Morris Croll,* J. M. Patrick, R. O. Evans, J. M. Wallace, & J. R. Schoeck (eds.), Princeton University Press, Princeton, New Jersey, 1966.
3. D. Bush, *English Literature in the Earlier Seventeenth Century* (2nd ed.), Cambridge University Press, New York, 1962.
4. G. R. Elton, *England, 1200-1640,* Cornell University Press, Ithaca, New York, 1969.
5. J. H. Fisher, *The Emergence of Standard English,* University Press of Kentucky, Lexington, 1996.
6. M. Richardson, The Distamen and the Influence on Fifteenth Century English Prose, *Rhetorica, 2,* pp. 207-226, 1983.
7. M. Richardson, *Middle-Class Writing in Late Medieval London,* Pickering Press, London. 2011.
8. A. C. Baugh, *A History of the English Language,* Appleton-Century-Crofts, New York, 1957.
9. M. Elsky, *Authorizing Words: Speech, Writing, and Print in the English Renaissance,* Cornell University Press, Ithaca, New York, 1989.
10. M. T. Clanchy, *From Memory to Written Word, England, 1066-1307,* Basil Blackwell, London, 1993.
11. E. M. Miller, *The Professional Writer in Elizabethan England,* Harvard University Press, Cambridge, Massachusetts, 1959.
12. E. Hallem and A. Prescott, *The British Inheritance: A Treasure of Historic Documents,* University of California Press, Berkeley, 1999.
13. D. C. Douglas and G. W. Greenway (eds.), *English Historical Documents 1500-1700* (2nd ed.), Routledge, London, 1981.
14. B. Cusack, *Everyday English 1500-1700*, University of Michigan Press, Ann Arbor, Michigan, 1998.
15. M. Prestwich, English Government Records, 1250-1330, in *Pragmatic Literacy East and West, 1200-1330,* R. Britnell (ed.), Boydell, St. Edmunds, United Kingdom, pp. 95-106, 1997.
16. C. Johnson and H. Jenkinson, *English Court Hand A.D. 1066 to 1500* (Vol. 2), Ungar, New York, 1967.
17. H. S. Bennett, *English Books & Readers, 1475-1557,* Cambridge University Press, Cambridge, United Kingdom, 1952.
18. R. Hirsch, *Printing, Selling and Reading, 1489-1550*, Harrassowitz, Wiesbaden, 1975.

19. F. J. Furnivall (ed.), *Child-Marriages, Divorces, and Ratifications etc in the Diocese of Chester, A.D. 1561-65,* Early English Text Society, No. 108, Kraus Reprint, Milwood, New York, 1973.
20. J. M. Mueller, *The Native Longue and the Word: Developments in English Prose Style 1380-1580,* University of Chicago Press, Chicago, 1984.
21. R. W. Chambers, *On the Continuity of English Prose from Alfred and His School,* Early English Text Society, No. 186, Oxford University Press, Oxford, United Kingdom, 1950.
22. K. Alfred, *The Whole Works of King Alfred the Great. An English Translation of King Alfred's Anglo-Saxon Version of the Historian Orosius* (Vol. 2), J. A. Giles (Gen. ed.), AMS Press, New York, 1969.
23. D. Freeborn, *From Old English to Standard English* (2nd ed.), University of Ottawa Press, Canada, 1998.
24. K. D. Bulbring (ed.), *The Earliest Complete English Prose Psalter,* Early English Text Society, No. 97, Kegan Paul, Trench, the Netherlands, 1891.
25. T. Austin, *Two Fifteenth-Century Cookery-Books,* Early English Text Society, No. 91, Oxford University Press, Oxford, United Kingdom, 1888.
26. T. Smith (ed.), *English Gilds: The Ordinances of More than One Hundred Early English Gilds,* The Early English Text Society, No. 40, Oxford University Press, United Kingdom, 1963.
27. E. Lamond (ed.), *Walter of Henley's Husbandry,* Longmans, Green, London, 1890.
28. W. F. Bolton, *A Living Language: The History and Structure of English,* Random House, New York, 1979.
29. E. Tebeaux, Books of Secrets: Authors and Their Perception of Audience in Procedure Writing of the English Renaissance, *Issues in Writing, 3,* 41-67, 1990.
30. E. Tebeaux, Ramus, Visual Rhetoric, and the Emergence of Page Design in Medical Writing of the English Renaissance: Tracking the Evolution of Readable Documents, *Written Communication, 8,* pp. 411-445, 1991.
31. T. Elyot, *The Castle of Health,* Woods, London, 1534/1535.
32. T. Gale, *Certaine Workcs Of Chirurgerie* (Four Parts), Hall, London, 1563.
33. P. Miller, *The New England Mind: From Colony to Province,* Harvard University Press, Cambridge, Massachusetts, 1954.
34. W. Leybourn, *The Compleat Surveyor, Containing the Whole Art of Surveying of Land,* Sawbridge, London, 1656.
35. N. Grew, *The Anatomy of Plants, With an Idea of a Philosophical History of Plants, and Several Other Lectures,* Read Before The Royal Society, W. Rawlins, London, 1682.
36. J. H. Fisher, Chancery and the Emergence of Standard Written English in the Fifteenth Century, *Speculum, 52,* pp. 870-899, 1977.
37. H. MacDonald, Another Aspect of Seventeenth-Century Prose, *Review of English Studies, 19,* pp. 33-43, 1943.
38. E. Tebeaux, The Voices of English Women Technical Writers, 1641-1700: Imprints in the Evolution of Modern English Prose Style, *Technical Communication Quarterly, 7,* pp. 125-152, 1998.
39. J. Glanvill, *The Vanity of Dogmatizing: Or Confidence in Opinions. Manifested in a Discourse of the Shortness and Uncertainty of Our Knowledge, and Its Causes;*

With Some Reflexions on Peripateticism; And an Apology for Philosophy, E.C. for Henry Eversden at the Grey-Hound in St. Pauls-Church-Yard, London, 1661.

40. J. Glanvill, *Scepsis scientifica, or, Confest ignorance, the way to science in an essay of The vanity of dogmatizing, and confident opinion: with a reply to the exceptions of the learned Thomas Albius*, Printed by E. Cotes, for Henry Eversden, London, 1665
41. J. Glanvill, *Essays on several important subjects in philosophy and religion by Joseph Glanvill . . .*, printed by J. D. for John Baker . . . and Henry Mortlock . . . , London, 1676
42. T. Spratt, *The History of the Royal Society,* J. I. Cope and H. W. Jones (eds.), Washington University Press, St. Louis, Missouri, 1958.
43. H. Lyons (ed.), *The Records of the Royal Society of London* (4th ed.), Royal Society, London, 1940.
44. R. Gunning, *The Technique of Clear Writing,* McGraw-Hill, New York, 1963.
45. R. Flesch, *The Technique of Clear Writing,* McGraw-Hill, New York, 1952.
46. R. Lanham, *Revising Business Prose,* Allyn & Bacon, Boston, Massachusetts, 2000.
47. J. M. Williams, *Style: Ten Lessons in Clarity and Grac*e (6th ed.), Longman, New York, 2000.

CHAPTER 3

Technical Writing and the Development of the English Paragraph 1473–1700*

As I argued in the previous chapter, the influence of technical or practical writing on the evolution of English prose has long been ignored [e.g., 1, 2]. Scholars who have traced the development of English have focused on humanist writing, the canon of works of most interest to literary historians [e.g., 3, 4]. Similarly, the influence of technical writing on the development of the English paragraph has never been considered, even though the first printed English technical books (1490) illustrate concepts of the paragraph first described by Alexander Bain in 1866. Paragraphs can be found in 12th century English how-to manuscripts, but tracking the emergence of the paragraph in the first printed how-to books provides a method of examining a wide range of books from 1473 onward. The fruits of the printing press provide a story of the development of English prose from multiple genres.

Exclusion of a broad range of writing to describe the early development of English prose has created a distorted history, even though, as the venerable Douglas Bush noted, "Plain prose was the natural medium for most kinds of utilitarian writing, and most writing was utilitarian" [5, p. 192]. Therefore, in this chapter, I have three goals:

1. To show, with the use of examples from the first printed technical books, that paragraphs as we know them today existed in England in the late 15th century and continued to occur and develop during the 16th and 17th centuries;
2. To broaden the history of the English paragraph by describing the role of technical writing in early English paragraphs; and

*Originally published in the *Journal of Technical Writing and Communication, 41*(3), pp. 219–253, 2011.

3. To show the influence of Ramus' rhetoric on the paragraph, not just technical writing in general. Ramus' influence produced attractive, organized, printed texts exhibiting careful format and page design.

Paragraphs were linked and structured to reveal organization and to improve readability. Ramist-influenced technical works (1550–1650) cemented many of the early characteristics of paragraphs sustained in late 16th century English printed how-to books [6].

THE HISTORY OF THE ENGLISH PARAGRAPH—AN INCOMPLETE ASSESSMENT

The history of the English paragraph has received little attention since the late 19th century, even though the paragraph remains fundamental to English composition instruction. Studies of the rise of the English paragraph have relied principally on the work of Alexander Bain, whose *English Composition and Rhetoric* (1866) remains the first comprehensive work on the English paragraph. Bain described the paragraph as "a division of discourse next higher than the sentence . . . a collection of sentences with unity of purpose" [7, p. 108]. From Bain emerged the rules of the paragraph [7, pp. 92–134]. Erica Lindemann provides a modern parsing of Bain, and I refine those further:

1. The paragraph must equal the sum of the sentences.
2. Each sentence of a paragraph should clearly relate/connect with the sentence before it.
3. When several consecutive sentences state or illustrate the same idea, they should, as far as possible, be formed alike (parallel structure).
4. The opening sentence, unless it serves as a preparatory sentence, should indicate what concept the paragraph will present.
5. Unity in a paragraph implies a sustained purpose and forbids digressions not connected with the topic sentence.
6. Arrangement of sentences could be consecutive, i.e., related topics should be kept close together.
7. Principle and subordinate statements should have their relative importance clearly indicated. [8, p. 147]

The first composition books that dealt with the paragraph were largely derivative of Bain.[1] Writers such as A. D. Hepburn, an American, further defined the

[1] See also, A. S. Hill, *Foundations of Rhetoric* and *Principles of Rhetoric*; D. J. Hill, *Elements of Rhetoric* and *Science of Rhetoric*; J. S. Clark, *Practical Rhetoric*; T. W. Hunt, *Principles of Written Discourse*; G. R. Carpenter, *Exercises in Rhetoric*; Barrett Wendell, *English Composition*; J. G. R. McElroy, *Structure of English Prose.* Scott and Denney, according to Rogers, compiled the first bibliography of paragraph studies, which they include in *Paragraph-Writing*, 1898, p. 106.

paragraph as a "connected series of sentences containing the development of a single topic" *Manual of Rhetoric* (1875) [9]. John Genung (1886) [10] developed Bain's concepts, followed by Scott and Denney (1896) [11], who wrote the first composition book focusing on the paragraph and defined it as "A unit of discourse developing a single idea. It consists of a group or series of sentences closely related to one another and to the thought expressed by the whole group or series" [11, p. 5]. Their second text, *Paragraph-Writing; A Text for Colleges* (1896), the third edition [12], focused on the paragraph as the foundation of the essay and is the first book to do so. The legacy of these early compositionists yielded the three canons of paragraph structure of the early 20th century: unity, coherence, and emphasis.

As Paul C. Rogers noted in his 1965 article, *Alexander Bain and the Rise of the Organic Paragraph* [13], early English rhetoric books (16th, 17th, and 18th centuries) did not discuss the paragraph, which usually focused on oral delivery and application of schemes and tropes to oral discourse. At that time, rhetoric was generally the province of oral delivery. However, in the 19th century, oral delivery separated from rhetoric. The separation was critical to the teaching of composition: courses emerged that focused on belletristic, philosophical, or historical prose as examples with no inclusion of oratory. Yet the first American composition textbooks, such as J. R. Boyds' *Elements of Rhetoric and Literary Criticism* (1844) [14] and H. N. Day's, *Elements of the Art of Rhetoric* (1850) [15], included no discussion of the paragraph. A. D. Hepburn's *Manual of Rhetoric* (1875) [9] launched the discussion of the paragraph and its organic nature in the composition. Following Hepburn's lead, other books followed with only slightly varying approaches to the paragraph.

The paragraph, nonexistent in spoken discourse, quickly assumed a central function in the study of composition: "to do for the eye what vocal pauses and stress do for the ear—to show what parts of a composition belong together, and among those parts to indicate the most significant" [16, p, 38]. In short, analysis of the paragraph began to emerge when composition emerged as a separate study from speech. Wendell [16], and then Scott and Denny (1896) [12], approached the paragraph from the visual aspect of the printed page, which they believed was important in helping the reader move through the text.

Another driving force was the decreasing length of the English sentence, a change that occurred with emergence of middle-class English readers in the late English Renaissance. The paragraph demanded attention when shorter sentences replaced the Latinate sentence, often several hundred words of "clause heaps," a phrase used by Lewis [17, p. 46] via George Saintsbury [18]. By the early decades of the 19th century, when composition became a course by itself, student essays, composed of numerous short sentences, exemplified incoherent, illogical, rambling, pointless text. Bain was the first to address the issue with depth, although Joseph Angus, in his 1862 *Handbook of the English Tongue,* devoted 14 pages to the paragraph [19, p. 111]. For unknown reasons, Scott and Denney omitted this work in their bibliography of composition texts that discussed the paragraph [11, p. 106]. Lewis mentions Angus, but focuses on Bain [17, p. 20].

Studying Early English Paragraphs: The Need for Adequately Reflective Illustrative Examples

When we study the example works used in the first books dealing with the paragraph and then its place in the English composition, we find no references to or examples of practical discourse. This omission is particularly noticeable in Edwin Lewis's *History of the English paragraph* (1894) [17], a dissertation that remains the first and only attempt to present the history of the English paragraph. Lewis analyzes paragraphs from humanist, philosophical, and historical works. Similarly, Bain and writers of the first composition texts published after Bain, excerpt examples from similar works. Lewis believed that the modern paragraph was first exemplified in Sir William Temple (1686) [17, pp. 42, 44].

While Lewis noted the rise of literacy on the evolution of the paragraph, he lacked the research of Henry Stanley Bennett's three-volume set, *English Books and Readers* [20–22], Louis Wright's *Middle-Class Culture in Elizabethan England* [23], and Thomas Laquer's influential article on the rise of popular literacy in England, 1500–1850 [24]. These works, mentioned in these modern publications, along with many others, would likely have reshaped both Bain and Lewis' thinking in that they exemplify 20th century assessments of the impact of English middle-class readers in the 17th century. These readers sought how-to and self-help books in a wide range of fields. However, as Lewis did observe, these readers helped diminish the presence of the Latinate sentence structure in favor of short, loose sentences [17, pp. 44–46]. Both authors and printers were more than happy to meet the demands of this new group of readers for practical how-to books. Early printers, like modern publishers, could not afford to print what would not sell, a point noted by both Bennett and Wright [20, p. 17; 23, pp. 82–83].

Thus, while Lewis' work provides an important beginning for understanding the history of the English paragraph, Bain, a humanist, ignored writing outside the pale of humanist and university studies. Rogers noted, as late as 1965, that

> the sentence is as old as language and sentence theory as old as Greco-Roman rhetoric, but the paragraph as we know it today did not begin to emerge until the late seventeenth century, did not attain full development till the eighteenth, and was consistently ignored by rhetoricians until the mid-nineteenth. [13, p. 399]

Rogers is correct in dating the age of the sentence; but he and the compositionists of the 18th and 19th centuries erred (1) in ignoring the quantity of practical writing printed and used by English men and women prior to the 19th century; and (2) in recognizing its role in the rise of the paragraph. Rhetoricians and compositionists of the 20th century hadn't questioned the origins of the English paragraph since Lewis' dissertation and focused their attention on modern and contemporary issues in composition. Thus, this study provides an additional 400 years of history to our knowledge of how the English paragraph developed, as shown by examining English Renaissance how-to texts.

The Intrinsic Nature of the Paragraph

The lack of mention of paragraphs in 16th and 17th century English rhetoric books, such as those by Whately [25], Blount [26], and Rainolde [27], makes us wonder about the paucity of analysis of historical paragraph development. However, English grammar schools in the 16th and 17th centuries focused on rhetoric in oratory, and writing instruction often focused on imitation of classical literature. Even a cursory reading of later medieval incunabula reveals that the concept of the paragraph existed. As Naomi Barron notes,

> Division of written text into argument-sized chunks dates back to the second century BC, when a new paragraph or capitulum (as in caput, or "head") marked a change of topic or argument in a text (Parkes, 1993, p. 65). In early insular manuscripts, the beginning of a new capitulum was set off by the notation ".K.". By the twelfth century, the K had been replaced by a C. With the addition of a vertical line to indicate a *littera notabilior*, this C evolved into the paragraph symbol used in contemporary text editing. Rubricators who developed and began using the symbol by the end of the twelfth century typically colored it red. [28, p. 37]

The prevalence of the paragraph in incunabula suggests that writers could not think of complete texts that did not have segments or partitions that combined to form the complete idea. In addition, readers require segmented, or coherent text segments that move toward completion of a whole idea. Wendell and then Scott and Denney, as I previously noted, saw the visual value of paragraphing for the reader: Paragraphed text appears more attractive, and "the indented lines serve as landmarks for the reader's eye, enabling him to find his place again if he should happen to turn aside for a moment" [16, p. 93]. Scribes preparing Latin manuscripts apparently also saw the need to partition the text at junctures where ideas shifted. In, "The External Form of the Paragraph" [12, pp. 5–29], Scott and Denney discuss lowered readability when indentations are omitted. Most of the early compositionists, however, saw the necessary division of an idea as it is developed to support a central topic: As Scott and Denny noted in *Paragraph-Writing*, the essay has an organic character, and the paragraph,

> as a constituent element of the essay, can be neither arbitrary nor accidental. It must be part of the essay finding a reason for existence in the peculiar function which it performs. It must play a definite part in the structure of the

> whole organism. Whatever peculiarities of function or structure a paragraph possesses must be explainable by its relation to the function and structure of the whole composition. [11, p. 96]

As we will see in the examples that follow, early printed technical books exemplify the role of the paragraph that Scott and Denny and others described.

In his dissertation, Lewis began by stating that the paragraph is the oldest mark of punctuation in Greek manuscripts [17, p. 9]. How these partitions occur and are marked to alert readers emerged in manuscripts and then changed in printed English documents until the early 17th century, when indentation became the main indication of paragraphing. As Lewis noted [17, pp. 10–15], many versions of the "P" mark emerged and changed design, along with other devices—three dots (∴) and the darkened reverse **P** **C** minus a tail on the P—by Caxton's time. Another early method, exemplified in the first technical writing example, uses several blank spaces at the end of paragraph plus a **C** to note the end of one paragraph and the beginning of another, a technique shown in *Gouernayle of helth* "The Species of exercise," discussed in the next section. Examination of paragraphs allows researchers to see how writers partitioned and then presented their concepts, along with the logic of the division.

Paragraphs in 15th Century Printed Technical Books

Many of the first printed English technical or how-to books dealt with health and can provide a convenient starting point for examining examples of early English paragraphs in printed how-to texts. One of the first, *In this treatyse that is called Gouernayle of Helth*, written by Joannes de Burgundia and published by Caxton in 1490 [29], exemplifies an opening paragraph that begins with a topic sentence and then lists chapters and their content that will follow. This paragraph, which appears below, anticipates a modern opening, or introductory, paragraph that prepares readers for what will follow. As Scott and Denney wrote in *Paragraph-Writing* (1896, some 400 years later than the *Gouernale of Helth*),

> The object of an introductory paragraph is to segregate the idea of the composition in hand from all other ideas. As this is nearly always apparent from a mere statement of the theme, the introduction usually needs to do little more than state the theme, and indicate briefly the line of development to be followed. [11, pp. 61–62]

This description fits the *Gouernayle*, as well as other how-to books (such as Alexander Read's book on tumors (1635) [30], shown in Figure 3-1).

Perhaps the introductory paragraph emerged because this book, like many manuscripts and early printed books, had no title page and no table of contents. The paragraph gives the theme, stated in a topic sentence. The following sentences use parallel structure in describing the content of the eight chapters of the

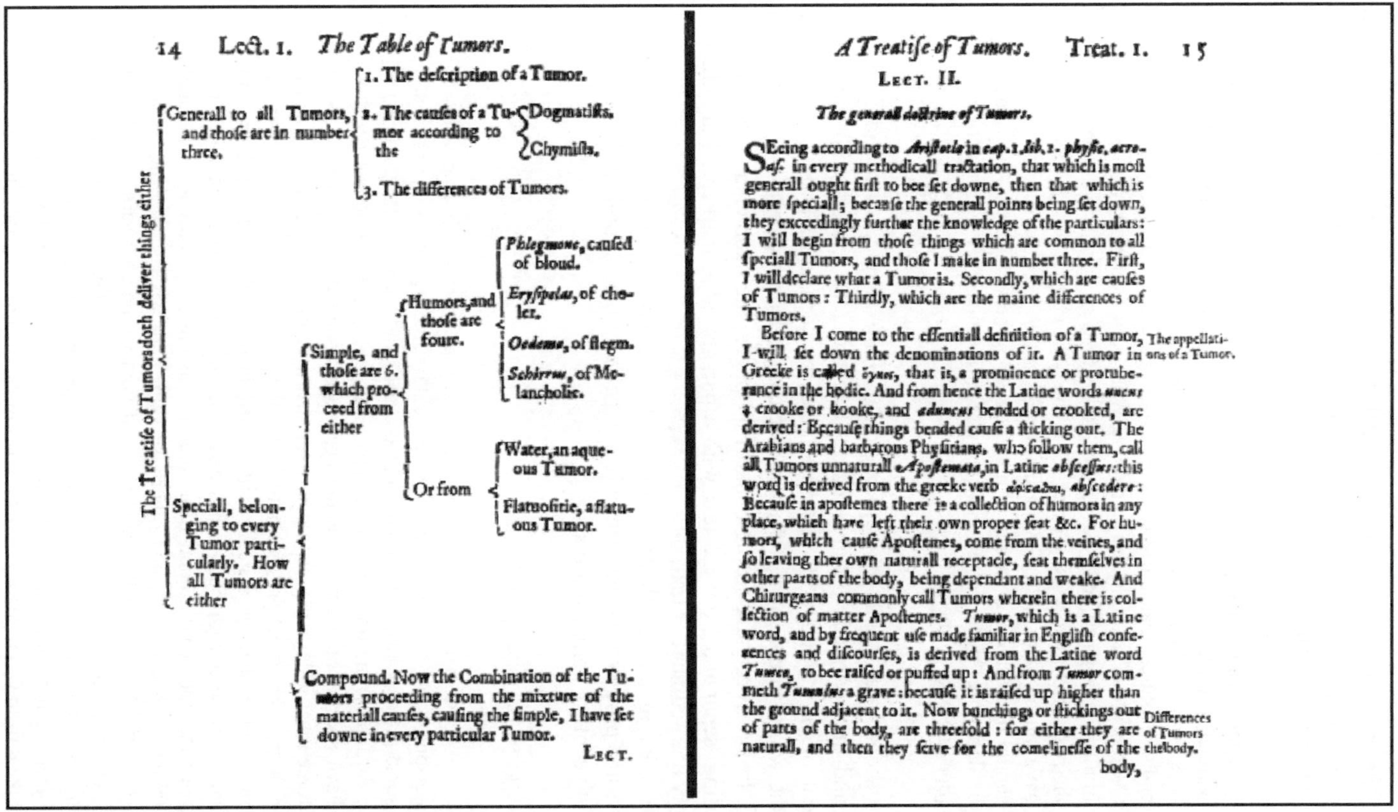

14 Lect. 1. *The Table of Tumors.*

The Treatise of Tumors doth deliver things either

Generall to all Tumors, and those are in number three.
1. The description of a Tumor.
2. The causes of a Tumor according to the Dogmatists. / Chymists.
3. The differences of Tumors.

Speciall, belonging to every Tumor particularly. How all Tumors are either

Simple, and those are 6. which proceed from either

Humors, and those are foure.
Phlegmone, caused of bloud.
Erysipelas, of choler.
Oedema, of flegm.
Schirrus, of Melancholie.

Or from
Water, an aqueous Tumor.
Flatuositie, a flatuous Tumor.

Compound. Now the Combination of the Tumors proceeding from the mixture of the materiall causes, causing the simple, I have set downe in every particular Tumor.

Lect.

A Treatise of Tumors. Treat. 1. 15

Lect. II.

The generall doctrine of Tumors.

SEing according to *Aristotle* in *cap.1.lib.1. physic. acroas.* in every methodicall tractation, that which is most generall ought first to bee set downe, then that which is more speciall; because the generall points being set down, they exceedingly further the knowledge of the particulars: I will begin from those things which are common to all speciall Tumors, and those I make in number three. First, I will declare what a Tumor is. Secondly, which are causes of Tumors: Thirdly, which are the maine differences of Tumors.

Before I come to the essentiall definition of a Tumor, I will set down the denominations of it. A Tumor in Greeke is called ὄγκος, that is, a prominence or protuberance in the bodie. And from hence the Latine words *uncus* a crooke or hooke, and *aduncus* bended or crooked, are derived: Because things bended cause a sticking out. The Arabians and barbarous Physitians, who follow them, call all Tumors unnaturall *Apostemata*, in Latine *abscessus*: this word is derived from the greeke verb ἀφίστασθαι, *abscedere*: Because in apostemes there is a collection of humors in any place, which have left their own proper seat &c. For humors, which cause Apostemes, come from the veines, and so leaving their own naturall receptacle, seat themselves in other parts of the body, being dependant and weake. And Chirurgeans commonly call Tumors wherein there is collection of matter Apostemes. *Tumor*, which is a Latine word, and by frequent use made familiar in English conferences and discourses, is derived from the Latine word *Tumeo*, to bee raised or puffed up: And from *Tumor* commeth *Tumulus* a grave: because it is raised up higher than the ground adjacent to it. Now bunchings or stickings out of parts of the body, are threefold: for either they are naturall, and then they serve for the comelinesse of the body,

The appellations of a Tumor.

Differences of Tumors the body.

Figure 3-1. Ramist Bracketed Division & Opening Paragraphs, Read's *The chirurgial lectures of tumors and ulcers* (1635).

book. The paragraph, clearly unified, equals the sum of its parts. The order of the sentences is bound by the order of the chapters that comprise the book. Thus, the sentences are clearly consecutive, as Bain required:

> In this treatise that is called Gouernale of helthe: What is to be sayd with Chryst's help of some thynges that belong to bodily helth, had and to be kept on to bodily helth. Lost and to be recovered and is divided into eight chaptures, that is to say In the fyrste chaptyre of the profyte of good Gouernayle of helth, In the ii chaptyre What is first on morrow to be done, In the iii chapter of bodily excersyce that is to says, begins to his profit, In the iv chapter of species of exercise, In the fythe chapter how a man should have him in mete and in eating his metes: In the sixth chapter how a man should have him in drynking of his drynker, In the vii chapter what should be done after mete, In the viii chapytre of the noyse of evil gouernance. [opening page of the text]

The second paragraph of *Gouernayle of Helth* provides a rationale for the topic and the approach, again a feature that provides readers with the second segment of an introduction, that anticipates modern introductions: It explains why the work should interest readers, a characteristic required by Genung [10, p. 47]. In De Burgundia's little book of 39 pages, the main chapters begin with a heading, separated from the text with a paragraphing marker. Many chapters are only one segment of text.

The opening paragraph of Chapter IV, The Species of exercise, reads as follows and provides another example of an effective paragraph as defined by the late 19th century compositionists:

> ℭ The Species of exercise
>
> Species of exercise be there well many as there be diverse states of persons, some be strong and some be feeble, some rich and some poor, and some prelates large and some subjects small, and some weather is fair and clear, and some time not so but darke and rainy, and therefore it needs to have species of exercise for the best species and the first is to walk to force meat in high places to clear. Another species is to ride and this is for rich men, but great prelates must have other manners of exercise for when in chamber shall be a great cord knitted in the end and hanged up, and take that cord with both hands and stand upright so that you touch not the earth and stand a good while then run as much as you may hither and thither with that cord, and otherwise skip, if this please thee not have a stone of 38 weight or thereabouts and if your would afterward often heave that stone from one side of that house to that other side, or a long while hold up that stone or bear it about thy neck or between thy hands, and so of other manner until you faint or thus holde a staff in thy hand and let another take it from you if he may with even strain, or you close a penny in your hand

> a let another take it if he may or hold your breath as long as thou mayst do and then puff if out as hard as thou mayst do, And this manner of exercise is full profitable to put out noises superfluous, for why many superfluous in sleep be withholden therefore if thou have no other means of exercise holding of breath helps much. ℂ Yet, there be other manners of exercise for young men that be lusty as to run or wrestle to leap, to cast the stone.

From our modern perspective, we might say that the paragraph is too prolix in its use of conversational English, but the content coheres: The paragraph begins with a topic sentence that prepares readers for a diverse presentation of exercise. The key phrase, "species of exercise," describes all species; all sentences contribute an idea to the topic—diverse species of exercise for diverse persons. Thus, the whole equals the sum of its parts. The types of exercises appear in a logical order; and the paragraph provides a progression of ideas that describe types of exercise to ensure good health. Note, too, that the sentences use subject-verb arrangement, short phrases, which begin with a subject-verb arrangement. While the "sentences" exemplify short independent clauses linked to phrases, the text does not use sentences that begin with dependent clauses. Content occurs in a "loose" style of additive clauses and phrases. Many lack a / (ferule) or comma to show the end of one clause or phrase. From Ong's perspective [31], this paragraph—and many others of the 15th century—exemplify oral tradition: conjunctive connectives, a loose style, which would become the foundation of plain English prose of the Royal Society. Note too that the next paragraph of *Gouernayle of Helth* begins with "Yet," a modern transition word.

Paragraphs in 16th Century Technical Books

Other works that can be defined as technical writing also illustrate effective paragraphing. The first printed English book on how to care for horses, *Here begynneth the Proprytees and medycynes for hors* (1502) [32], printed by Wynkyn de Word, contains a table of contents that lists topics on choosing horses, training them, and treating four dozen remedies for various conditions and ailments. Each topic begins with a heading and a number, coordinated with the table of contents. The heading is repeated in the opening sentence of the content for that segment, which exemplifies a paragraph. For example, segment 31 uses a topic phrase, followed for a logically ordered set of instructions for how to treat splints in horses. Again, we have logical order, parallel phrases (imperative mood), unity, sequential relationship among instructions, and a topic theme.

> ¶For the splint. xxxi.
>
> For the splint that growth under the kne. Take therefore an onyen and pyke out the core and put therein a sponefull of unflecked lyme and four peniweight of vertegrece/ a halfe a pound of launder sede/ and roft the onion till it be soft and then kytte the skynne a lytle that the medicine may enter to

> frete the malady out/ But let your medycyne lye styll two dayes and two nyghtes. But let the hors stande styll. But let your medicine lye still two days and two nights, and he shall be hole for ever.

This work also exemplifies paragraphs of one sentence, a topic extensively discussed by Bain and his followers, who sought to determine if a paragraph could be only one sentence. For example, in the following section of the opening chapter, the paragraph begins with a topic sentence that provides a division or direction that the paragraph would go. Each explanatory segment is then presented with a ¶ mark. Thus, the complete paragraph is, essentially, a topic statement with supporting sentences demarcated with the ¶. This paragraph illustrates unity, coherence, parallelism, sequencing, and even a concluding statement that reiterates the opening sentence. We can see how this paragraph could become a list with ¶ as bullets, a format device which did indeed appear within two decades.

> ¶ The .xv proprytees of a good hors. ii.
>
> There belongeth to a good hors xv. manner and properties that is to say three of a man/three of a woman/three of an hare/ three of a fox & three of an asse. ¶ The first three of a man is to be proude strong and hardy. ¶ The iii. of a woman is to have a fayre breste and a fayre traset and easy to lep upon. ¶ The three of a hare is to have a lean head/grete eyes/and therewith well runnynge. ¶The three of a fox is to have a fair tail and short ears and therewith a grete trot. ¶ The three of an asse is to have a tight chin/ a flat legg. And therewith a gode horse.

Perhaps the first printed surgical book, translated from German into English, Hieronymus Von Brunschweig's *The noble experience of the handy work of surgeri* (1525) [33], provides sustained examples of paragraphs that begin with a topic sentence that are also unified, sequential, and coherent. Many supporting sentences within the paragraphs also use parallel structure, as required by Bain. Figure 3-2 places the translation and the original material side by side to illustrate the development.[2]

A third example, Guy's *Questyonary of Chryugeons*, first published in 1542 [34], is divided into two sections. The first, composed of questions and answers that medical students would be expected to know in preparing for their oral examination, is furthered divided into paragraphs: the question in one paragraph,

[2] Subsequent works by Braunschweig also illustrate effective paragraphs and complete instructional essays. For example, A most excellent and perfecte homish apothecarye or homely physic booke, for all the grefes and diseases of the body. *Translated out the Almaine speche into English by Ihon Hollybush*, Imprinted at Collen: By [the heirs of] Arnold Birckman, in the yeare of our Lord M.D.LXI. [1561]

¶ The Anatomie in general

The skynne is a coveringe of the whole universal body woven out of the veins and sinews/made for the defence of the witts of man/ and the same skin is twayne indysseret. The one doth cover the uttermost member and is named the very skynne/ The other covereth the innermost parte of the members and it Is named panniculus or pannycle /and that is as a fleece/ lyke as by the sells or chambers of the brain/ and pyrocraneum is the thin flece aboue the brain pan/ and that panicle of the ribs is name Pleura and Siphac/ and the pannycle of the harte and other etc. ¶ Hereafter foloweth the fatnesses /some/is within the skynne and thatt is named greace/ some is within the belly about the kidneys and that is named tallow. ¶ Then commeth the flesh which be in iii. maners/ some is pure simple fleshe/ as is only the yarde and the gums of the teeth/ that other is as carneliys/like the flesh of the codds and of the breast; the iii. is comprehending almost over all the hole body and is named caroplena lacertis. ¶ Then be= ginneth the sinews to take their effect of the brains and of the mary of the backe/ After these the veins taketh their effect of the liver and of the heart. ¶ Then follows the bones of the body/ and of them be in number CC.xlviii beside the small bones that be called ossa cisanina And the bone laude/wheron the tongue is founded. ¶The navles be made on the uttermost parte of the members because they should and stronger group and hold all manner of things. ¶ The hear is ordained for sayrnes and to purify the manyfold parts of the body.

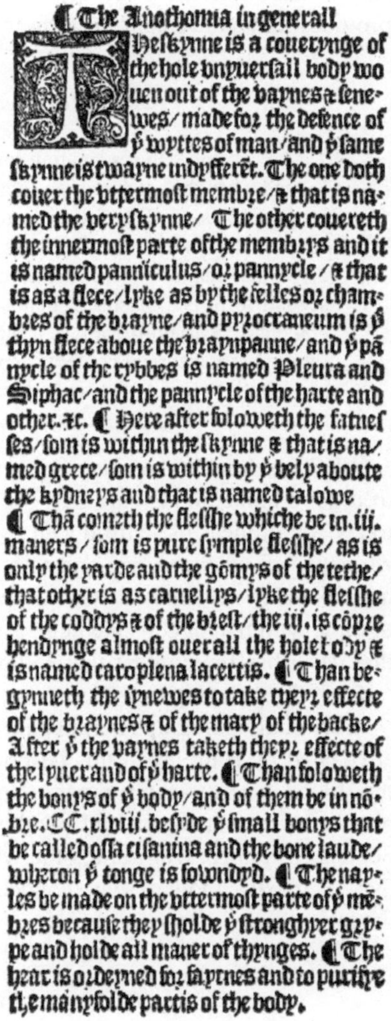
¶ The Anothomia in generall
The skynne is a coverynge of
the hole unyuersall body wo
uen out of the vaynes ⁊ sene-
wes/ made for the defence of
yͤ wyttes of man/ and yͤ same
skynne is twayne indyfferēt. The one doth
couer the uttermost membre/ ⁊ that is na-
med the very skynne/ The other couereth
the innermost parte of the membrys and it
is named panniculus/ or pannycle/ ⁊ that
is as a flece/ lyke as by the celles or cham-
bres of the brayne/ and pyrocraneum is yͤ
thyn flece aboue the braynpanne/ and yͤ pā
nycle of the rybbes is named Pleura and
Siphac/ and the pannycle of the harte and
other. ⁊c. ¶ Here after foloweth the fatnes
ses/ som is within the skynne ⁊ that is na/
med grece/ som is within by yͤ bely aboute
the kydneys and that is named talowe
¶ Thā cometh the flesshe whiche be in. iij.
maners/ som is pure symple flesshe/ as is
only the yarde and the gōmys of the tethe/
that other is as carnellys/ lyke the flesshe
of the coddys ⁊ of the brest/ the iij. is cōpre
hendynge almost ouer all the hole body ⁊
is named caroplena lacertis. ¶ Than be-
gynneth the synewes to take theyr effecte
of the braynes ⁊ of the mary of the backe/
After yͤ the vaynes taketh theyr effecte of
the lyuer and of yͤ harte. ¶ Than foloweth
the bonys of yͤ body/ and of them be in nō-
bre. CC. xlviij. besyde yͤ small bonys that
be called ossa cisanina and the bone laude/
wheron yͤ tonge is fowndyd. ¶ The nay-
les be made on the uttermost parte of yͤ mē-
bres because they sholde yͤ stronghyer gry-
pe and holde all maner of thynges. ¶ The
hear is ordeyned for fayrnes and to purifye
the manyfolde partis of the body.

Figure 3-2. Hieronymus Von Brunschweig, *The Noble Experyence of the Vertuous Warke of Surgerii* (1525).

and the answer in the other. The second part of the book, on medical formularies, provides tight paragraphs describing various medicines and their applications:

¶The cure of the exitures. [running abscesses]

The cure of exitures has four intentions. The first is to ripe the matter. The second after the maduracyon [coming to a head] to open it. The third is to purge the place. The fourth is to drye it after the opening and cleaning. ¶The first intention accomplishes four forms. The first form is tetrafarmacu of.G. that makes it with meale of wheate, saffron, water, and oyle. ¶The second form is a plaster of malows and is made thus. [latin pharmaceutical contents] ¶The third form is a plaster of Tebes, that is proper to our form. [latin pharmaceutical contents and quantities] ¶The fourth form is dyaguilon and basiticon and to the same intention is [latin medications] aboue said.

What we should notice here echoes the previous example on the characteristics of a good horse and Figure 3-1: the text uses ¶ marks to create what we would today call a linear list, introduced by a topic sentence that unifies the paragraph: we expect four intentions, and that is precisely what we get. The sentences all have parallel structure. In short, the paragraph denoted then, as now, a change of topic within a longer statement. Each section of this book opens with a heading that describes the content that follows. Thus, headings served to announce the topic of the paragraph.

Individual paragraphs illustrate the same carefully crafted paragraphs advocated by Bain, as in this description of how to "bleed" a patient and what the blood tells the physician. Note that the procedure creates sequential phrases, coherent ideas, and parallel structure (use of imperative commands). Sentences connect with "and, otherwise, another way, yet," all words and phrases that anticipate modern day connective devices. The first sentence contains the topic idea.

¶The maner to let them blode, and
to wash and strain it.

Floyne says, that there must be a great opening in the vein when they beletting blood because the thick blood shall not remain and the thin only come out. And when it is drawn, consider the substance and the color of it because as is above said, and than washe it, and pass it through a sair white cloth/ and ten loke on the fleshe that abides in the cloth/ and if it be graneous and troublous it is a great token. Otherwise take salt and mix it in the blood, and if it melt soon. Another way, take his urine and vinegar and loke if they will mingle together. Yet do thus/ and some of the blode in to a basin full of water/ and if it go down to the bottom like meals it is a token that he is a leperous.

Figure 3-3 from this work exemplifies paragraphs used to describe rude woodcut drawings of surgical tools. This method of linking text with visuals and using a text wrapping style is used in modern works to integrate text and visual. This example appears to be the first English printed technical work that uses this technique.

HUMANISTS AS TECHNICAL WRITERS—CONTRASTS IN PARAGRAPHING: THE 16TH CENTURY

Differences in paragraphing, as these affect content, presentation, and coherence, can be seen in writers who authored both humanistic discourse and practical discourse. One excellent example is Sir Thomas Elyot, who wrote *The Boke named the Gouernour* (1557) [35], a treatise describing the proper education of leaders, and *The Castle of Helth* (1561) [36], a book modeled on other self-help medical books in the 16th century. When we compare these works, we find that *The Gouernour* follows the style of much humanistic discourse of the English Renaissance. Elyot's plan for his work is not specifically laid out in the opening paragraphs, and subsequent paragraphs are often compilations of information on a specific topic. The work also exemplifies indentions that do not reflect any sort of paragraph, as Bain required. Elyot does link his paragraphs through concluding comments, which he then begins to develop in the succeeding paragraph, a strategy often used in orations. Many of his paragraphs, complete with side notes, lack conciseness and directness as he presents his central topic, the requirements of the person who will lead others, that is, the "Governour." Elyot, an icon of English humanism, was steeped in the Latin tradition. This work in English exemplifies what Lewis defined as the Latin influence on much English writing. The style would have been appropriate for readers of the English upper classes who had attended grammar school, where Latin was taught, and possibly the universities.

In contrast, Elyot's paragraphs and presentation in *The Castle of Helth* differ markedly from those in *The Gouernour*. Elyot uses lists, many within paragraphs or combined with short paragraphs, to present health advice. Figure 3-4 from *The Castle* exemplifies this strategy, clearly designed for easy reference and easy reading. His paragraphs describe the various humors that affect health. He lists the qualities of the melancholic, accentuated with the use of a large {. Next he adds additional information in one paragraph, which opens with a topic sentence. Note that Elyot uses sentences with parallel construction to list characteristics (one of Bain's requirements).

Throughout *The Castle*, Elyot uses simple headings to organize the contents. The following excerpt describes the virtues of bread and exemplifies the concise paragraphs he uses in *The Castle*.

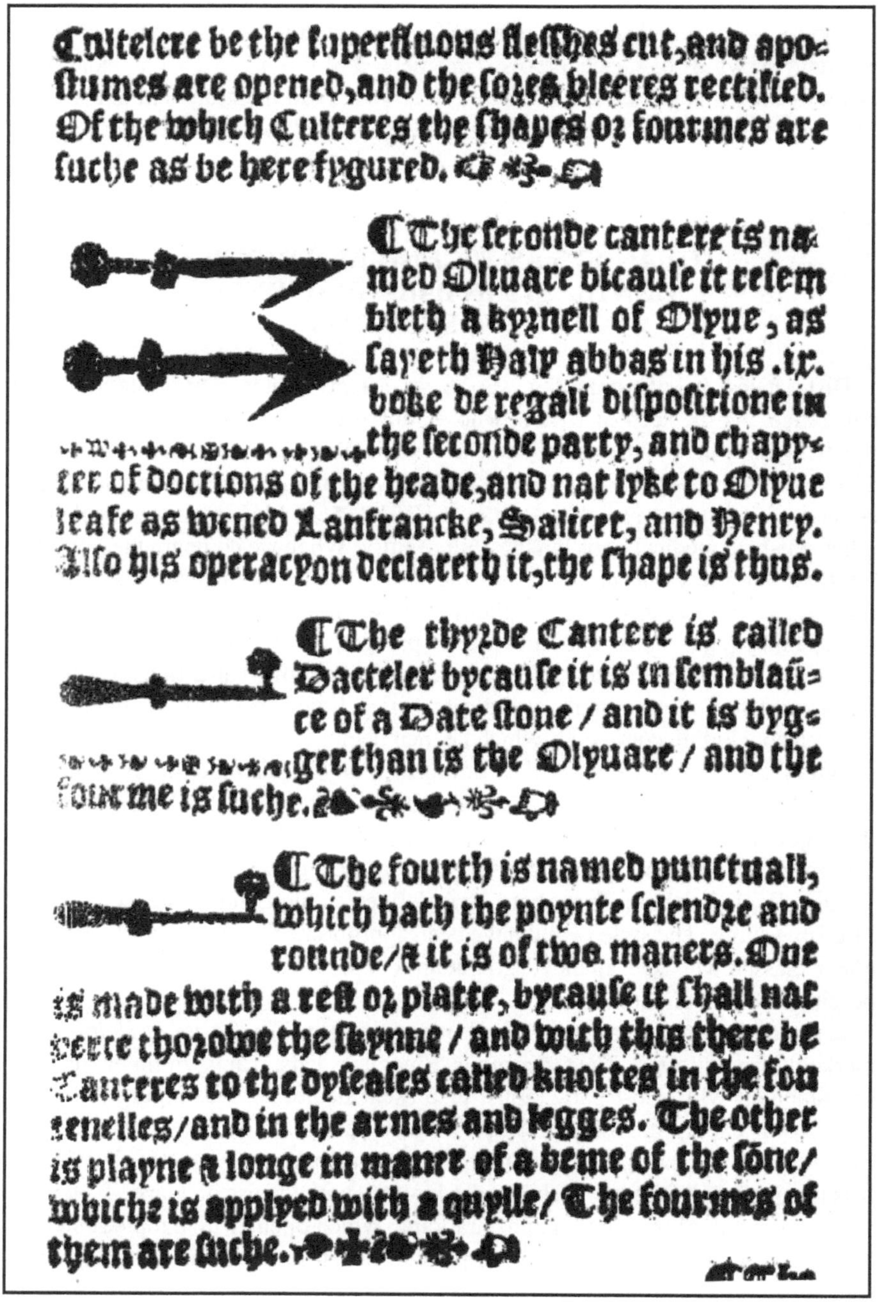

Cultelere be the superfluous flesshes cut, and apostumes are opened, and the sores ulceres rectified. Of the which Culteres the shapes or fourmes are suche as be here fygured.

The seconde cantere is named Oliuare bicause it resembleth a kyrnell of Olyue, as sayeth Haly abbas in his .ix. boke de regali dispositione in the seconde party, and chapyter of doctions of the heade, and nat lyke to Olyue leafe as wened Lanfrancke, Salicet, and Henry. Also his operacyon declareth it, the shape is thus.

The thyrde Cantere is called Dacteler bycause it is in semblaũce of a Date stone / and it is bygger than is the Olyuare / and the fourme is suche.

The fourth is named punctuall, which hath the poynte sclendre and rounde/ & it is of two maners. One is made with a rest or platte, bycause it shall nat perce thorowe the skynne / and with this there be Canteres to the dyseases called knottes in the fontenelles/and in the armes and legges. The other is playne & longe in maner of a beme of the sonne/ which is applyed with a quylle/ The fourmes of them are suche.

Figure 3-3. Graphics introduced and integrated with paragraphs from *The Questyonary of Cyurgeyons* (1490).

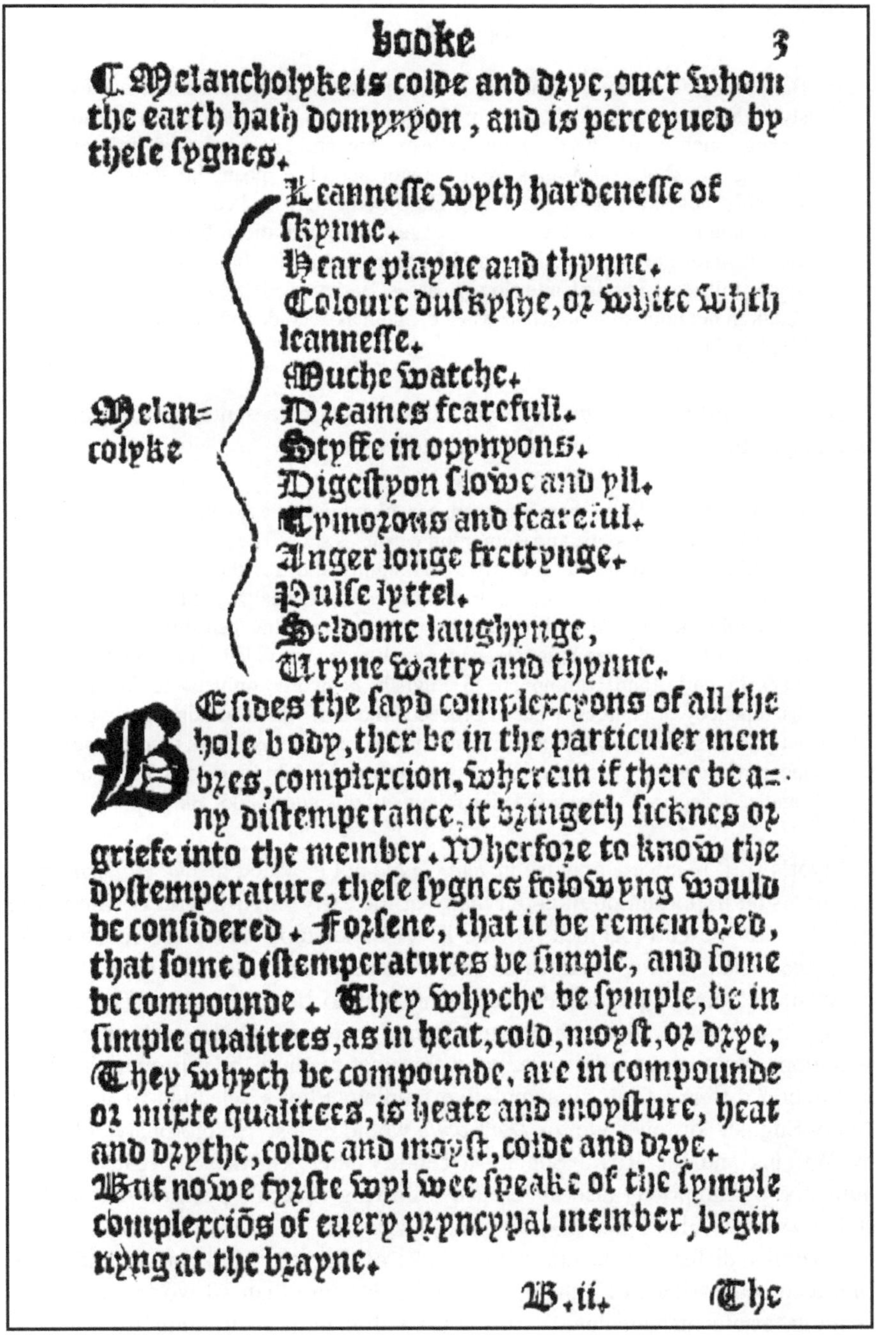
booke 3

¶ Melancholyke is colde and drye, ouer whom the earth hath domynyon, and is perceyued by these sygnes.

Melan-colyke

Leannesse wyth hardenesse of skynne.
Heare playne and thynne.
Colour duskyshe, or whyte wyth leannesse.
Muche watche.
Dreames fearefull.
Styffe in opynyons.
Digestyon slowe and yll.
Tymorous and fearefull.
Anger longe frettynge.
Pulse lyttel.
Seldome laughynge,
Uryne watry and thynne.

Besides the sayd complexcyons of all the hole body, ther be in the particuler membres, complexcion, wherein if there be any distemperance, it bryngeth sicknes or griefe into the member. Wherfore to know the dystemperature, these sygnes folowyng would be considered. Forsene, that it be remembred, that some distemperatures be simple, and some be compounde. They whyche be symple, be in simple qualitees, as in heat, cold, moyst, or drye, They whych be compounde, are in compounde or mixte qualitees, is heate and moysture, heat and drythe, colde and moyst, colde and drye.
But nowe fyrste wyl wee speake of the symple complexciōs of euery pryncypal member, begin nyng at the brayne.

B.ii. The

Figure 3-4. Thomas Elyot, *The Castle of health* (1561).

¶ *Of breadde. Cap. 7.*

> Bred of fine flour of wheat, having no leaven, is slow of digestion, and makes slymy humors, but nourishes much: if it be leavened, it digests sooner: bread having much bran, fills the belly with excrements, and nourished little or nothing, but shortly descends from the stomach: The meane between both suffienly levened, well moulded, and moderately baked is the most wholsome to every age, the greatest loaves do nourish most, for as much as the fire hath not exhausted the moisturer of them. Hot bred, much eaten, makes fullness and thirst, and slowly passes. Barley brdde cleanses the body, and doth not nourish so much as wheate, and makes colder juyce in the body. [36, Ch. 7]

However, in *The Gouernor*, Elyot chose a Latinate sentence structure and paragraphs that use a highly ornate, elevated style:

> ℭ And consequently there may appear like diversitie to be in English, between a publike weale and a common weale, as should be in Latin between Republica, and Replebia. And after that signification, if there should be a common weale, either the commoners only must be wealthy, and the gentle and noble men needy and miserable: or else excluding gentility all men must be of one degree and sort, and a new name provided. Forasmuch as Plebs in Latin, and commoners in English, be words only made for the discrepencey of degrees, whet of proceedeth order: which in things, as well naturall or supernaturall hath ever had such a preeminence, there thereby the incomprehensible majestie of God, as it were by a bright flame of a Torch or candle, is declared to the blinde inhabitants of this world. [35, Bk. 1, p. 2]

Elyot's shift in style—concise in *The Castle* but elevated in *The Gouernour*—illustrates an important point about many of the early English writers who wrote both humanistic and technical books. Elyot understood that those reading *The Gouernour* had been well educated and were accustomed to a Latinate style. In contrast, he sensed that readers accessing how-to books, such as *The Castle*, needed a different style: a direct, unadorned, concise, subject-verb-object presentation. Elyot apparently understood that that reading for information or for instruction differed from contemplative reading. Eliot's selection of a kind of "plain English" or low style for *The Castle* would be a style vigorously defended by Wilkins and the Royal Society a century later. Or Elyot, because of his humanist training, may also have been following Cicero's position that a plain style was most suitable for instruction.

A similar difference occurs in the works of Gervase Markham, perhaps the first technical writer in England, who authored more than 40 works on horses, farming, estate management, animal husbandry, military training for soldiers, as well as various literary pieces, such as *Arcadia* [37]. In this dream vision, we find a languid sentence structure, and paragraphs are often little more than indentations, perhaps in keeping with the dreamlike quality of this work. The opening of the *Arcadia* (1607) follows:

> At such time as the flowers appearing upon the earth, had summoned the ayrie choristers to entertain the first Ambassadors of the Spring, and that Nature (delivered through the barraine wombe of Winter) had showed her selfe livelie as the morning, faire as the nightes Gouenesse, pure as the sun, and as almighty as an army of invisible fortune. The unhappy and forlorn Shepard *Credula* being come to the foot of the mountain *Tagetus*, from whose late distributed skirts, ran an even and well leveled plain through which the silvering flowing *Ermanthus* had made many curious and enterchangeable windings, till she delivered the abundance of her tribute into the bottom of the Ocean's unruely treasurie. [37, p. 1]

Is this a paragraph? Probably not, and most "paragraphs" throughout *Arcadia* are not by Bain's standards. However, Markham uses a different style and a different approach to paragraphing in his how-to books, as the following excerpt from *Country Contentments: Or The Husbandmans Recreations* (1649) [38], one of Markham's popular books on sports written for new country property owners:

> Know then, that if you Angle in any Pond or standing water, you shall before you fall to your businesse, with your Plumbe, found the water in divers places, and where you find it deepest, blakest, and least transparent, there you shall stand to Angle, placing your selfe under the banke, and if it be possible, so as your shadow may bee carried from the water: For you must at som time, if you can chuse, let your shadow lye upon the water: and although in these deep laces your standing open or close, are either of them reckoned indifferently, because the waters depth is a sufficient concealement, yet the closer you stand is accounted amongst Angler most handsome and artificial.
>
> But if you go to angle at the River, then the best place to cast in your Line, is where it is deepest and clearest, so as you may behold the Sand or Gravell at the bottome: and in these places you shall strive to conceale your selfe as much as possible, as standing behind Poplars, Oziers, or other Trees, or under the cover of some Banke, Rocke, or other ruines at the side of the River: also in covert places, where are many Weekes, rootes of Trees, and other rubbish, is good Angling, but very troublesome, for Fish lying there warm and in safety, will have a great resort thereunto, freely, so that the Angler must be careful in the putting in of his hooke, and very deliberate in striking, lease doing any thing rashly, he breake his Line and Hooke, being never so little intangled. [38, Bk. 1, p. 75].

These paragraphs exemplify Markham's paragraphs, formed from coherent, unified groups of sentences, in which content occurs in sequence. He may or may not use topic sentences—in these two paragraphs he does not. He also used parallelism as shown in Figure 3-5, paragraph 2. The chapter title and the marginal headings reveal the sequence of the paragraph. The first paragraph

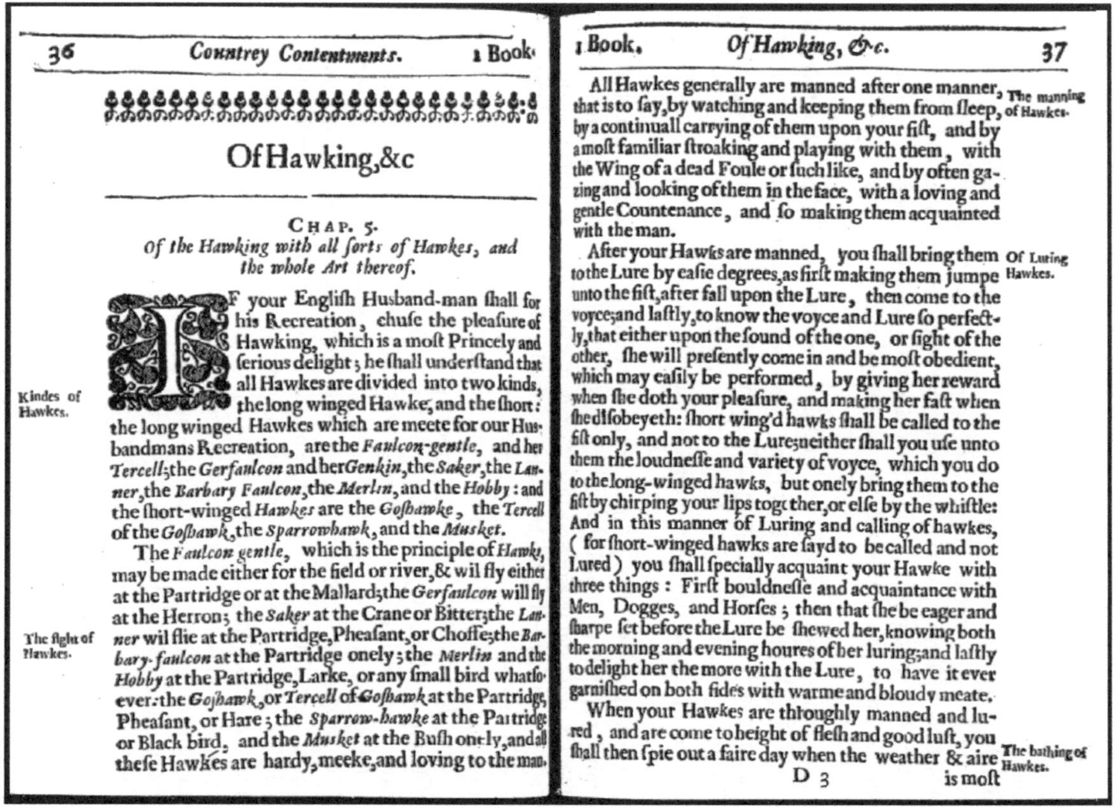

36 *Countrey Contentments.* 1 Book.

Of Hawking, &c

CHAP. 5.

Of the Hawking with all ſorts of Hawkes, and the whole Art thereof.

IF your Engliſh Husband-man ſhall for his Recreation, chuſe the pleaſure of Hawking, which is a moſt Princely and ſerious delight; he ſhall underſtand that all Hawkes are divided into two kinds, the long winged Hawke, and the ſhort: the long winged Hawkes which are meete for our Husbandmans Recreation, are the *Faulcon-gentle*, and her *Tercell*; the *Gerfaulcon* and her *Genkin*, the *Saker*, the *Lanner*, the *Barbary Faulcon*, the *Merlin*, and the *Hobby*: and the ſhort-winged *Hawkes* are the *Goſhawke*, the *Tercell* of the *Goſhawk*, the *Sparrowhawk*, and the *Musket*. — Kindes of Hawkes.

The *Faulcon gentle*, which is the principle of *Hawks*, may be made either for the field or river, & wil fly either at the Partridge or at the Mallard; the *Gerfaulcon* will fly at the Herron; the *Saker* at the Crane or Bitter; the *Lanner* wil flie at the Partridge, Pheaſant, or Choffe; the *Barbary-faulcon* at the Partridge onely; the *Merlin* and the *Hobby* at the Partridge, Larke, or any ſmall bird whatſoever: the *Goſhawk*, or *Tercell* of *Goſhawk* at the Partridge, Pheaſant, or Hare; the *Sparrow-hawke* at the Partridge or Black bird, and the *Musket* at the Buſh onely, and all theſe Hawkes are hardy, meeke, and loving to the man. — The flight of Hawkes.

1 Book. *Of Hawking, &c.* 37

All Hawkes generally are manned after one manner, that is to ſay, by watching and keeping them from ſleep, by a continuall carrying of them upon your fiſt, and by a moſt familiar ſtroaking and playing with them, with the Wing of a dead Foule or ſuch like, and by often gazing and looking of them in the face, with a loving and gentle Countenance, and ſo making them acquainted with the man. — The manning of Hawkes.

After your Hawks are manned, you ſhall bring them to the Lure by eaſie degrees, as firſt making them jumpe unto the fiſt, after fall upon the Lure, then come to the voyce; and laſtly, to know the voyce and Lure ſo perfectly, that either upon the ſound of the one, or ſight of the other, ſhe will preſently come in and be moſt obedient, which may eaſily be performed, by giving her reward when ſhe doth your pleaſure, and making her faſt when ſhe diſobeyeth: ſhort wing'd hawks ſhall be called to the fiſt only, and not to the Lure; neither ſhall you uſe unto them the loudneſſe and variety of voyce, which you do to the long-winged hawks, but onely bring them to the fiſt by chirping your lips together, or elſe by the whiſtle: And in this manner of Luring and calling of hawkes, (for ſhort-winged hawks are ſayd to be called and not Lured) you ſhall ſpecially acquaint your Hawke with three things: Firſt bouldneſſe and acquaintance with Men, Dogges, and Horſes; then that ſhe be eager and ſharpe ſet before the Lure be ſhewed her, knowing both the morning and evening houres of her luring; and laſtly to delight her the more with the Lure, to have it ever garniſhed on both ſides with warme and bloudy meate. — Of Luring Hawkes.

When your Hawkes are throughly manned and lured, and are come to height of fleſh and good luſt, you ſhall then ſpie out a faire day when the weather & aire is moſt — The bathing of Hawkes.

D 3

Figure 3-5. Gervase Markham, *Country Contentments: Or the Husbandman's Recreation* (1649).

describes the varieties of hawks. The Faulcon-gentle, as announced in this paragraph, becomes the topic sentence of the next paragraph. The sequence of training hawks then begins, as illustrated in the third and fourth paragraphs and the beginning of the fifth. Markham uses a modern strategy of linking the paragraphs via the phrasing of the topic sentences.

RAMISTIC RHETORIC AND ITS EFFECT ON THE ENGLISH PARAGRAPH

As the previous examples of technical writing illustrate, the English paragraph was fully operative by the end of the 16th century and the early decades of the 17th century. These paragraphs occur within a framework dictated by a heading announcing the topic followed by a paragraph characterized by completeness, coherence, logic, unity, and in many cases internal parallel structure of supporting sentences. As shown in Markham, paragraphs without topic sentences nevertheless often exemplify coherence, unity, sequence, and completeness of thought.

At this point, we may ask, why would technical writing nurture development of the English paragraph, as defined by Bain and entrenched in modern paragraph theory? The answer to that question links to the next major philosophic influence on English prose—the rhetoric of Peter Ramus, whose principles heavily shaped technical writing during 1550–1670. Ramus' theory was particularly palatable to technical writers in numerous fields because technical writing usually involves procedures, instructions, technical description, and process description/analysis. In each of these genres, sequence is critical to the content, as are unity, coherence, and completeness. The development of typography, the demand for books by middle-class readers, and the spread of literacy worked to improve ways of presenting information.

As I have shown in other studies [6], Ramist rhetoric, used in numerous technical books of the English Renaissance, reshaped not only complete documents but also elements of documents: introductions, overviews, and paragraphs. In addition, Ramist rhetoric fostered use of visuals integrated into well-structured works that exhibited readable format and page design. Ramist principles, when applied to the arrangement of text on the page, exemplifies the power of print technology to present readable content. Truth could be efficiently displayed, visualized, understood, and remembered. For followers, remembering text had critical importance.

Ramus' logic, although it was replaced by Cartesian logic in the latter 17th century, needed less than a century to leave its indelible mark on the consciousness or writers who learned to use technology and visual display of information as a rhetorical strategy. To revitalize Oxford's curriculum, which was languishing from the effects of degenerate scholasticism, Henry VIII decreed in 1530 that Agricola's dialectic be taught along with the works of Aristotle. Ramus incorporated Agricola's dialectic to attack the medieval logicians' interpretation of Aristotle, which Ramus believed was confusing and useless. It gave readers, particularly students, no practical understanding of the concepts to which it was applied. With its endless categories and enumeration of disconnected predicables, which existed only for the sake of disputation, logic had lost its practicality to help students learn, understand, and remember subject matter. Ramus' goal was to make the truth of any concept accessible and memorable by visual organization based on the principles of partition and division.

Agricola and ultimately Ramus' logic used a schematic arrangement of logical terms. By first defining and then dividing concepts into increasingly discrete entities and then using brackets to display these dichotomies, Agricola and Ramus created a schematic diagram of the components of the idea to be presented. Ramistic logic became a method of "laying things out in series" in a descending order of generality. Material from any discipline could be "sown asunder" by this kind of logic. Thus, to Ramus and his followers, logic became a picture of reality. Visual display became an important catalyst in strengthening organization of text and making the presentation more evident, particularly in practical books.

Invention, for Ramus, became synonymous with arrangement, that is, finding the appropriate dichotomies or partitions within a concept. For Ramus, invention discovers, or lays open to view, that dichotomous or clearly partitioned arguments such as these exist. Format and organization—of complete works, sections, and paragraphs within these works—reveal the truth of the concept and allow truth to be seen and understood. To minds tired of degenerate scholasticism that extracted one proposition from another with no goal beyond disputation for the sake of disputation, Ramus' method offered refreshing relief.

Ramus' influence on many important technical books from 1550 to 1650 was enormous. Even if brackets were not used (and many times they were), rigid organization and then partition of ideas occurred in major medical works, such as *Certaine workes of chirurgerie*, by Thomas Gale (1563) [39], *A most excellent and Learned Woorke of Chirurgerie*, by John Halle, a Chirurgiun (1565) [40]; *Tables of Svrgerie*, by Robert Caldwell (1585) [41], and a host of others. Each of these well-educated physicians sought to improve the knowledge of medicine or surgery for English physicians who lacked access to medical education outside England. Each also wrote to teach medical students, so clarity of presentation was essential.

Ramism defines the works of Alexander Read, a well-known English surgeon. Read's *The chirurgical lectures of tumors and ulcers delivered in the Churgeons Hall* [30] combined text developed according to Ramist method with bracketed tables. The lectures opened with a bracketed summary of the content of the work. Then a bracketed summary precedes each discourse section, as shown in Figure 3-1. The work shows Read's painstaking effort to produce a text that would be easy for students to follow during lectures and during individual study. The content of each chapter is immediately evident, as also shown in Figure 3-1.

The work as a whole, as well as each individual page, was designed for efficient reading and memorization.

Works emphasizing text rather than bracket divisions nevertheless used a similarly rigorous division of content. For example, *A Worthy Treatise of the eyes* (1587) [42] by Jacques Guillemeau begins every paragraph with a topic sentence, usually a definition. Most of these define the point to be discussed followed by a sequential description of the process and cure, a strategy that persists in this work (see Figure 3-6). Guillemeau's approach is echoed in John Banister's *A needefull, new, and necessarie treatise of chyrurgerie* (1575) [43].

As Ong (1982) states, Ramus was one of the humanist educators who "were pupil-oriented, interested in the kind of person whom they were producing quite as much as in the professional attitudes towards the subjects that they were teaching." These educators subscribed to methods "that show the overpowering tendency to view knowledge in terms of those to whom it was purveyed" [31, p. 173]. However, as I have noted, writers such as Sir Thomas Elyot apparently anticipated the importance of readability and used a visual method of displaying content that anticipated Agricola and Ramus, although Elyot may well have known about Agricola before he published his first edition of *The Castle* in 1539 [36]. That design, which emerges from the desire to teach clearly, definitely affected how Vesalius developed his anatomical drawings in the 16th century, which I argue in Chapter 4, had a power to impact on the design of all anatomical works in England in the 17th century.

Paragraphing in Technical Description in the 17th Century

When truth became associated with visual clarity, as it did with Robert Hooke, author of *Micrographia*, [44] Nehemiah Grew, a disciple of Hooke, used Ramist strategy in developing *The Anatomy of Plants with an Idea of a Philosophical History of Plants* (1682) [45]. The effects in paragraphing and integration of visuals and text were impressive. As Ong observed, Ramist philosophy, combined with typography, "heightened the value of the visual imagination and the visual memory over the auditory imagination and the auditory memory" [46, p. 167], a major step away from orality toward textuality in an English society in which literacy became critical. The result was a change in sensibility driven by a philosophy approving the use of typography to diagram and visually organize ideas. Grew's work clearly illustrates this point. As we will see in Grew's *The Anatomy of Plants with an Idea of the Philosophical History of Plants* [45], Ramist dialectic and rhetoric used schematic arrangement of texts and visuals to present technical description.

Nehemiah Grew, one of the first botanists to use a microscope to study plant morphology, was a disciple of Robert Hooke. Both knew each other's work and the role of microscopy in opening a new world of scientific study. Grew explored

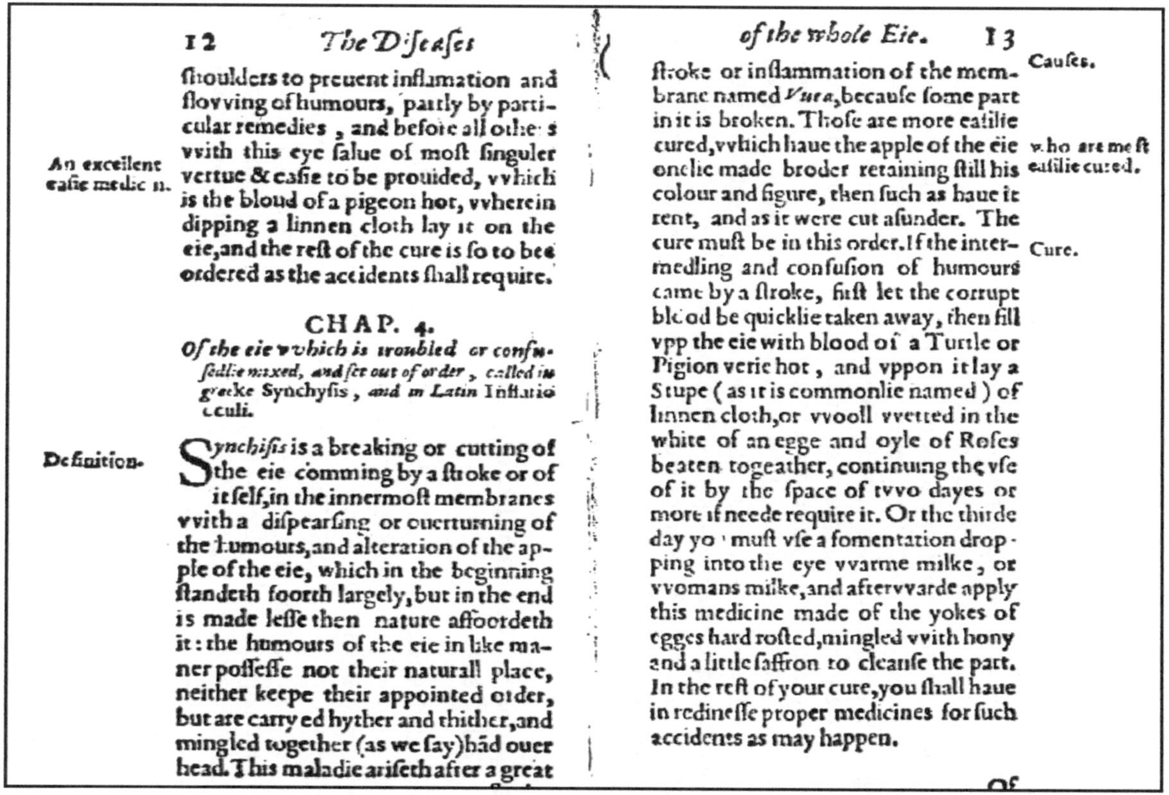

12 *The Diſeaſes*

ſhoulders to preuent inflamation and flovving of humours, partly by particular remedies, and before all othe s vvith this eye ſalue of moſt ſinguler vertue & eaſie to be prouided, vvhich is the bloud of a pigeon hot, vvherein dipping a linnen cloth lay it on the eie, and the reſt of the cure is ſo to bee ordered as the accidents ſhall require.

An excellent eaſie medicin.

CHAP. 4.

Of the eie vvhich is troubled or confuſedlie mixed, and ſet out of order, called in greeke Synchyſis, *and in Latin* Inſtatio oculi.

Definition.

Synchiſis is a breaking or cutting of the eie comming by a ſtroke or of it ſelf, in the innermoſt membranes vvith a diſpearſing or ouerturning of the humours, and alteration of the apple of the eie, which in the beginning ſtandeth foorth largely, but in the end is made leſſe then nature affoordeth it: the humours of the eie in like maner poſſeſſe not their naturall place, neither keepe their appointed order, but are carryed hyther and thither, and mingled together (as we ſay) hād ouer head. This maladie ariſeth after a great

of the whole Eie. 13

ſtroke or inflammation of the membrane named *Vuea*, becauſe ſome part in it is broken. Thoſe are more eaſilie cured, vvhich haue the apple of the eie onelie made broder retaining ſtill his colour and figure, then ſuch as haue it rent, and as it were cut aſunder. The cure muſt be in this order. If the intermedling and confuſion of humours came by a ſtroke, firſt let the corrupt blood be quicklie taken away, then fill vpp the eie with blood of a Turtle or Pigion verie hot, and vppon it lay a Stupe (as it is commonlie named) of linnen cloth, or vvooll vvetted in the white of an egge and oyle of Roſes beaten togeather, continuing the vſe of it by the ſpace of tvvo dayes or more if neede require it. Or the thirde day you muſt vſe a fomentation dropping into the eye vvarme milke, or vvomans milke, and afterwarde apply this medicine made of the yokes of egges hard roſted, mingled vvith hony and a little ſaffron to cleanſe the part. In the reſt of your cure, you ſhall haue in redineſſe proper medicines for ſuch accidents as may happen.

Cauſes.

vvho are moſt eaſilie cured.

Cure.

Of

Figure 3-6. Example Paragraphs from Jacques Guillemeau, *A Worthy Treatise of the eyes* (1587).

plant physiology and described it in his *Anatomy of Plants* [45], first published in 1674. From the perspective of paragraph structure and development, *The Anatomy of Plants* offers a plethora of well-designed paragraphs.

For example, the introduction, "An Idea of a Philosophical History of Plants," is an essay composed of 63 numbered paragraphs [45]. Each paragraph is indented before the number. An early Latin paragraph mark occurs before the topic sentence. Like Hooke's *Micrographia* [44], the *Anatomy* effectively links paragraphs, often with transitional words, a method often used today. As readers, we may scan this essay via the topic sentences for each paragraph and observe the content. Paragraphs 26–29 link and are introduced by the topic sentence of paragraph 26:

> 26. § Likewise their Colours, Smells, and Tastes: The general and particular Kinds of all which should be noted. And to what Contained Parts, and in what Variety, they appertain. So most Resinous Gumms are Tinctur's, some not; as that which drops from the Domestick Pine, is as clear as Rock-water. The Milks of some Plants are Paler, as in Burdock; of other Whiter, as in Dandelyon, Scorzonera; Citrine, as in the Root of Tracheliumk Angelica; Yellow, as in Lovage. In some plants, Odorous, as in Umbelliserous; in other now, as in Cichoraceojs. That of Little Maple, Tasteless; of Garden Chervil, Sweet; of Fenil, Hot; of Scorzonera, Astringent; of Dandelion, Bitter; and generally, in other plants; but with many degrees of Strength, and in conjunction with other Tasts. But most Mucilages, have little either Colour, Taste, or Smell; and the like. Here also the same Qualities are to be inquired into, as, in general speaking, they are said to belong to a Vegetable. Since it is more than probable, that all Colours (excepting White, which is sometimes common both to Containing and Contained Parts) all Odours, and Tastes, which are more immediately, and without a resolution of their Essential Principles, perceptible in a Plant; are not ascribable either to the Organical, or Containing Parts, but only to Those, contained in them; as some diverse reasons hereafter may appear.

The remaining topic sentences of paragraphs 27–29 appear as follows:

> 27. And first, their Colours; where, with respect to several Plants and Parts, they are more Changeable; as Red, in Flowers; or Constant, as Green, in Leaves.
>
> 28. Next their Odours: what may be their principal Seat; whether one or divers seats in the same Plant.
>
> 29. But especially their Tastes, which it much importeth us more precisely to distinguish. First, by their general Kinds; for the number, even of these may be coputer greater than usually it is. [45, pp. 12–13]

Figure 3-7 illustrates how carefully Grew builds his paragraphs. He opens Book III "Of the Trunk," by referring to the bean (Book I) and the root (Book II). He then begins his discussion on trunks by another series of carefully structured and linked paragraphs. Figure 3-8 shows the drawings of these descriptions, to which readers are alerted by the marginal notes. Each visual, which allows readers to see the part of the plant he describes, is carefully integrated into the text.

Grew develops his technical description of plant parts, the bulk of *The Anatomy* [45], as carefully as he does his introduction to the work, which is basically a longer version of the same structure we examined in *The Gouernayle of Helthe* (1490) [29].

TOWARD A NEW THEORY OF THE EMERGENCE OF THE ENGLISH PARAGRAPH

Examples shown in this analysis of the early history of the English paragraph show that paragraphs, as they were defined by Bain and his followers at the end of the 19th century, existed at least 400 years earlier. While the scope of this study focuses only on paragraphs in the first printed technical books, these examples show that the writers were adept at developing paragraphs. The use of various paragraph marks in these works, some dating from the 10th century, according to Lewis' dissertation [17, p. 11], further illustrate that the paragraph was familiar to the writers/translators of these works. Examples of well-developed paragraphs can be found in several 12h century classic English works, such as *The Peterborough Chronicle* and *Ancrene Wisse*; but we can also find distinct paragraphing in some of the earliest estate management manuscripts, such as *The Seneschaucy*, *Walter of Henley*, and *The Rules of Robert Grosseteste* [47]. In short, the history of the English paragraph remains far from complete. Yet when it is written, history must take into account practical writing in addition to liturgical, humanist, and historical texts. Students of the paragraph must recognize that work written and published before the 19th century has much to offer in understanding how early English writers used the paragraph; how they viewed it in their development process.

Why would early technical books exemplify the distinctive qualities of the paragraph, as described and affirmed by the first paragraph compositionists? Is the paragraph inextricable from the construction of English prose? Could humanists such as Bain have been unconsciously influenced by many of the how-to books published throughout England? Did these philologists simply assume that writing outside the pale of academic study was not germane? Whatever the answers (and these may never be known), the paragraphs I have presented in this chapter illustrate the following:

CHAP. III.

Of the TRUNK.

HAVING thus declar'd the degrees of *Vegetation* in the *Root*; the continuance hereof in the *Trunk* ſhall next be ſhew'd: in order to which, the Parts whereof this likewiſe is compounded, we ſhall firſt obſerve.

1. §. That which without diſſection ſhews it ſelf, is the *Coarctture:* I cannot ſay of the *Root*, nor of the *Trunk*; but what I chooſe here to mention, as ſtanding betwixt them, and ſo being common to them both; all their *Parts* being here bound in cloſer together, as in the tops of the grown *Roots* of very many *Plants*, is apparent.

2. §. Of the *Parts* of the *Trunk*, the firſt occurring is its *Skin:* The Formation whereof, is not from the Air, but in the *Seed*, from whence it is originated; being the production of the *Cuticle*, there inveſting the two *Lobes* and *Plume*.

3. §. The next *Part* is the *Cortical Body*; which here in the *Trunk* is no new ſubſtantial Formation; but, as is that of the *Root*, originated from the *Parenchyma* of the *Plume* in the *Seed*; and is only the increaſe and augmentation thereof. The *Skin*, this *Cortical Body* *Tab.3.f.1,* or *Parenchyma*, and (for the moſt part) ſome Fibers of the *Lignous* *& 4.* mixedherewith, alltogether make the *Barque*.

4. §. Next, the *Lignous Body*, which, whether it be viſibly divided into many ſofter *Fibers* or ſmall *Threads*, as in the *Bean*, *Fennel*, and moſt *Herbs*; or that its Parts ſtand more compact and cloſe, *Tab.3.f.1.* ſhewing one hard, firm and ſolid piece, as in *Trees*; it is, in all, one and the ſame *Body*; and that not formed originally in the *Trunk*, but in the *Seed*; being nothing elſe but the prolongation of the *Seminal Root* diſtributed in the *Lobes* and *Plume* thereof.

5. §. Laſtly, The *Inſertions* and *Pith* are here originated likewiſe from the *Plume*, as the ſame in the *Root*, from the *Radicle:* So that as to their *Subſtantial Parts*, the *Lobes* of the *Seed*, the *Radicle* and *Plume*, the *Root* and *Trunk* are all one.

6. §. Yet ſome things are more fairly obſervable in the *Trunk*. Firſt, the *Latitudinal* ſhootings of the *Lignous Body*, which in *Trunks* of ſeveral years growth, are apparent in ſo many *Rings*, as is commonly known. For ſeveral young Fibers of the *Lignous Body*, as in the *Tab.3.f.5,* *Root*, ſo here, ſhooting in the *Cortical* one year, and the ſpaces be- *& 8.* twixt them being after fill'd up with more (I think not till) the next, at length they become altogether a firm compact *Ring*; the *Perfection* of one *Ring*, and the *Ground-work* of another, being thus made concomitantly.

Figure 3-7. From Chapter on Trunks, Grew's *The Anatomy of Plants* (1682).

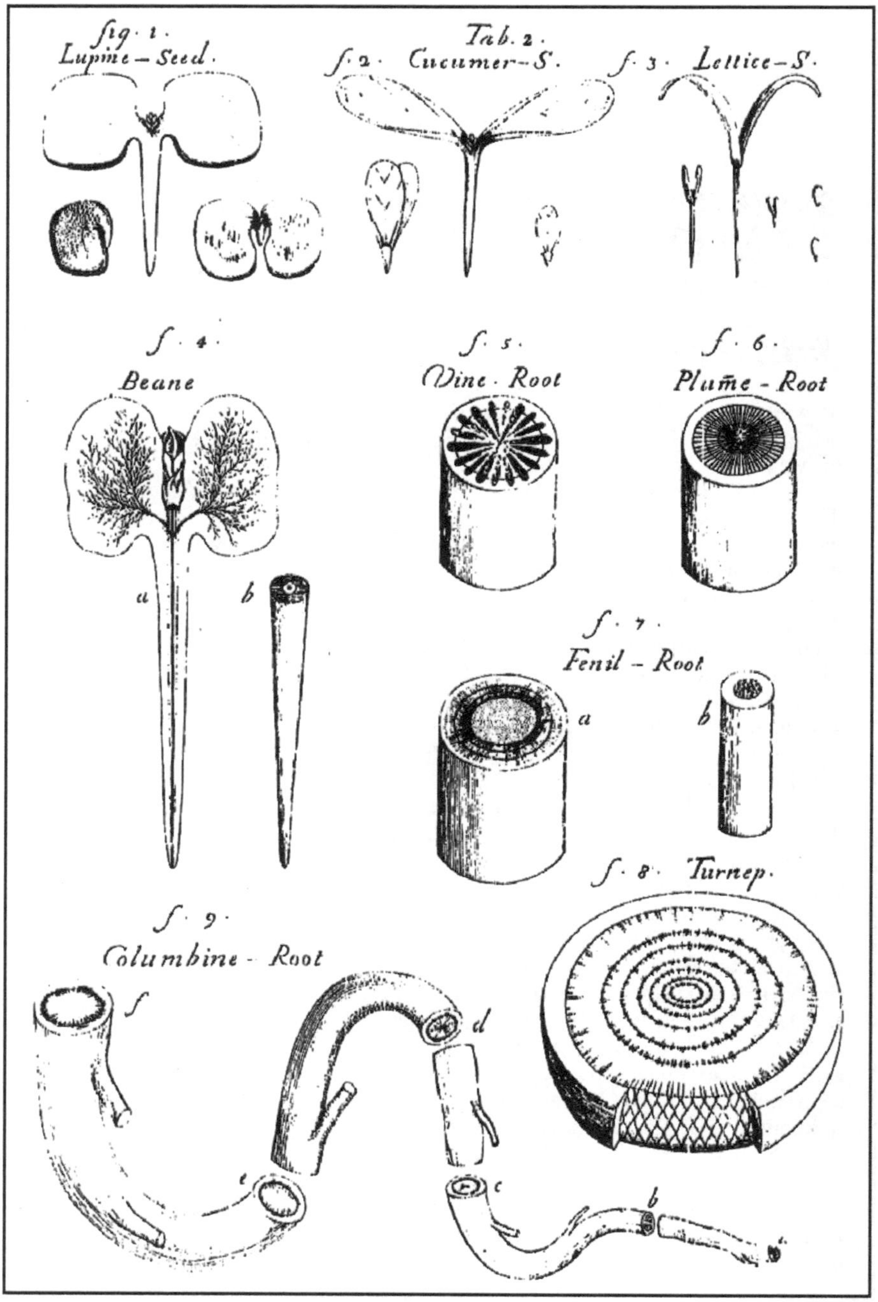

Figure 3-8. Drawings to Accompany Figure 3-7: Grew
The Anatomy of Plants.

1. English technical writing books, from the beginning of printing in England, included headings that helped readers locate information. Even printed texts that began with linear text moved to formatted text. The use of headings containing descriptors of the content that followed may have nurtured the development of topic sentences.
2. In contrast to religious, philosophical, and historical published writing of the English Renaissance, how-to books used a direct, concrete style. Writers who wrote both technical and philosophical/literary works seemed to understand that readers of technical books needed a different style as they sought to follow instructions. Headings provided one of the distinctive differences between humanist and technical books. In addition, technical books generally showed better paragraphing than humanist books. Ease of access seems to have been important to early technical writers.
3. Technical instructions were carefully organized, sequenced, unified, and coherent, as instructions must be to help readers follow them. Thus, paragraphs in technical books printed as early as the 15th century show unity, organization, and coherence, along with either topic sentences/phrases or opening sentences that contained "topic" ideas that echoed the "topic" words of the heading.
4. Paragraphs in technical books tended to be deductive, or "loose," rather than inductive or periodical, perhaps to aid readability. Both conciseness and readability appear have been important qualities to the first writers/printers of these English printed books.
5. The early printed technical books did not reflect Latinate style—long, periodical sentences composed of "clause heaps." Rather, they exemplified Anglo-Saxon sentence structures: subject-verb-object sentences and phrases designed to be direct and easy to process.

In his excellent essay, "Whatever Happened to the Paragraph?" Mike Duncan [19], like many other scholars, does not question the origin of the English paragraph. Instead he relies on Bain's essay and Lewis' dissertation as an adequate discussion of the paragraph's history. However, Duncan does conclude that a theory or reexamination of the paragraph—what it is, how it works, how it should be designed—should be attempted. As Duncan notes, the requirements for paragraphs remain unsettled, but "few would contest that robust paragraph formation is essential to effective writing" [19, p. 124].

While I am not recommending a theory of the paragraph, I do believe that the history of the English paragraph should be considered in discussing paragraph formation and effectiveness. Inclusion of how-to documents in the history of the paragraph suggest that it is clearly indigenous to English composition, and many Renaissance English technical writers used well-designed paragraphs to present content. A preponderance of paragraphs that I have examined over the past two decades of research on the history of English technical writing shows that Bain's requirements were prevalent in English Renaissance books, but technical writing seems to have been the first printed English texts to exemplify completeness, sequence, unity, coherence, and even conscious use of internal parallelism—the fundamental elements of the paragraph.

As I discussed in 1990 [48], much English Renaissance technical writing illustrates careful attention to the needs of readers. Many Renaissance technical writers apparently understood that how-to books should be easy to read, that "reading to learn to do" [49] required a different format and style than contemplative reading, and adapted Cicero's directives on style to the technical book. Many English printed technical books exemplify many of the same format and page design strategies still used today, and many of these can be traced to the influence of the rhetoric of Peter Ramus. Readability of text is enhanced by use of topic sentences, lists, headings, subheadings, white space, and a variety of page design strategies. Thus, the cognitive aspect of paragraphing, discussed by Duncan, clearly has roots in the 16th century.

Additional studies of the history of the paragraph in nonhumanist texts are needed. A study of the first scientific papers, many from the early decades of the 18th century, will provide a sequel to my observations. Similarly, analysis of manuscripts of medieval how-to works will likely illuminate the indigenous nature of the English paragraph and tell us how writers 500 to 700 years ago viewed the paragraph and its role in the construction of extended texts.

In short, the history of the paragraph, ignored since 1866, as compositionists looked forward rather than backward, provides a research area that, when more fully developed, will tell students of the history of English more about how authors in all periods of English constructed their texts. Perhaps the paragraph as we know it today is not only inextricable but also indigenous to how our language works.

REFERENCES

1. E. Tebeaux, *The Emergence of a Tradition: Technical Writing in the English Renaissance, 1475-1640,* Baywood's Technical Communication Series, Baywood, Amityville, New York, 1997.
2. E. Tebeaux, Pillaging the Tombs of Noncanonical Texts: Technical Writing and the Evolution of English Style, *Journal of Business and Technical Communication, 18,* pp. 165-197, 2004.
3. R. Adolph, *The Rise of Modern Prose Style,* MIT Press, Cambridge, Massachusetts, 1968.
4. J. Mueller, *The Native Tongue and the Word: Developments in English Prose Style 1380-1580,* University of Chicago Press, Chicago, Illinois, 1984.
5. D. Bush, *English Literature in the Earlier Seventeenth Century* (2nd ed.), Cambridge University Press, New York, 1962.

6. E. Tebeaux, Ramus, Visual Rhetoric, and the Emergence of Page Design in Medical Writing of the English Renaissance: Tracking the Evolution of Readable Documents, *Written Communication, 8*:4, pp. 411-445, 1991.
7. A. Bain, *English Composition and Rhetoric,* Longmans, Green, and Co., London, 1877.
8. E. Lindemann, A Rhetoric for Writing Teachers (4th ed.), Oxford University Press, New York, 2001.
9. A. D. Hepburn, *Manual of Rhetoric,* Wilson, Hinkle & Co., 1875.
10. J. H. Genung, *The Practical Elements of Rhetoric, with Illustrative Examples,* Ginn & Company, Boston, Massachusetts, 1886.
11. F. Scott and J. Denney, *Composition and Rhetoric,* Allyn and Bacon, Boston, Massachusetts, 1898.
12. F. Scott and J. Denney, *Paragraph-Writing* (3rd ed.), Allyn and Bacon, Boston, Massachusetts, 1896.
13. P. Rodgers, Alexander Bain and the Rise of the Organic Paragraph, *Quarterly Journal of Speech, 51*:4, pp. 339–408, 1965.
14. J. R. Boyd, *Elements of Rhetoric and Literary Criticism,* Harper and Brothers, New York, 1844.
15. H. N. Day, *Elements of the Art of Rhetoric,* A. S. Barnes & Co., New York, 1850.
16. B. Wendell, *English Composition: Eight Lectures Given at the Lowell Institute, 1905,* F. Ungar Pub. Co., New York, 1963.
17. E. Lewis, *The History of the English Paragraph,* University of Chicago Press, Chicago, Illinois, 1894.
18. G. Saintsbury, *Specimens of English Prose Style from Malory to Macaulay, Selected and Annotated,* with an introductory essay by G. Saintsbury, K. Paul, Trench & co., London, p. xix, 1885.
19. M. Duncan, *Whatever Happened to the Paragraph? Style in Rhetoric and Composition: A Critical Sourcebook*, P. Butler (ed.), Bedford/St. Martins, Boston and New York, pp. 108-131, 2010.
20. H. S. Bennett, *English Books & Readers, 1475-1557,* Cambridge University Press, Cambridge, England, 1952.
21. H. S. Bennett, *English Books & Readers 1558-1603,* Cambridge University Press, Cambridge, England, 1965.
22. H. S. Bennett, *English Books & Readers, 1603-1640,* Cambridge University Press, Cambridge, England, 1970.
23. L. Wright, *Middle-Class Culture in Elizabethan England,* Cornell University Press, Ithaca, New York, 1935.
24. T. Laquer, The Cultural Origins of Popular Literacy In England, 1550-1850, *Oxford Review of Education, 2,* pp. 255-275, 1976.
25. R. Whately, *The Elements of Rhetoric, 1850: A Facsimile Reproduction,* Scholar's Facsimiles & Reprints, Delman, New York, 1991.
26. T. Blount, *The academie of eloquence. Containing a compleat English rhetorique, exemplified, with common-places, and formes, digested into an easie and methodical way to speak and write fluently, according to the mode of the present times,* London. Printed by T. N. for Humphrey Moseley, at the Princes-Arm's in S. Paul's Churchyard, 1654.

27. R. Rainolde, *The Foundacion of Rhetorike,* Scholars' Facsimiles & Reprints, Delman, New York, 1543.
28. N. Barron and S. Naomi, Commas and Canaries: The Role of Punctuation in Speech and Writing, *Language Sciences, 23*:1, pp. 15-67, 2001.
29. J. De Burgundia, *In this treatyse that is called Gouernayle of Helth,* Caxton, Westminster, 1490.
30. A. Read, *The Chirurgical Lectures of Tumors and Ulcers Delivered in the Churgeons Hall*, London, Pauls Churchyard, 1635.
31. W. S. J. Ong, *Orality and Literacy: The Technologizing of the Word,* Methuen, New York, 1982.
32. *Here begynneth the proprytees and medycynes for hors,* [S.l.: W. de Worde, 1502?
33. H. Von Brunschweig, *The Noble experience of the Virtuous Handy Warke of Surgeri* Imprynted at London In Southwarke by Petrus Treueris, 1525.
34. D. Guy, de Chauliac, *The Questyonary of Cyrurgyens,* printed by Robert Wyer, London, 1542.
35. T. Elyot, *The boke Named the Gouernour,* Thomas Marsh, London, 1557.
36. T. Elyot, *The Castle of Helth,* Thomas Marshe, London, 1561.
37. G. Markham, *The English Arcadia alluding his beginning from Sir Philip Sydneys ending, By Iaruis Markham.* London : Printed by Edward Allde, and are to bee solde by Henrie Rocket, at his shop vnder Saint Mildreds Church in the Poultrie. 1607.
38. G. Markham, *Country Contentments: Or the Husbandman's Recreations,* William Wilson, St. Paul's, 1649.
39. T. Gale, *Certaine works of chirurgerie* (4 parts), R. Hall, London, 1563.
40. J. Halle, *A most excellent and Learned Woorke of Chirurgerie,* Thomas Marshe, London, 1565.
41. R. Caldwell, *Tables of Svrgerie, Brieflie Comprehending the whole art and Practise Therof In a Maruelous GoodMmethod,* H. Denham, London, 1585.
42. J. Guillemeau, *A Worthy Treatise of the Eyes, Contauning the Knowledge and the Cure of One Hundredth and Thirteen Diseases, Incidens vntoTthem,* Robert Waldegrane, London, 1587.
43. J. Banister, *A Needefull, New, and Necessarie treatise of Chyrurgerie,* Thomas Marshe, London, 1575.
44. R. Hooke, *Micrographia: Or Some Physiological Descriptions of Minute Bodies Made by Magnifying Glasses,* J. Martyn and J. Allestry, London, 1665.
45. N. Grew, *The Anatomy of Plants with an Idea of a Philosophical History of Plants,* London, 1682.
46. W. S. J. Ong, *Ramus, Method, and the decay of Dialogue,* Harvard University Press, Cambridge, Massachusetts, 1958.
47. D. Oschinsky, *Walter of Henley and Other Treatises on Estate Management and Accounting*, Clarendon Press, Oxford, United Kingdom, 1971.
48. E. Tebeaux, Books of Secrets—Authors and Their Perception of Audience in Procedure Writing of the English Renaissance, *Issues in Writing, 3*:1, pp. 41-67, 1990.
49. J. Redish, Reading to Learn To Do, *The Technical Writing Teacher, 15,* pp. 223-233 1988.

CHAPTER 4

Format, Page Design, and the Visual Display of Information: The Contribution of Renaissance English Technical Writing

As Chapters 1, 2, and 3 discussed and illustrated, the English plain style sentence and paragraph contributed to the organization and clarity of English Renaissance and 17th century technical writing, renowned for its use of visuals to allow readers to "see the text." But book technology and its relationship to the development of technical writing before 1700 have not been extensively discussed. However, as I have shown in *Emergence of a Tradition*, technical writing—most often occurring as instructional writing or descriptions of processes and objects frequently illustrated distinctive page format, document design, and illustrations.

When I began my study of early English technical writing, I read Pollard and Redgrave's *Short-Title Catalog*, as indicated in Chapter 1, and made a list of works that seemed, according to their titles, like technical writing. By examining nearly all the items on my list either as facsimiles or via microfilm, I was struck by the many examples of format and page design in "technical writing" that differed from religious, humanistic, and popular media printed during the period. Printers, authors, or perhaps both viewed much practical writing in a new way, as a form that needed to be accessed easily and read quickly. While Louis Wright suggested that simplified prose occurred because middle-class readers needed a less academic prose [1], we now know that many of these how-to works were written for use by a wide range of readers who used these texts to learn how to perform work.

Thus, style, page design, visuals, and format—enabled by typography, continued to improve throughout the period—set technical and how-to writing apart from other genres of English Renaissance and 17th century printed books. As a result of my three decades of research, I have concluded that many early authors, translators, and printers of these books knew that technical writing differed from literary, religious, historical, and other humanistic texts in one important way: Traditional texts would be carefully and often meditatively read. In contrast, technical writing helped readers perform tasks. Visuals that supported text and in some cases supplanted text enabled readers to follow descriptions and instructions more easily than linear text by itself. A number of small books composed only of visuals show that authors themselves knew which tasks could be conveyed with little or no text.[1] These "picture books" also suggest to modern readers what work was taught via oral instruction and which could (and needed to) be written, to enable readers to learn and apply new steps. In pursuing the Royal Society transactions and publications from 1665 to 1700, I found numerous illustrations, such as those discussed in Chapter 1. Numerous Royal Society Fellows found the value of visuals, drawings, and tabular display in explaining their research.

After 1500, printed English technical books, many translations of technical works printed in Germany, Italy, or France, included illustrations used without alteration from the original work. These illustrations were woodcuts—many carefully crafted—or copperplate engravings borrowed from their European sources. These borrowed illustrations show the expertise of Continental printers. In this chapter, I will demonstrate, via excerpts from a range of Renaissance printed technical books, how early technical writing pushed the boundaries of print from 1500 to 1700. In contrast to many humanistic and religious writers who continued to focus on illuminated or decorative books, technical writers and printers during this period found innovative ways to present content that would be easy for readers to understand and use. By the closing decades of the 16th century, technical writing exhibited distinctive visual characteristics enabled by print technology, which continued to develop throughout the 17th century.

This chapter will survey major examples of visual design that emerged in Renaissance and then 17th century technical writing and show their value to the development of technical writing in the topics discussed.

TECHNICAL WRITING AND THE VISUALIZATION OF CONTENT FURTHERED KNOWLEDGE

As the examples I include here show, print became a powerful means of sharing new knowledge in fields such as medicine and anatomy, mathematics, military science, navigation, shipbuilding, science, and botany—topics central to

[1] See, for example, Walter Gedde, A Book of Sundry Draughts (1615), which is composed of 100 pages of designs for windows. Grevase Markham, *Excellent and New Invented Knots and Mazes*, presents designs for use in planting maze hedges; *A Book of Engraved Lace Patterns* (1605), which provides a collection of designs for lace patterns with no commentary; and *A Booke of Curious and strange Inventions, called the first part of Needleworkes* (1596), which lacked prose and was composed only of pages of exquisite drawings of lace patterns.

the development of English commerce and middle-class industrialists in addition to the work of the Royal Society. Drawings, maps, diagrams, tables, and charts became critical to the conveyance of technical information, as shown in Chapter 1. Increasing knowledge made transmission of knowledge via oral methods inadequate: written instructions became essential. Seminal thinkers in these expanding fields saw the potential of printed texts to share their discoveries. The requirements of texts for these fields—innovative formats, page design, and visuals needed to insure usable, easily accessible text—produced works that had major impact. The medium was the message. Visual design became the hallmark of many how-to works that used print technology in new, impressive ways that linear text by itself did not.

RENAISSANCE TECHNICAL BOOKS AND THEIR DESIGN

Before examining early Renaissance and then 17th century technical works, we need to remember again that technical, or utilitarian, writing was not neatly partitioned into specific disciplines as it is today, that is, scientific writing, medical writing, and computer writing. Technical writing, as a field, did not exist. Showing its existence has been the product of modern technical communication history researchers. Technical writing, which appeared most often as how-to writing and technical description, first emerged in anatomy works and then in books on science, navigation, shipbuilding, military science, and agriculture. Without the groundbreaking work in anatomy, which I discuss first, much of what transpired in English technical writing would have been delayed. Anatomy books opened the door to new ways to present technical information. English technical writers of the 16th and 17th centuries were deeply indebted to works in countries beyond English shores. This indebtedness begins with anatomical works.

Andreas Vesalius—Anatomical Woodcuts as Art

The most important Renaissance challenge to print technology can be found in the work of Andreas Vesalius of Brussels, whose work was translated from Latin into English by Thomas Geminius (1553) [2] and then Thomas Willis (1654) [3]. These translations allowed English physicians access to Vesalius' groundbreaking revision of Galenic medicine. Because many English physicians studied outside England and brought home their knowledge, Vesalius' anatomical works influenced English anatomy as powerfully as the original Latin version influenced Continental anatomy. Vesalius' three works, *Tabulea Sex* (1537) [4], followed by *De Humani Corpis Fabrica* (1543) [5], and its companion volume, *Epitome* (1543) [6], all written and designed by Vesalius, stand as landmark works in establishing anatomy as a science. However, the knowledge presented by Vesalius would not have had the impact it had without print combined with a creative blend of format, typography, and illustration. As noted by Singer and Rabin, Vesalius' works mark

> the abrupt intrusion into a non-scientific renaissance society, the intellectual interests of which were centered on the ancient classics, . . . The *Fabrica* introduced a new method of representation to the printed book. . . of an immense and highly finished monograph in quite a new manner. Nothing of the sort or class had been printed before. [4, p. iii]

Or, as noted by Saunders and O'Malley, in both *Fabrica,* known "as one of the most magnificent volumes in the history of printing" [7, p. 9], and *Epitome*, a summary of *Fabrica,* illustration, text, and typography blend to achieve an unsurpassed work of art. In these two works, Vesalius captured human anatomy in exquisite detail that embodied the spirit of Renaissance art [7, p. 19].

The Visual Display of Information—A New Form of Communication

Ironically, however, before Vesalius' groundbreaking works, the use of illustrations for biological works was deemed vulgar, a perspective likely based on the perspective of humanistic scholars whose works advocated literary form—linear text with illuminated drawings [4, p. iv], wherein text assumed the primary position of the works. However, in his three works, Vesalius deemphasized text in favor of anatomical drawings with parts described in succinct phrases, as shown in *Fabrica* (see Figure 4-1) and *Epitome* (see Figure 4-2). Vesalius developed his illustrations as teaching tools or mnemonics for both physicians and students. He wanted those who used his illustrations to be able to move back and forth, from picture to the brief descriptors of each part. Text, combined with visuals, became the hallmark of technical writing throughout the period discussed here. However, without Vesalius' work, which highlighted visuals over text, many of the other technical books described in this chapter would likely not have incorporated or reflected visual-based book design launched by Vesalius.

Singer and Rabin note that Vesalius intended *Tabulae Sex* as a summary. The work included six single sheets that provided three illustrations of the portal and artery systems and three of the skeleton. Their size suggests that he intended the sheets to be suspended on a wall for use during demonstration during dissection lectures [4, p. viii]. In *Tabulae Sex*, the descriptors of the anatomical parts occur around the perimeter of each sheet, as they are in *Epitome* (Figure 4-2) which can "stand alone" as a teaching tool. When he prepared *Tabulae Sex,* Vesalius was already planning the *Fabrica*, which would provide more illustrations of the human body but in full color. The *Tabulae* remains critical in the development of printed material for one major reason: Vesalius believed that individual sheets, rather than printed books (a method used frequently in the 17th century), provided better teaching tools.

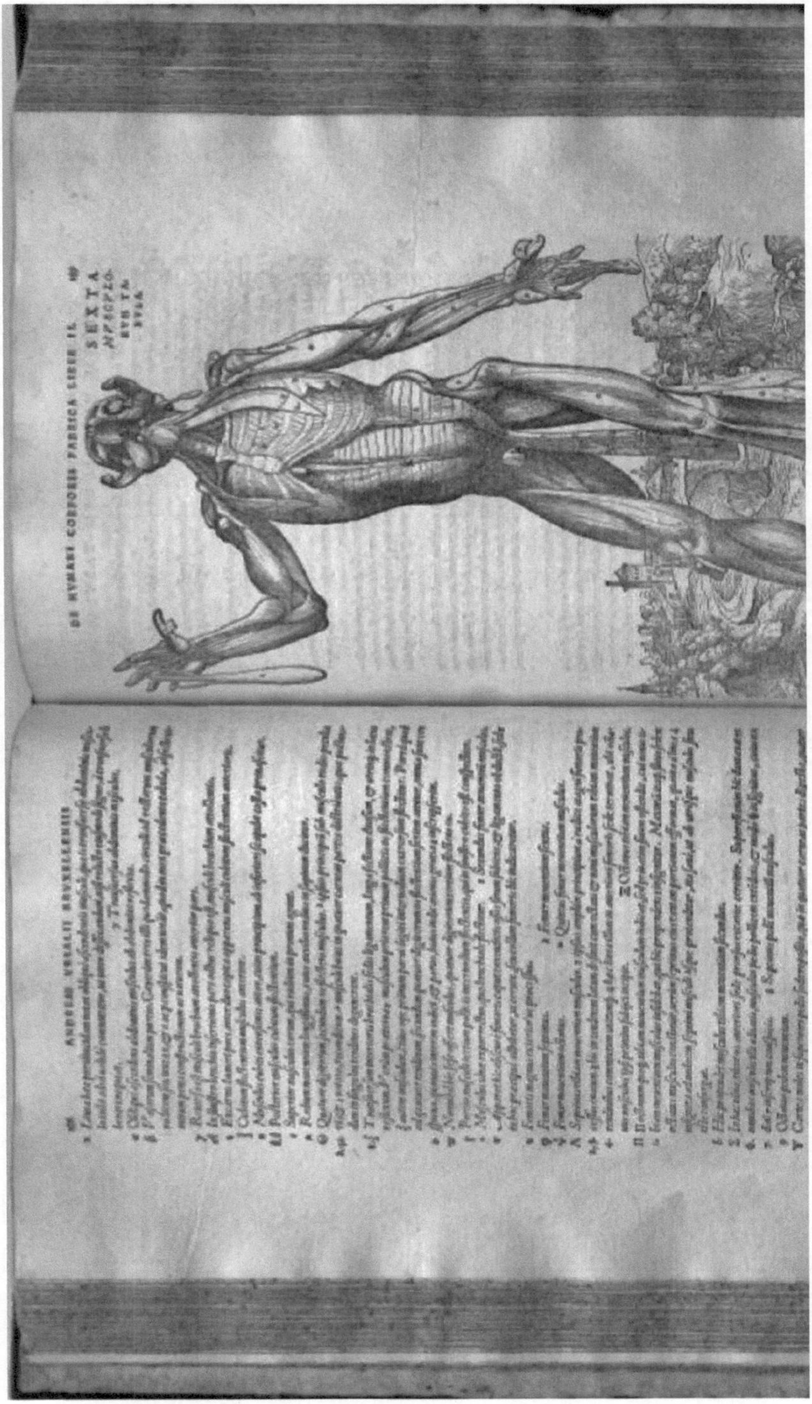

Figure 4-1. Page from *Fabrica*. Vesalius Image credit Octavo Corp. and the Warnock Library.

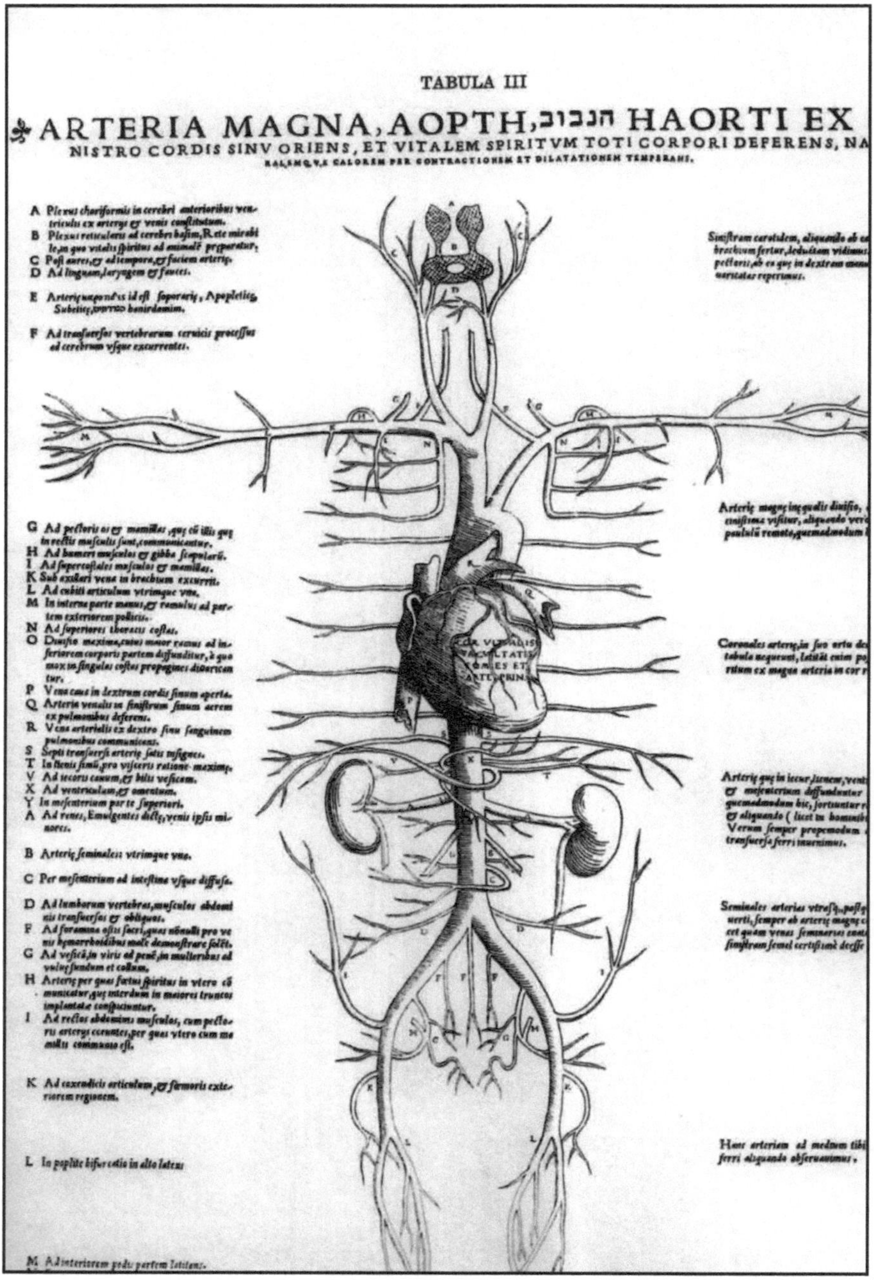

Figure 4-2. Page from *Epitome*.
De Humani Corporis Fabrica Librorum Epitome.
Used by permission of University of Glasgow Library, Special Collections.

Tabula Sex, Fabrica, and *Epitome*—Icons of the New Frontier for Technical Representation

The *Tabulae Sex* and then *Fabrica* are the largest works ever developed by Vesalius and are likely some of the largest examples of technical writing produced in the Renaissance. Despite integration of format, page design, and type in these works, the illustrations directed the size of *Tabulae*, *Fabrica*, and *Epitome*. The *Tabulae* measured 19 × 13.5 inches [4, p. ix]. In all three works, succinct descriptions of anatomical parts appear at the top and on the sides of the pages; the figures, created with woodcuts, take center stage. Only a few copies of the original woodcut *Fabrica* remain, and most copies show the use of copperplate engravings with the descriptive text across the page, as shown in Figure 4-1. Nevertheless, in all three works, Vesalius abandoned linear text in favor of brief descriptive phrases coordinated with the figures. The size of the illustrations was needed to capture the size ratio of the human body and its parts.

Many versions of the *Fabrica* were 19 × 13.5 inches [4, p. 25]. The *Epitome* consisted of 14 folio pages measuring 21 × 16 inches. These pages too were among the largest woodcut illustrations ever printed in Venice. The *Epitome* contained 11 plates made from wood blocks, showing the bones, muscles, external parts, nerves, veins, and arteries. Some portions of the plates were intended to be cut out and attached to their proper places upon the drawings of the body or skeleton [6, p. xxii]. When the cutouts were lifted or moved aside, the organs could be observed in sequence from the surface to the innermost parts. Vesalius also had copies of the *Epitome* printed so that the figures were already prepared and colored by hand.

The *Epitome* stands as a triumph of summation. This guide, or brief manual, provides an index to the parts of the body. As Vesalius states in his dedicatory introductory, the *Epitome* provides a compendium of anatomy for physicians and may be treated as an index to the *Fabrica.* In essence, it may have served as a summary in the true modern use of the term.

While the text of both *Fabrica* and *Epitome* illustrate the textual conciseness and organization of a master teacher, the illustrations have captured the imagination of readers for nearly five centuries. The dramatic postures of the figures, which emanated from the workshop of the master painter Titian, combined with the postures of death of Holbein's pictures reflecting the desolation of plagues and massacres [7, p. 9, xxii]. Yet the illustrations, with their system of cross-references between text and illustrations, made *Fabrica* unique in the history and development of the printed book as medium for the communication of a descriptive science. Vesalius wove text and picture into an integrated whole; illustration delimited the need for excessive text. Thus, in the final product, void of heavy roman type, Vesalius had his printer substitute a delicate font reminiscent of Plantin [7, p. 22] another feature that adds to the artistic quality of the massive pages [8]. This typeface can be seen in Figures 4-1 and 4-2.

TECHNICAL VISUALIZATION AND RENAISSANCE ART

Vesalius' creative interplay of format, design, and typography were nurtured by Renaissance art that demanded a direct and faithful representation of natural phenomena. The artist had to acquaint himself with the structure and physical properties of natural phenomena to insure objectivity within the rules of perspective and mathematics to obtain representational correctness. Thus the large page size of *Tabulae Sex*, *Fabrica*, and *Epitome* not only preserved the proportions of human anatomy [7, p. 22] but also ensured readability and usability for both teacher and practitioner.

The appearance of *Tabulae Sex*, and then *Fabrica* and *Epitome,* launched factual anatomy and the beginning of rejection of humoral-based medicine and Galen's anatomy, which contained no figures. While Vesalius initially sought to portray Galen's concepts, Vesalius actually corrected more than 200 anatomical errors contained in Galen's work. Both *Fabrica* and *Epitome*, with their visual distinctiveness, captured the minds and the imaginations of physicians and scientists. The faithful representation, demanded by art of the time and then by science and mathematics, was fulfilled in visuals that largely supplanted text and thus established a new direction in printed works, a technique that would appear in other technical books of the 17th century.

The First Printed English Surgical Books

Thomas Geminius' 1553 translation of *Fabrica* into English [2], 15 years after the *Fabrica* appeared in Italy, nurtured trends in English medical books that continued to improve as print technology improved.

For example, Guy de Chuliac's *The Questyonary of cyrurgyens* [9], first published in 1542 by Robert Wyer, appears to be the first printed English surgical book. The book, divided into two main sections, was prepared for physicians who had to pass oral surgical examinations. Section two features rude woodcut drawings of surgical instruments. A brief description of how the instruments were to be used accompanied these drawings. Figure 4-3 displays an example of tongs. These crude drawings contrast sharply with the woodcuts of Vesalius, printed a year later in Italy and illustrated in Figures 4-1 and 4-2. The difference illustrates the advanced print technology in Italy by the early 16th century and the reason English printers and technical writers became eager to adapt the most up-to-date print technology to their own work. A full page with descriptive paragraphs appears in Chapter 3, Figure 3-3.

Another evidence of the quality of European woodcuts can be seen in Von Brunschweig, *The noble experience of the virtuous handy warke of surgeri,* printed in London in 1525 [10] (see Figure 4-4). This translation contained much higher quality woodcuts than those prepared in England but served a similar

Figure 4-3. Early illustration of surgical instruments (1542) in *Questyonary of cyrurgyens.*

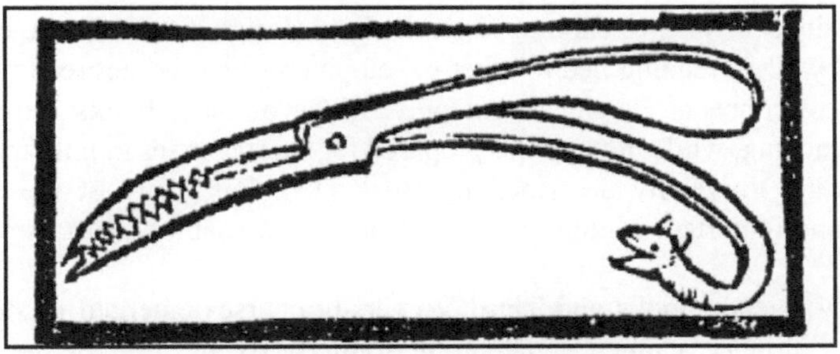

Figure 4-4. Woodcut of surgical tool illustration in a 1525 English translation of a German work, *The noble experience of the virtuous handy warke of surgeri* [10].

purpose, that is, to provide a visual example of the surgical tools described. These drawings in this work show that the power of visuals to communicate what would become increasingly important in all kinds of how-to books. After Thomas Geminius' English translation of Vesalius [2] appeared, English printers began imitating European examples incorporating high quality woodcuts and then copperplate engravings into a variety of technical books.

Euclid's Geometry: The First English Translation

But improved visuals also emerged in other forms to enhance readers' understanding. The first English edition of Euclid's *Geometry*, dated 1570 [11], exemplifies an astounding feature in print technology. To help students of geometry visualize the differences between solid objects as these incorporated principle concepts of plane geometry, this first edition provides patterns of solid geometric forms: cubes, pyramids, polyhedrons, and polypipedons. The user could then trace the patterns on thick paper, cut and fold these, glue them to the appropriate sites in the text, and "see" (as well as feel) a solid geometrical object. Figure 4-5 shows a photograph of a page from one of these first editions. We can see three kinds of pyramids: three-sided, four-sided, and five-sided. Over 400 years have passed since the book was printed, but these pyramids, which fold down neatly when the book is closed, can still be unfolded and stand erect to illustrate the three pyramidic forms. I can only guess at the wonder that many of the first English readers must have experienced in their effort to cut, fold, and view many of these solid geometric objects. Perhaps this first edition of Euclid in English was the first pop-up book in English!

Renaissance Military Science Books

Innovations in print technology, particularly design features that enhanced learning by silent reading, led to other developments that did not require readers to visualize concepts via linear text only. Many of these books were printed for self-teaching, while others were prepared for short courses in military science at the Third Vniversity described by Buck [12]. For many of these books, size became important in improving readability and usability, just as Vesalius' works did.

Military science books and Third Vniversity courses emerged to strengthen England's military acumen and to train soldiers. By the turn of the century, books such as Robert Barrett's *The Theoricke and Practice of Moderne Warres* (1598) [13], printed a half-century after *Fabrica*, appeared as large folios. The descriptive text includes integrated block drawings of the arrangement of regiments during the battle, one large foldout drawing of a military encampment, and 40 tables showing how to position soldiers, depending on the fronts shown by the enemy. The tables, with their broken horizontal and vertical separation lines, illustrate the printer's challenge in fitting numerical information into the 9 × 12 pages. The need for clarity and readability may have dictated the large folio design for many of these works. Carefully drawn foldouts, combined with precisely articulated verbal descriptions, again show the Renaissance commitment to detailed, accurate verbal description as well as visual portrayal. We may also assume that these large foldouts could have been removed for display during war strategy planning.

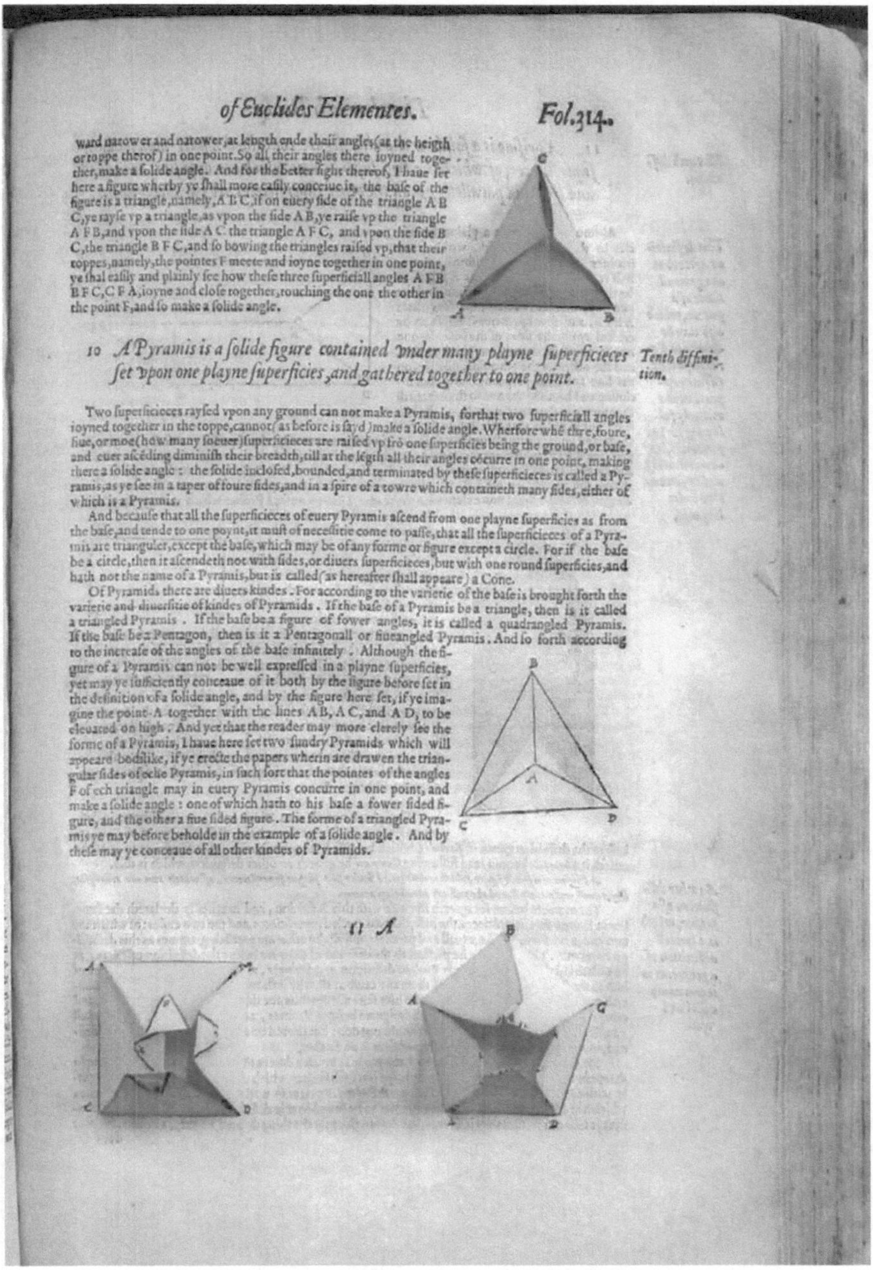

of Euclides Elementes. Fol.314.

ward narower and narower, at length ende their angles (at the heigth or toppe therof) in one point. So all their angles there ioyned together, make a ſolide angle. And for the better ſight thereof, I haue ſet here a figure wherby ye ſhall more eaſily conceiue it, the baſe of the figure is a triangle, namely, A B C, if on euery ſide of the triangle A B C, ye rayſe vp a triangle, as vpon the ſide A B, ye raiſe vp the triangle A F B, and vpon the ſide A C the triangle A F C, and vpon the ſide B C, the triangle B F C, and ſo bowing the triangles raiſed vp, that their toppes, namely, the pointes F meete and ioyne together in one point, ye ſhal eaſily and plainly ſee how theſe three ſuperficiall angles A F B B F C, C F A, ioyne and cloſe together, touching the one the other in the point F, and ſo make a ſolide angle.

10 *A Pyramis is a ſolide figure contained vnder many playne ſuperficieces ſet vpon one playne ſuperficies, and gathered together to one point.* Tenth diffinition.

Two ſuperficieces rayſed vpon any ground can not make a Pyramis, forthat two ſuperficiall angles ioyned together in the toppe, cannot (as before is ſayd) make a ſolide angle. Wherfore whē thre, foure, fiue, or moe (how many ſoeuer) ſuperficieces are raiſed vp frō one ſuperficies being the ground, or baſe, and euer aſcēding diminiſh their breadth, till at the lēgth all their angles cōcurre in one point, making there a ſolide angle: the ſolide inclosed, bounded, and terminated by theſe ſuperficieces is called a Pyramis, as ye ſee in a taper of foure ſides, and in a ſpire of a towre which containeth many ſides, either of which is a Pyramis.

And becauſe that all the ſuperficieces of euery Pyramis aſcend from one playne ſuperficies as from the baſe, and tende to one poynt, it muſt of neceſſitie come to paſſe, that all the ſuperficieces of a Pyramis are trianguler, except the baſe, which may be of any forme or figure except a circle. For if the baſe be a circle, then it aſcendeth not with ſides, or diuers ſuperficieces, but with one round ſuperficies, and hath not the name of a Pyramis, but is called (as hereafter ſhall appeare) a Cone.

Of Pyramids there are diuers kindes. For according to the varietie of the baſe is brought forth the varietie and diuerſitie of kindes of Pyramids. If the baſe of a Pyramis be a triangle, then is it called a triangled Pyramis. If the baſe be a figure of fower angles, it is called a quadrangled Pyramis. If the baſe be a Pentagon, then is it a Pentagonall or fiueangled Pyramis. And ſo forth according to the increaſe of the angles of the baſe infinitely. Although the figure of a Pyramis can not be well expreſſed in a playne ſuperficies, yet may ye ſufficiently conceaue of it both by the figure before ſet in the definition of a ſolide angle, and by the figure here ſet, if ye imagine the point A together with the lines A B, A C, and A D, to be eleuated on high. And yet that the reader may more clerely ſee the forme of a Pyramis, I haue here ſet two ſundry Pyramids which will appeare bodilike, if ye erecte the papers wherin are drawen the triangular ſides of eche Pyramis, in ſuch ſort that the pointes of the angles F of ech triangle may in euery Pyramis concurre in one point, and make a ſolide angle: one of which hath to his baſe a fower ſided figure, and the other a fiue ſided figure. The forme of a triangled Pyramis ye may before beholde in the example of a ſolide angle. And by theſe may ye conceaue of all other kindes of Pyramids.

Figure 4-5. [Pop-up view of pyramids] Euclild. *The Elements of Geometrie.*
University of Oklahoma Libraries, History of Science Collection
http://129.15.14.63/galleries/01Ancient/Euclid/1570/

Even more striking is Captain John Bingham's *The Art of Embattailing an Army* (1629) [14], which uses copperplate illustrations, six large foldout illustrations of the arrangement of soldiers. The drawings are bound so that they appear across the page from the text (as Geminius did in his translation of *Fabrica* [2]), allowing the reader to move easily from text to illustration. The drawings appear at the appropriate point of the text. Each is labeled with a title and the chapter to which it refers. As in Vesalius' works and many 17th century how-to books, visuals and page design combined with text heighten the communicative quality of the book.

Another military science book, Richard Norwood's *Fortification or Architecture Military* (1639) [15], likely served as both a text book and a reference work (see Figure 4-6). Norwood provided a numerical list of "axioms observed in fortification," but he used instructional problems to teach students how to design ramparts. The verbal description that follows, integrated with the drawing, illustrates the precision required in constructing fortifications. His work exemplifies a Renaissance technical description. This small book, approximately 5 × 7 inches, could have been easily carried in a military officer's saddlebag. The drawings were woodcuts, likely to save printing costs.

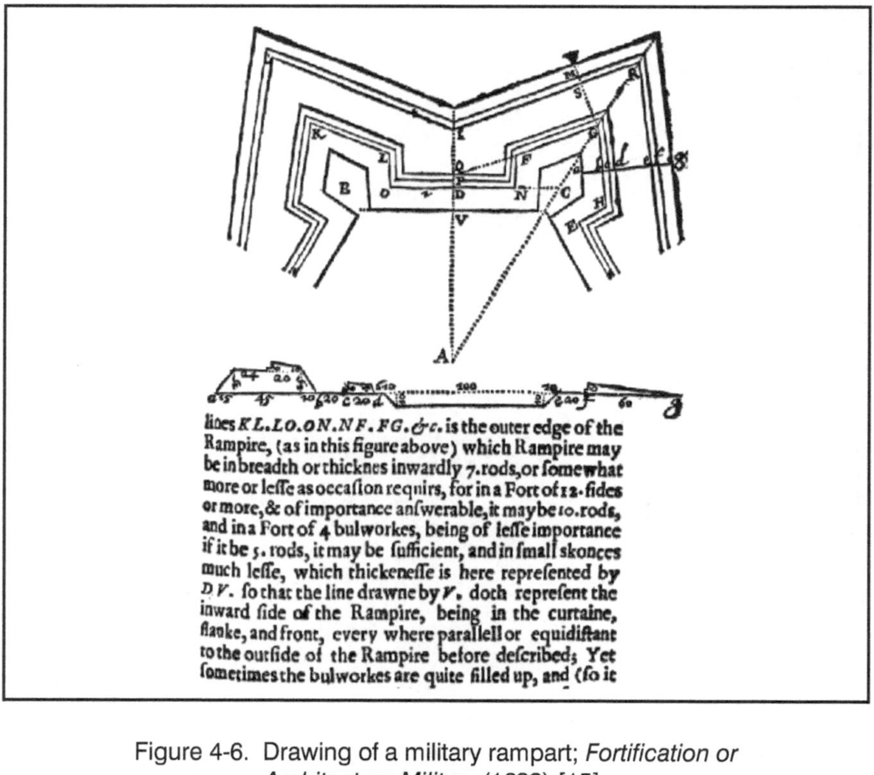
lines *KL.LO.ON.NF.FG.&c.* is the outer edge of the Rampire, (as in this figure above) which Rampire may be in breadth or thicknes inwardly 7.rods, or ſomewhat more or leſſe as occaſion requirs, for in a Fort of 12. ſides or more, & of importance anſwerable, it may be 10. rods, and in a Fort of 4 bulworkes, being of leſſe importance if it be 5. rods, it may be ſufficient, and in ſmall skonces much leſſe, which thickeneſſe is here repreſented by *DV*. ſo that the line drawne by *V*. doth repreſent the inward ſide of the Rampire, being in the curtaine, flanke, and front, every where parallell or equidiſtant to the outſide of the Rampire before deſcribed; Yet ſometimes the bulworkes are quite filled up, and (ſo it

Figure 4-6. Drawing of a military rampart; *Fortification or Architecture Military* (1639) [15].

Another small book of 36 pages, designed for use on the battlefield, contains concise text accompanied by drawings to show how a company should march [16]. Achesone's book also contains descriptions of commands used on the field and to teach soldiers how to perform during military exercises. The little book, a training manual, also gives descriptions of duties of captains, lieutenants, sergeants, and corporals. Figure 4-7a shows one formation used during military exercises:

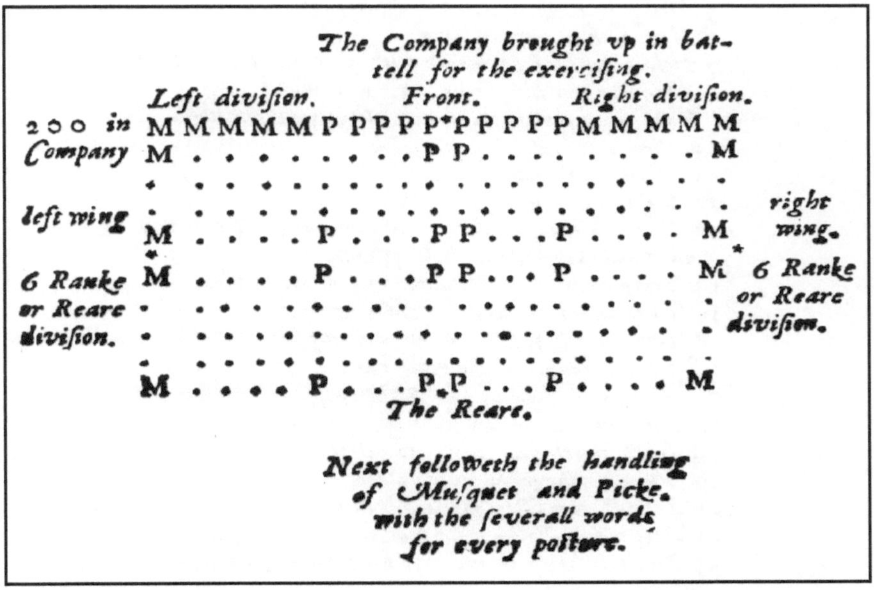

Figure 4-7a. Visual portraying how a company should march.

As Figure 4-7b states, the next page will give the verbal commands for soldiers in handling muskets and picks.

Norwood's *Fortification of Architecture Military* used spacing to enhance accessibility and included a foldout table, shown in Figure 4-8 that did not compromise the accessibility of the material in his little book.

English Shipbuilding Books

Books on shipbuilding illustrate the importance of books that could be easily transported as well as used by shipwrights, but they also created a host of design problems for printers. English shipbuilding books in the 16th and 17th centuries never contained detailed instruction. Instead, they included some text and a large number of tables and drawings to assist designers and builders (see Figure 4-9).

The po-ſturs of the Muſquet. 6. *The Military Garden.*

The words for the Muſquet.
The Muſquet ſhouldered.

Charge. 1. *Sinke your Muſquet.*
2 *Handell your Muſquet with your right hand.*
3 *Unſhoulder your Muſquet and hold her vp.*
4 *Fall backe with your right leg and hand.*
5 *Bring your reſt to your Muſquet.*
6 *Ioyne both in your left hand.*
7 *Hold your Muſquet mouth over your leaders right ſhoulder.*
8 *Open your pan with right finger and thumb.*
9 *Take your pryming wyre and clenſe your touchholl.*
10 *Blow your pan.*
11 *Morſe with powder.*
12 *Cloſe your pan.*
13 *Grip the but-end of your Muſquet and reſt with your right hand and ſhake off the looſe powder.*
14 *Bring about your Muſquet to your left ſyde.*
15 *Diſcharge your meaſure into your barrell.*
16 *Draw foorth your ſcrow or rammingſticke with the middle finger and thumb of the right hand.*
17 *Shorten the ſame at your right pappe within a handfull.*
18 *Ram in your powlder, furring, and bullet.*
19 *Draw foorth your ſcrow and ſhorten it as before.*
20 *Put vp your ſcrow in its place.*
21 *Recover your Muſquet and hold it vp with your left hand.*
22 *Take it bee the butt hard at the pan and ſhoulder it.*
23 *Your reſt being in your left hand with your thumb vpon it above the butt.*

Diſcharge. 24 *Sinke your Muſquet.*
25 *Handell your Muſquet.*
26 *Unſhoulder your Muſquet.*
27 *Hold vp your Muſquet.*
28 *Bring your reſt to your Muſquet.*

Figure 4-7b. Verbal commands to accompany Figure 4-7a.

These books reflect the 15th century Venetian works that described a design system based on geometrical and proportional relationships and a small number of fundamental measurements. The manuals included some linear text, but they focused on visuals and visual relationships rather than on textual description. Until the mid-17th century, shipbuilding, both in design and construction, remained orally transmitted. To deter design theft, shipwrights wished to keep their design knowledge secret and confined to use within shipyards [17, p. 10].

A Table of the dimenſions of any Regular Fortification from the Tetragon to the Dodecagon; the flanked angle being halfe the angle of the Poligon, and 15. degrees.

Poligons the number of their ſides	4	5	6	7	8	9	10	11	12
	degrees.	degrees.	degrees.	degrees.	degrees.	degrees.	degrees.	degrees.	degrees.
The angle of the Poligon — BCE.	90	108	120	128 4/7	135	140	144	147 3/11	150
The flanked angle of the bulworke — FGH.	60	69	75	79 2/7	82 1/2	85	87	88 7/11	90
The angle of the ſhoulder — NFG.	105	109 1/2	112 1/2	114 9/14	116 1/4	117 1/2	118 1/2	119 1/11	120
The inward flanking angle — SGF.	15	19 1/2	22 1/2	24 9/14	26 1/4	27 1/2	28 1/2	29 1/11	30
The outward flanking angle — KMG.	150	141	135	130 5/7	127 1/2	125	123	121 9/11	120
The angle forming the flanke — FCN.	40	40	40	40	40	40	40	40	40

	4 rod.	4 cent	5 rod.	5 cent	6 rod.	6 cent	7 rod.	7 cent	8 rod.	8 cent	9 rod.	9 cent	10 rod.	10 cent	11 rod.	11 cent	12 rod.	12 cent
The Curtaine — ON.	42	00	42	00	42	00	42	00	42	00	42	00	42	00	42	00	42	00
The front of the bulworke — FG.	28	00	28	00	28	00	28	00	28	00	28	00	28	00	28	00	28	00
The Gorge-line — NC.	10	77	12	18	13	26	14	22	14	83	15	42	15	92	16	36	16	73
The Semidiameter of the inner Poligon — AC.	44	92	56	45	68	52	80	94	93	63	106	49	119	49	132	57	45	80
The ſide of the inner Poligon — BC.	63	53	66	36	68	52	70	24	71	66	72	84	73	85	74	71	75	47
The perpendicular of the inner poligon — AD.	31	76	45	67	59	34	72	90	86	50	100	06	113	64	127	20	140	83
The Semidiameter of the outer Poligon — AG.	67	95	80	63	93	74	107	05	120	50	124	02	147	59	161	16	174	83
The ſide of the outer Poligon — KG.	96	09	94	79	93	74	92	90	92	22	91	67	91	21	90	83	90	50
The perpendicular of the outer Poligon — AI.	48	04	65	23	81	28	96	41	111	32	125	93	140	37	154	63	168	87
The diſtance of the poligons — DI.	16	28	19	57	21	84	23	52	24	83	25	87	26	72	27	43	28	04
The flanke — FN.	9	03	10	22	11	13	11	85	12	44	12	94	13	36	13	72	14	04
The head-line — CG.	23	02	24	18	25	22	26	11	26	87	27	53	28	10	28	59	29	03
The ſhoulder from the Center — FC.	14	05	15	90	17	31	18	43	19	36	20	13	20	79	21	35	21	85
The ſecond flanke — OP.	8	29	13	12	15	13	16	14	16	77	17	14	17	39	17	56	17	68
The longeſt line of defence — QG.	70	94	71	14	71	30	71	43	71	55	71	67	71	77	71	86	71	94

Figure 4-8. Foldout table from Norwood [15].

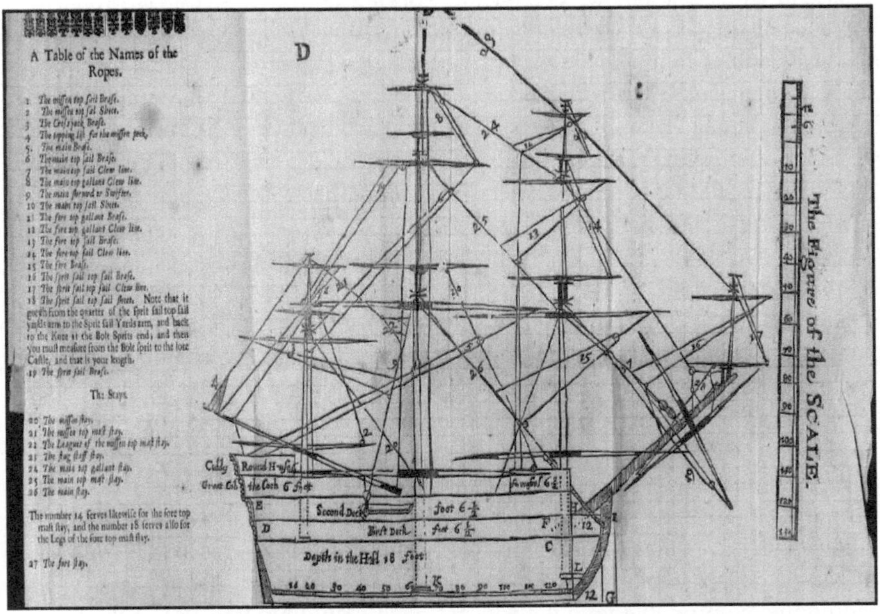

Figure 4-9. [Fold-out drawing of a ship to show proportions]
Miller, Thomas, *The Complete Small or Great,* William Fisher, London, 1676.
Print. *Modellist: Shewing he True and Exact Way of Raising the Model of Any Ship or Vessel.*

Tables became a mnemonic and a reference source for planning visual and proportional relationships.

Many of these books, like military science books, illustrate the ongoing challenges that writers and printers faced in designing and printing tables, a point evident in Norwood [15], who had no choice but to use a foldout table (Figure 4-8). Tabular format had yet to be defined. Books by two shipbuilders illustrate the preponderance of tables used in shipyards. First, Edward Hayward's *A Full and Perfect Account of the Sizes and Lengths of Riggins for All His Majesties Ships and Frigats* (1666) [18] provides lengths and numbers of ropes and sails. Six pages of tables comprise the majority of this work, with each table extending across the page of this quarto. Readability is excellent, and many tables, on the right side, have space for notes. This technique illustrates yet another effort to make tabular displays more readable and useful.

Edmund Bushnell, *The Complete Ship-wright* (1678) [19] another small book of "tables" (Figures 4-10a and 4-10b), shows the difficulties writers had in displaying in an easily accessible way information ship yard crews needed for their work [19]. Henry Bond's two books, *The Art of Apparelling and Fitting of Any Ship with Masts, Yards, and Cordage* (1655) [20] and *The Boate Swains Art* (1699) [21], give tables of the sizes of rigging required for a ship of 300 tons and explain a simple method for working out other sizes via proportion. Developed for use as a reference by builders, *The Boate Swain's Art* contains extensive tables followed by a short concluding summary for rigging a ship with known dimensions. The summary was also printed separately as a small booklet of only eight pages.

Other shipbuilding books showed how to build a model of a ship before launching a full shipbuilding project. Thomas Miller's *The Compleat Modellist: Shewing the True and Exact Way of Raising the Model of any Ship or Vessel, Small or Great* (1676) [22] illustrates the shift from orality to textuality in shipbuilding. This book textualizes the ancient practice of building small ship models and then building full-size ships from the models. This small 24-page pamphlet explains how to draw a rigging plan to use in determining lengths of various ropes. It has two plates, one a 16 × 14 foldout drawing of a fully rigged ship, shown in Figure 4-9, and tables of sizes and proportions, exemplified in the excerpt below from one of the pages.

Miller uses extensive tables and drawings to help readers understand the proportions. He includes a second foldout drawing that shows the rigging of a ship-with the names of each rope and then other tables-that relate rope sizes to masts. These books, directed to shipbuilders, assume that readers will be thoroughly comfortable with the concept of shipbuilding. Because of the limited literacy of many shipyard workers, the books assume that drawings and tables would also serve for those able to read shipyard nomenclature but not extensive linear paragraphs [15, p. 7] (see Figures 4-11a and 4-11b). Some times

4 *The Boat-Swain's Art.*

A Table of the Names, the Sizes, the Number, and the Lengths of each Rope, of Rigging, belonging to the fore-propounded Ship.

1. *Cordage of 1 Inch $\frac{1}{10}$ Parts.*

		Fa.	Fa.
2	Fore Top-gallant Braces	25	50
4	Fore Top-gallant Bowling Bridles	4	4
2	Fore Top-gallant Lifts	22	44
8	Main Top-gallant Lanniards	$1\frac{1}{2}$	12
4	Main Top-gallant Bowling Bridles	1	4
1	Main Flag Staff-ſtay	11	11
8	Mizen Top-maſt Lanniards	$1\frac{1}{2}$	12
1	Fall of the Mizen Top-ſail Cranlines	36	36
1	Mizen Top-ſail Bowling	$12\frac{1}{2}$	25
2	Mizen Top-ſail Braces	17	34
34			232

2. *Cordage of 1 Inch $\frac{1}{2}$ Parts.*

8	Lanniards of the Sprit-ſail Top maſt	$1\frac{1}{2}$	12
2	Falls of the Sprit-ſail Top-maſt Tackles	6	12
1	Fall of the Sprit-ſail Cranlines	18	18
1	Sprit-ſail Top-ſail Halliard	6	6
2	Sprit-ſail Top-ſail Lifts	5	10
2	Pennants of the Sprit-ſail Top-ſail Braces	2	2
2	Sprit-ſail Top-ſail Braces	10	20
2	Sprit-ſail Top-ſail Cluelines	10	20
2	Fore Top-maſt Tacklefalls	$13\frac{1}{2}$	27
2	Falls of the Fore Top-gallant Back-ſtays	18	36
1	Fore Top-gallant Halliard	32	32
2	Pennants of the Fore Top-gallant Braces	3	3
2	Fore Top-gallant Bowlings	24	48
2	Fore Top-gallant Cluelines	22	44
6	Fore Top-gallant Lanniards	$1\frac{1}{2}$	9
2	Falls of the Main Top-gallant Tackles	7	14
39			2 Falls

Figure 10a. A Bushnell table.

[26]

A Plain and Eaſie Rule *to* Rigge *any* SHIP *by the Length of his* Maſts *and* Yards, *without any further Trouble.*

For the Spritſail-Yard.

. *Halliards*	THree times the length of the Yard.
. *Lifts*	Three times the length of the Yard.
. *Clulines*	Two times the length of the Yard.
. *Braces*	Two times the length of the Fore-yard.
. *Buntlines*	Two times the length of the Yard.
Sheets	Three times the length of the Yard.
. *Pennants*	One third of the Yard.

For the Spritſail-Top-maſt.

. *Shroudes*	Moſt be the length of the Maſt from the Heel to the Croſtrees.
. *Halliards*	Three times the length of the Maſt.
. *Lifts*	Three times the length of the maſt.

Clulines

Figure 10b. Another example of Bushnell's tables.

Bigneſs of Miſſen-top-maſt-rigging.	Maſt of 34 inch.	Maſt of 32 inch.	Maſt of 30 inch.	Maſt of 29 inch.	Maſt of 28 inch.	Maſt of 20 inch.	Maſt of 26 inch.	Maſt of 24 inch.	Maſt of 23 inch.	Maſt of 19 inch.	Maſt of 13 inch.	Maſt of 12 inch.
	1 inch.	2 inch.	3 inch.	4 inch.	5 inch.	6 inch.	7 inch.	8 inch.	9 inch.	10 inch.	11 inch.	12 inch.
Penents of Tackles	3½	3	2½	2½								
Falls of Tackles	2½	2¼	2	2								
Shrowds	3	3	2½	2½	2	1½	1	2	2	2		
Laniards	2	1½	1½	1½	1½	1	1½	1	¼			
Puttocks	3½	3	2½	2½	2	1½	2	1½	1½	1		
Penents of braſes	2½	2½	2	1½	1½	1½	1	2	1½	1¼		
Braſes	2	2	1¼	1½	1	1	1	1	¼			
Bow-lines	2½	2	1¼	1½	1¼	1	1	1½	1	¼		
Bridles	2	2	1½	1¼	1	1	1	1	1	¼		
Clew-lines	2	2	¼	1½	1½	1½	1	2	1½	1½	1	1
Tye	3	3	3	2½	2½	2½	2	3	2	2	2	1½
Halliards	2½	2½	2	2	1½	1½	1¼	2	1½	1½	1½	1
Lifts	2	2	1¼	1½	1½	1½	1	1	1	1	¼	1
Parrel-rope	3	2½	2	1¼	1¼	1½	1½	2	1½	1½	1	1
Runner of the ſtay	2½	2½	2									
Penent of the ſtay	3	3	2½	2	1½							
Top-rope	4	3½	3½	2½	2	2	1½					
Parts of the ſtay	2	2	1½	1½	1	1	1					1
Penents for the back-ſtays	3	2½	2½									
Falls	2½	2	1¼									

Figure 11a. Portion of a table from *The Complete Modellist* (1676) [22].

The Table of minutes, which euery degree conteyneth in euery of the paralelles.

G	M	S	G	M	S	G	M	S	G	M	S	G	M	S	G	M	S
1	59	59	16	57	41	31	51	26	46	41	41	61	29	5	75	14	31
2	59	58	17	57	23	32	50	53	47	40	55	62	28	10	77	13	30
3	59	55	18	57	4	33	50	19	48	40	9	63	27	14	78	12	28
4	59	51	19	56	44	34	49	45	49	39	22	64	26	18	79	11	27
5	59	46	20	56	23	35	49	9	50	38	34	65	25	21	80	10	25
6	59	40	21	56	1	36	48	32	51	37	46	66	24	24	81	9	23
7	59	33	22	55	38	37	47	55	52	36	56	67	23	27	82	8	21
8	59	25	23	55	14	38	47	17	53	36	7	68	22	29	83	7	19
9	59	16	24	54	49	39	45	38	54	35	16	69	21	30	84	6	16
10	59	5	25	54	23	40	45	58	55	34	25	70	20	31	85	5	14
11	58	54	26	53	56	41	45	17	56	33	33	71	19	32	86	4	11
12	58	41	27	53	28	42	44	35	57	32	41	72	18	32	87	3	8
13	58	28	28	52	59	43	43	53	58	31	48	73	17	33	88	2	5
14	58	13	29	52	29	44	43	10	59	30	54	74	16	32	89	1	3
15	57	57	30	51	58	45	42	26	60	30	0	75	15	32	90	0	0

Figure 11b. Portion of a table from *The Compleat Modellist* the True and Exact Way of Raising the Model of Any Ship or Vessel, Small or Great (1676) [22].

Bushnell's and Miller's works were bound and sold together, a fact that shows their complementary information.

Navigation Books

Like military science and shipbuilding books, navigation books also presented challenges for printers. Some of the most challenging were visuals that had to use woodblock drawings. Several books with engraved visuals of navigation instruments had to be carefully constructed from instructions given in the text. Other drawings needed to show how to navigate, so numerous navigation tables were needed. Martin Cortes' *The Art of Navigation* (1589) [23] also illustrates the continued struggle that English printers experienced in creating tables, particularly in small duodecimos or quartos rather than in large folios. Despite the rude tables used in this book Cortes' book, translated into English by Richard Eden, was the most important navigation work that appeared in England in the 16th century. It was the first book to include a map of the New World in the same sheet with the western parts of Europe and Africa.

English navigation books of the 17th century that required extensive maps often emerged as rectangular books, bound on the short side, to allow maps to be printed across two adjoining pages to improve readability and usability. Other challenges included the need for drawings of instruments to accompany instructions on how to use and perhaps make these. Some books even included patterns that could be used to trace parts of these tools so that they could be constructed by a seaman or artisan. Richard Eden's translation of Cortes' *The Art of Navigation* (1615) [23], a large folio, features the first important published navigational map, a foldout drawing; additional pages of tables containing navigation information; and drawings of navigation instruments such as spheres, globes, and dials.

William Johnson's, *The Light of Navigation* (1612), printed as a 12 × 18 folio printed to be read lengthwise, included extensive tables and copperplate maps that covered two adjacent pages [24, p. 36]. The maps were followed by woodcut drawings of landmarks on these coastlines. The drawings, as exemplified in Figure 4-12, were designed to help ships sailing near the coastlines know the location of the ship in reference to the land. Drawings often included houses, trees, town skylines, and livestock.

Bulleted Lists and White Space

The need for readability often characterized much English Renaissance technical writing. Lists often appeared in books, usually as columns, and printers often employed white space combined with lists to improve the display of information. We find these first lists in English printed texts, such as the *Boke of St. Albans* by Julian Berners (1485) [25], which uses paragraph signs to introduce lists. The ¶ was the first "bullet" in English printed text. This list, Figure 4-13, with red ¶ marks, explains what may be used in designing a coat of arms.

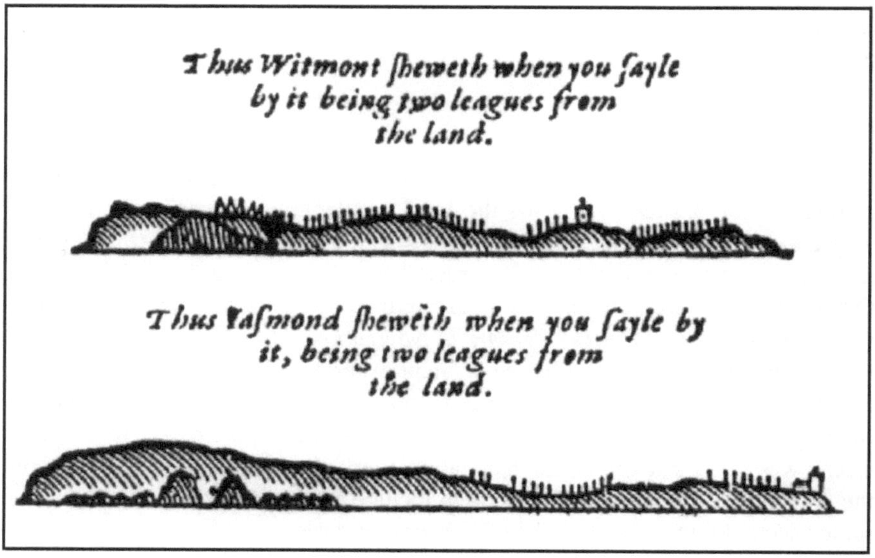

Figure 4-12. Example figure of a coastline, Johnson, *The Light of Nauvigation* (1612).

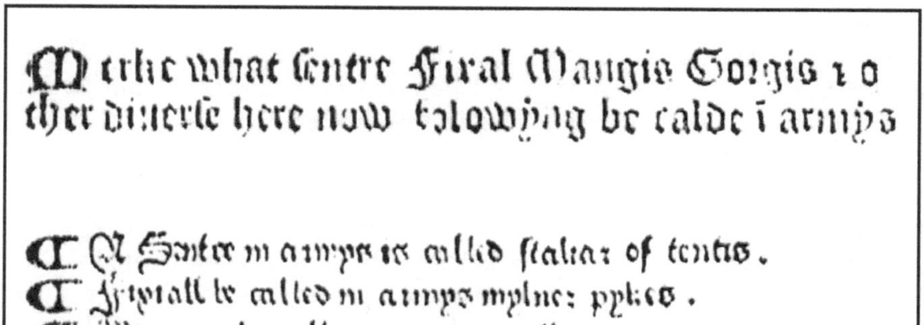

Merke what sautre Fyrmal Maungis Gorgis & o ther diuerse here now folowyng be calde i armys

¶ A Sautre in armys is callid stakar of tentis.
¶ Fyrmall be callid in armys mylner pykes.
¶ Maungys be called in armys a sleue.
¶ Gorgys be callid in armys water bulgets.
¶ Elynellis be calde in armys .iiij. quadrantis trunchelis.
¶ Oglys be calde in armys gonestonys.
¶ Tortellis be calde in armys Wastell.
¶ Diaclys be called in armys scoppellys.

Figure 4-13. First use of bulleted text in an English Book (1486).

About a half-century later, we find other kinds of bullets, such as urinals, used as bullets, in uroscopy books which explained, via lists, what different colors, textures, and densities of urine meant in terms of self-diagnosis. This excerpt comes from *Here begynneth the seyng of uriness* (1544) [26]. Some versions of these uroscopy books contained urinal shaped bullets with numbers inserted (see Figure 4-14).

Micrographia by Robert Hooke

In *The Nature of the Book: Print and Knowledge in the Making*, Johns states that *Micrographia* [27] was "by common consent the most impressive single document produced during the first decade of its existence" [27, p. 429]. Its influence came from its prescriptive practices for pursuing and recording natural philosophy, which became the New Science. But the power of the work emerged more from the engravings. Hooke sought to use his microscope to circumvent the imperfections of our own minds and our senses. Engravings that captured

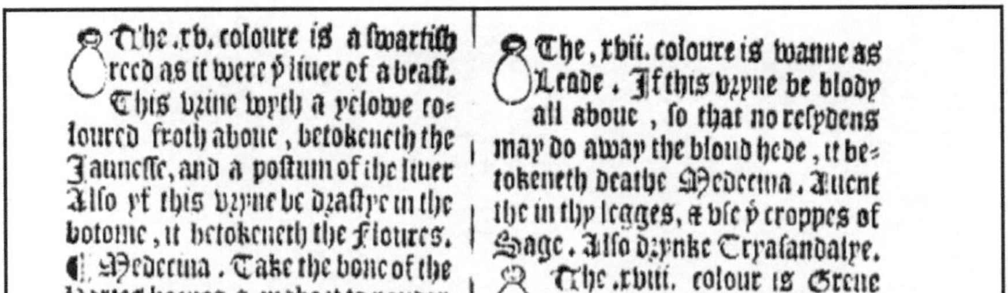
The .xv. coloure is a swartish reed as it were ẏ liuer of a beast. This urine wyth a yelowe coloured froth aboue, betokeneth the Jaunesse, and a postum of the liuer Also yf this uryne be drastye in the botome, it betokeneth the Floures. ¶ Medecina. Take the bone of the Hartes hornes & make it to pouder & drinke it. And use Diacameron.

The .xvi. colour is as blacke wine or rotten bloud. This urine in a lastynge Feuer betokeneth death. If it be troubled, fatti and stynkynge, as the uryne before ẏ was reed as a cole, and this uryne apere in a hole bodye, it betokenethe brustynge of a vayne in the raynes of skyppyng or lyftynge. Medecina. Take Osmonde, Sauyne and gromel, and boyle them wyth whyte wine, and wyth clarified Hony, and use early and late.

The

The .xvii. coloure is wanne as Leade. If this uryne be blody all aboue, so that no resydens may do away the bloud hede, it betokeneth deathe Medecina. Auent the in thy legges, & use ẏ croppes of Sage. Also drynke Tryasandalye.

The .xviii. colour is Grene as wortes. This uryne yf it be lytell in quantite, it betokeneth deathe.

The .xix. colour is blacke & shynynge as a Rauēs fether. This uryne muche in quantite & thin in substance betokeneth deliuerynge of the quartayne. Medecina. Take the ioyce of Tansye and of the reed Cole & boyle them with white wine and Hony, and use it early and late.

The .xx. coloure is blacke as a cole. This coloure in an agew litel fatty and stynkyng, if ẏ se thy face therin, than it betokenethe death. medecina. Drynke the ioyce of

Figure 4-14. Uroscopy used as bullets [23].

microscopic views of numerous common objects enabled readers to grasp the "true nature of things themselves" [28, p. 74]. Hooke's engravings support his point that the senses deceive us, and his microscopic engraving of frozen urine crystals (see Figure 4-15) illustrate Hooke's goal of showing the limits of human vision. As Harwood discussed, Hooke believed that his pictures supplied what verbal description could not, proof of a world that existed beyond words.

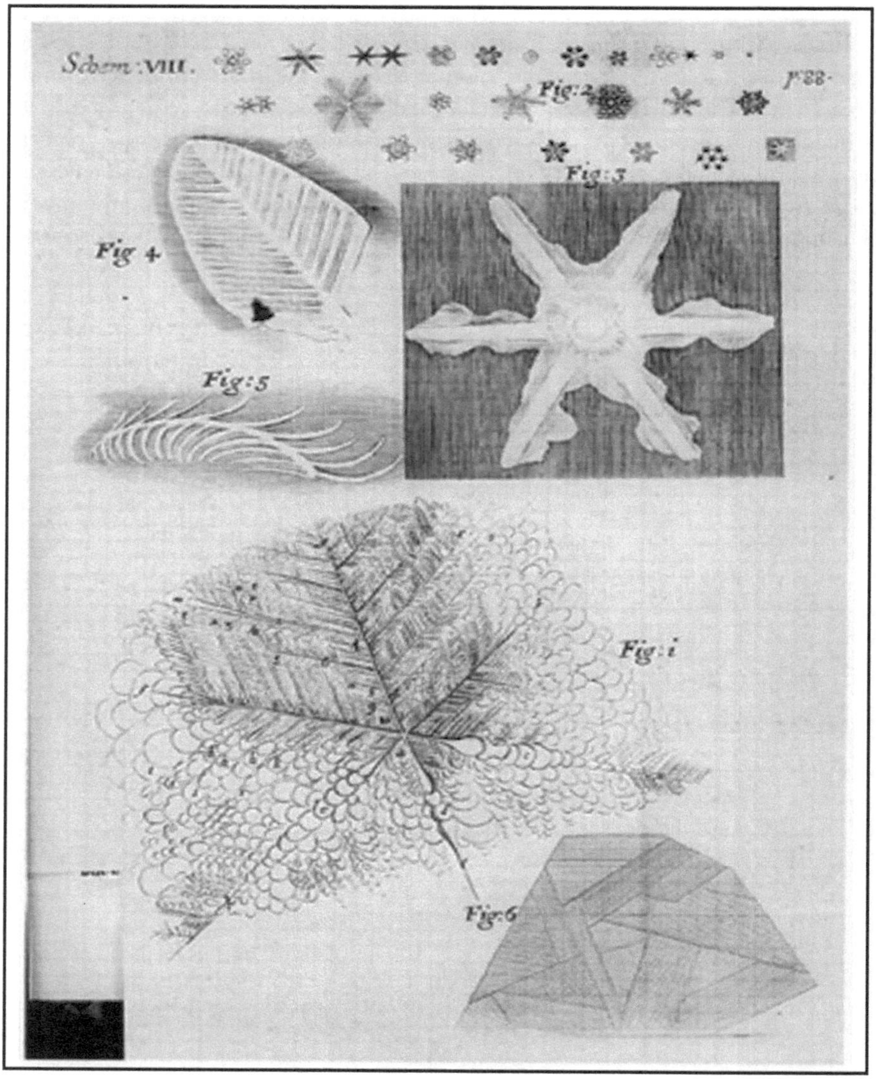

Figure 4-15. Frozen urine crystals.

As John Harwood also noted, Hooke's central thesis, the unreliability of the senses, attacked the foundation of scholastic disputation. The objects Hooke revealed with his microscope documented that (1) many "invisible" things actually exist, and that (2) seeing objects microscopically discloses radically new ways of describing reality. Hooke not only described by microscopically exploring and then drawing the world he saw with his microscope, but he also combined text and plates to support the new science [30]. *Micrographia* [29] provided copperplate drawings of a conglomeration of items, 60 in all (e.g., insects, the point of a needle, a razor's edge, silk, urine, shown in Figure 4-15, mould, sponges, sparks from steel, linen, cork). The result of this work was as dynamic as that of Vesalius: Hooke's text is important, but his visuals of the items he captures are as memorable of Vesalius' anatomical drawings. Visuals portrayed what words could not and created a new appreciation of the unseen world, the target of the Royal Society. As a result of Hooke's work, microscopy moved forward in becoming a major tool of science [30]. Hooke's student, Nehemiah Grew, used Hooke's strategies in *The comparative anatomy of trunks together with an account of their vegetation grounded thereupon, a version of his* Anatomy of Plants (167 4/5) [31], discussed in Chapter 3, and other works he wrote that revealed the unseen world of plant structure.

Nehemiah Grew

Nehemiah Grew, a student and disciple of Hooke, wrote a number of works on plants, principally technical descriptions of vegetables and plants in general. Each of these works incorporated careful technical description in addition to drawings of what Grew saw under his microscope of plant seeds, stems, roots, flowers, fruits, and leaves, particularly cross sections of these. Grew had no nomenclature for much of what he saw: He called plant cells "bladders;" but his works, published two decades after Hooke's *Micrographia* [29], furthered the impact of Hooke. These works appear in large folios, many with a foldout drawing of his view of plant taxonomy.

Grew's visuals portrayed, as Hooke's did and as Vesalius had done over a century earlier, a new perspective on a world familiar to most readers, while they continued to build the foundation for botany as a science. The small segment of a dandelion stem cut and then magnified provides some idea of the persuasive, visual power of what his visuals captured. Many transverse drawings cover two pages. Grew's books exemplify rigid organization and occur as large folios, which needed to contain his drawings. The original drawing of the dandelion stem and the microscopic display almost wordlessly conveyed the power of microscopy (see Figure 4-16, his microscopic view of the stem of a dandelion.

Humphrey Ridley—Late 17th Century Anatomist

The Anatomy of the Brain (1695) [32] by Humphrey Ridley culminates this sojourn into the importance of creative print design as it was required by various

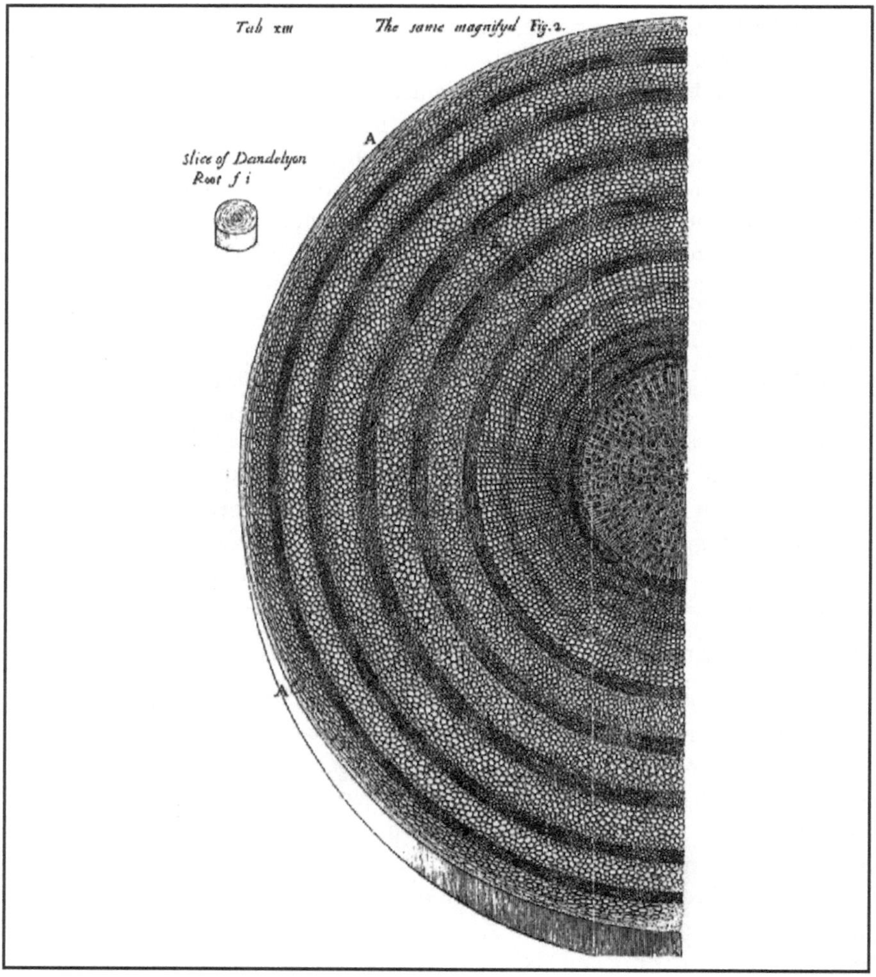

Figure 4-16. Example drawing from the *Anatomy of Plants*, 1682 [30].

forms of technical writing. Ridley provides a fitting closure to the power of visuals in technical writing in the 17th century for one important reason: Ridley's visuals stand as some of the finest examples of English copperplate illustrations developed and used in the 17th century. Ridley, held in high esteem by the Royal College of Physicians, was made fellow in 1692 and was granted the Gulstonian Lectureship in 1694. *The Anatomy of the Brain* was the first neuroanatomy published in the English.

As in many technical works that integrated visuals with verbal description and instruction, each plate occurs across from a page that provides a verbal description of the parts marked with letters. Figure 4-17 illustrates one of the seven figures developed for this work. Much of the effectiveness of Ridley's small book [32] emanates from the copperplate engravings, such as Figure 4-17 (Figures 6-7) of the brain. Figure 6 shows the Annular Protuberance; Figure 7 shows the Cerebellum. The visual provided both accurate and new information about the role of parts of the brain. Ridley was able to get bodies of freshly hanged criminals, which had engorged brains. He used wax and mercury injections to enable him to see (and his artists to outline and detail) with elegance the parts of the brain.

Ridley uses a clear, direct, first-person style to describe each part of the brain, as shown in the description of each labeled part of Figure 4-18. His concise description of the parts again reminds us of Vesalius, as shown in Figures 4-1 and 4-2.

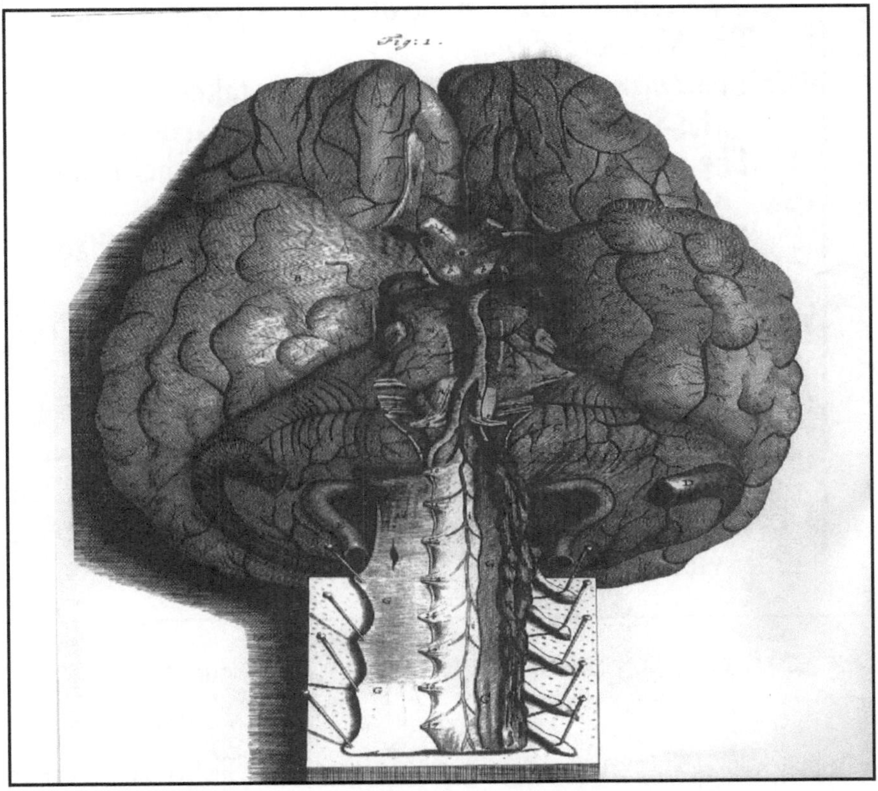

Figure 4-17. [The base of the brain] Ridley, Humphrey, *The Anatomy of the Brain.* The Classics of Neurology & Neurosurgery Library. Birmingham, Alabama, 1987. Print.

F I G. I.

Exhibits the Baſis of the Brain, with part of the Medulla Oblongata, *the Blood-veſſels being injected with Wax.*

AA The fore Lobes of the Brain.
BB The hinder Lobes.
CC The *Cerebellum.*
DD The lateral *Sinus's.*
EE The Vertebral Arteries as they paſs between the firſt *Vertebra* and the Bone of the *Occiput.*
F The Vertebral *Sinus.*
G,&c The *Dura Mater* on the right ſide taken off from the Spinal Marrow, and remaining on the left.
1,2,3 4,&c. The ten pair of Nerves belonging to the Brain, with ſeven of the Spinal Marrow.
a The *Foramen* that opens into the Pituitary Gland from the *Infundibulum.*
bb The two white Protuberances behind the *Infundibulum.*
cc The two Trunks of the Carotid Artery cut off where they begin to run betwixt the fore and hinder Lobes of the Brain.
dd The two Arteries joyning the Carotids with the Cervical Artery, called the Communicant Branches.

Figure 4-18. Description of the parts of Plate 1 (Figure 4-17).

Ridley's *Anatomy* [32] reemphasizes the impact of Vesalius' visual depictions of human anatomy, even though Ridley had access to 150 years of improvements in engraving technology. Ridley's foldout drawings in a quarto did not require a large printed folio but nevertheless captures the same detail that Vesalius' drawings did. Ridley's drawings likely derived from Vesalius' three major works.

The quality of the plates, as shown in Figure 4-19 illustrates the status of print typography by the close of the 17th century. Yet the visual power of Vesalius' artistic display, which continued in the significant writers discussed here, shows why the visual display of information became a necessary, powerful tool for the new world of knowledge that emerged from the Renaissance.

CONCLUSION

As I hope this brief journey into some of the first important technical works shows, technical writing made new demands on technology that other forms of writing did not. While linear text remained as important to technical writing as it did to humanistic, historical, or theological writing, the focus of technical writing led to the development of new ways of formatting works and designing individual pages for one important reason: Technical writing requires readers to understand text quickly and accurately. Those who needed and used technical

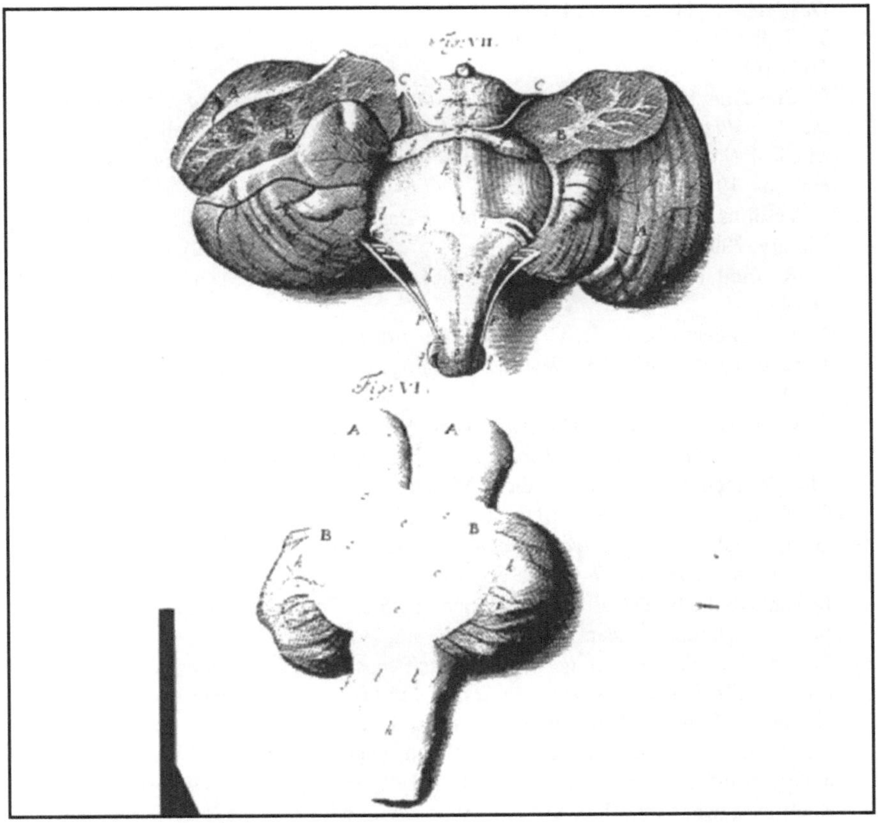

Figure 4-19. Cross-section of the brain. See Figure 4-18 for part names.

writing read and used these works in ways that differed from those who read other kinds of works.

As examples of the first technical writing shows, visuals are often imperative because they enable readers to "see the concept" presented in the text. As both Robert Boyle and Robert Hooke pointed out, the human mind, capable of perceptual errors, benefits from illustrations to enforce—present, understand, and use—correct concepts. English technical writers learned, most likely from the influence of Vesalius, that concise language correlated with visuals, typography, and page design, made writing easy to understand and use.

REFERENCES

1. L. B. Wright, *Middle-Class Culture in Elizabeth England*, Cornell University Press, Ithaca, New York, 1935.
2. T. Geminus, *Compendios a totius anatomie delineatio, aere exarata: per Thomam Geminum*, Nycholas Hyll, Londini, 1553.
3. T. Willis, *Five Treatises,* T. Dring, C. Harper, J. Leigh, and S. Martin, London, 1654/1681.
4. C. Singer and C. Rabin, *A Prelude to the History, Sources and Circumstances of the 'Tabulae Anatomicae Sex' of Vesalius*, Publications of the Wellcome Historical Medical Museum (New Series), Cambridge University Press, Cambridge, Massachusetts, 1946.
5. A. Vesalius, *DeHumani Corporis Fabrica,* Basel, 1543. Octavo edition, The Warnock Library, 2003. CD.
6. L. R. Lind (Trans.), *The Epitome of Andreas Vesalius,* Macmillan, New York, 1949.
7. J. B. Saunders and C. D. O'Malley, *The Illustrations from the Works of Andreas Vesalius of Brussels,* The World Publishing Company, Cleveland and New York, 1950.
8. A. Vesalius, *A Prelude to Modern Science, Being a Discussion of the History, Sources and Circumstances of the 'Tabulae Anatomicae Sex" of Vesalius,* Publications of The Wellcome Historical Medical Museum, No. 1. Cambridge University Press, Cambridge, Massachusetts, 1946.
9. G. de Chauliac, *The questyonary of cyrurgyens*, Robert Wyer, London, 1542.
10. H. Von Brunschweig, *The noble experience of the virtuous handy warke of surgeril,* In Southwarke by Petrus Treueris, London, 1525. The English Experience, Da Capo Press, Amsterdam: Theatrum Orbis Terrarum, 1973.
11. Euclid, *The elements of geometrie of the most auncient philosopher Euclide of Megara. Faithfully (now first) translated into the Englilshe toung* by H. Billingsley, citizen of London. Whereunto are annexed certaine scholies, annotations, and inuentions, of the best mathematiciens, both of time past, and in this our age. With a very fruitfull praeface made by M. I. Dee, specifying the cliefe mathematicall scie[n]ces, what they are, and wherunto commodious: where, also, are disclosed certaine new secrets mathematicall and mechanicall, vntill these out daies, greatly missed. Iohn Daye, 1570.

12. G. Buck, Third Vniversity of England. Or a Treatise of the Foundations Of All the Colleges, Ancient Schooles of Priviledge, and Of Houses of Learning, And Liberall Arts, Within And About the Most Famous Cittie London, in John Stowes, *Annales of a generall chronicle of England*, Thomas Dawson, London, p. 983, 1615.
13. R. Barret, *The Theorike and Practice of Moderne Warres*, London, 1598. The English Experience, Amsterdam and New York: Da Capo Press, Theatrvm Orbis Terrarvm Ltd. 1969.
14. J. Bingham (Trans.), *The Art of Embattailing an Army. Or, The Second Part of Aelians Tacticks.* London, 1629. Amsterdam and New York: DaCapo Press, Theatrvm Orbis Terrarvm Ltd., 1969.
15. R. Norwood, *Fortification of Architecture Military,* printed by Tho. Cotes, for Andrew Crooke, and are to be sold at the signe of the Beare in Pauls Church-yard, London, 1639. Amsterdam and New York: Da Capo Press, Theatrum Orbin Terrarum Ltd., 1973.
16. J. Achesone, *The Military Garden, or Instructions for All Young Souldiers*, John Wreighton, Edinburgh, 1629. Amsterdam and New York: Da Capo Press, Theatrum Orbin Terrarum Ltd., 1974.
17. E. Tebeaux, Technical Writing in English Renaissance Shipwrightery: Breaking the Shoals of Orality, *Journal of Technical Writing and Communication, 38*:1, pp. 3–25, 2008.
18. E. Hayward, *A Full and Perfect Account of the Sizes and Lengths of Riggins for All His Majesties Ships and Frigats.* Peter Cole, London, 1666.
19. E. Bushnell, *The complete ship-wright* (4th ed.), Peter Cole, London, 1660.
20. H. Bond, *The Art of Apparelling and Fitting of Any Ship with Masts, Yards, and Cordage,* F. R. Boydell, London, 1655.
21. H. Bond, *The Boate Swains Art: or the Compleat Boat-Swain,* Richard Mount, London, 1676.
22. T. Miller, *The Complete Modellist: Shewing he True and Exact Way of Raising the Model of Any Ship or Vessel, Small or Great,* William Fisher, London, 1976.
23. M. Cortes, *The Art of Navigation,* Richard Eden (Trans.), London, 1561, 1572, 1579, 1584, 1589, 1596, 1615.
24. W. Johnson, *The Light of Navigation*, Amsterdam, 1612. Amsterdam and New York: Da Capo Press, Theatreum Oorbis Terrarium, Ltd., 1964.
25. J. Berners, *Here in thys boke afore ar contenyt the bokkys of haukyng and hunting with Other plesuris dyuerse as in the boke apperis and also of cootarmuris a nobull werke*, Facsimile, St. Albans, 1486. Amsterdam and New York: Da Capo Press, Theatrvm Orbis Terrarvm Ltd., 1964.
26. *Here begynneth the seynge of urines of all the courlores that urynes be of.* Wyllyam Myddlyton, London, 1544.
27. A. Johns, *The Nature of the Book: Print and Knowledge in the Making,* University of Chicago Press, Chicago and London, 1998.
28. C. Pyle, Art as Science: Scientific Illustration, 1490-1670, In Drawing, Woodcut, and Copper Plate. *Endeavor, 24*:2, pp. 69–75, 2000.
29. R. Hooke, *Micrographia: Or Some Physiological Descriptions of Minute Bodies Made By Magnifying Glasses,* St. Paul's Church-yard, London, 1665.

30. J. T. Harwood, *Rhetoric and Graphics in Micrographia,* Michael Hunter and Simon Schaffer (eds.), Robert Hooke, New Studies, Woodbridge, Massachusetts, pp. 118–147, 1989.
31. N. Grew, *The Anatomy of Plants*, With An Idea Of A Philosophical History Of Plants, And Several Other Lectures, Read Before The Royal Society, W. Rawlins, London, 1682.
32. H. Ridley, The Anatomy of the Brain, *containing its mechanism and physiology: together with some new discoveries and corrections of ancient and modern authors upon that subject: to which is annex'd a particular account of animal functions and muscular motion: the whole illustrated with elegant sculptures after the life by H. Ridle,* printed for Sam. Smith and Benj. Walford, London, 1695.

CHAPTER 5

Instructions: The Oldest Form of Technical Writing—Tracking the Shift From Oral to Textual Instructions in English Estate Management Documents, 1200–1700

Instructions and procedures represent a dominant form of modern technical writing, as suggested in Chapter 1; but instructions are not a recent development, a point emphasized in Chapter 4. As shown in the *Short-Title Catalogue*, numerous examples of instructions appear in published works of the English Renaissance (1475–1600), but how this important type of utilitarian discourse evolved before the 18th century has yet to be written. In this chapter, I sketch the history of technical instructions with two goals for this initial study: (1) to describe the development of practical instructions, as they shifted from oral to textual transmission, and (2) to exemplify this shift in early how-to manuscripts and books on agriculture and estate management.

These were topics of sustained importance in England until well into the 18th century. As much of my research has shown, practical instructions existed throughout the medieval period, the Renaissance, and the 17th century in England in a range of fields, such as medicine, military science, navigation, and early Royal Society documents, in addition to agriculture and estate management. These documents anticipate characteristics of modern technical instructions. What we today call "good style" in technical instructions emerged as practical discourse shifted from academic oration style to plain, or "low" style, exemplified in Chapter 2, and shed much of its oratorical residue by the close of the 17th century. As I will show, the first instruction writers seemed aware that practical texts required a style much different from the style of humanistic works.

My larger goal in this historical sketch is to encourage others to add, revise, and challenge my theories. How instructions developed varies among disciplines, as I discussed in both Chapter 1 and next in Chapter 6. That is, military science instructions developed in a different way from agriculture, as did accounting instructions and medical and midwifery instructions. Each offers a unique research opportunity. Of one point I am sure: The study of instructions, whatever the focus, allows historians to explore the world in which practical writing existed and to unearth the history of our field.

The history of instructions, taken as a broad genre, must be inferred from extant scholarship and historical work, particularly that which tracks the shift from orality to textuality. This shift, which marks the foundation of text, enables us to apply theory to the first writing, which was practical writing. Studies of the evolution of text, such as those by Goody [1, 2], as well as Goody and Watt [3], for example, describe the evolution of orality to textuality in early societies. Studies by Ong [4, 5] examine elements and implications of oral residue in text in general. Studies by Stock [6, 7] and Glen [8] illuminate the oral/textual shift in humanistic works. A host of other fine studies could be cited, but all these authors provide useful information and perspective in suggesting how English practical instructional texts may have developed before the 18th century. Practical writing developed alongside humanistic and literary writing and in larger quantities than both, a fact we must never forget [9, pp. 60–61; 10, p. 162].

LISTS—THE FIRST PRACTICAL TEXTS

Emergence of English text echoes the emergence of text in ancient societies. The first texts that emerged from Sumeria, Assyria, and ancient Mesopotamia were lists that led to the development of ledgers. As Goody remarked, "It is not literary works, then, but administrative lists that dominate the uses of writing in ancient Mesopotamia":

> These lists can take a whole variety of forms, receipts of tribute, itemization of war booty on the income side, distribution of rations, payments to officials, among the expenditures. A recording of these transactions is especially important in a bureaucratic system of the Mesopotamian kind, whose economic activities were based upon the movement of personnel and goods . . . through the channels of bureaucracy under the supervision of personally responsible officials who serve for a definite term of office. [2, p. 82]

Early Syrian writing, such as that found on the Ugarit tablets of the 14th century BCE, used administrative lists extensively. Of the 508 documents recovered, two-thirds consist of lists rather than consecutive prose or poetry. These lists include quotas, inventories, receipts, guild lists, census records, land grants, statements of cost, purchases, loans, indications of ownership [2, pp. 85–86]. Technical communication faculty and practitioners can visualize how people in these early cultures may have used these lists, not only for recordkeeping but also for mnemonics and implementation guidelines. Oral instructions most likely surrounded these lists, which today can suggest the oral business activities embedded in these documents.

I believe that we can use Goody's work to build on this idea—the textual primacy of lists. Goody believes that "writing therefore provides a locational sorting device," perhaps a framework for instructions on what to do with these items and how the associated administrative task should be done [2, p. 75]. As Goody states, words become enduring objects that suggest evanescent aural signals. This transformation means that communications over time and space alter in significant ways. At the same time, the materialization of the speech act in writing enables it to be inspected, manipulated, and reordered in a variety of ways [2, p. 76]. These first texts—lists—therefore anticipate the potential of what full-blown text would become.

ORALITY, LITERACY, AND TEXTUALITY—FOUNDATIONAL CONCEPTS IN THE DEVELOPMENT OF INSTRUCTIONS

Orality preceded both literacy (the ability to read) and textuality (the ability to write), but the relationship among these three and the impetus for literacy and text plays a significant role in the development of instructional text. I wish to apply three observations that show how the interplay of orality, literacy, and textuality, excerpted from current scholarship, create an initial theory of how instructional text emerged.

First, people in early societies who could not read could be "literate" in the sense that they could remember and apply spoken text. When texts written in the rhetorical tradition were read aloud, listeners could partake of the rhetorical tradition via sound. In *Listening for the Text,* Stock states that "the text did not have to be written: oral record, memory, and reperformance sufficed" [7, p. 37]. The strong oral component of medieval popular literacy meant that "the masses of people read by means of the ear rather than the eye, by hearing others read or recite rather than by reading to themselves" [11, p. 88]. Glenn effectively augments this point: Those who could read print could share the content to the *illitterati.* The use of memory, the persistent habit of reading aloud, and the preference, even among the educated, for listening to a statement rather than scrutinizing it in script, comprised popular literacy practices. Thus,

> both lay men and women alike heard about texts they probably had not read, listened to texts they perhaps could not read, composed texts they probably could not themselves write, and talked about texts they all had in some measure committed to memory. [8, p. 498]

Second, while text thus coexisted with oral rendering, during the 11th and 12th centuries, texts became the governing segment. Ong [4, p. 123] and Harvey [12, p. 114], believe that lists, kept for accounting purposes and prepared on estates, exist as the first English texts. Then, as Stock explains, rapproachment between oral and written began to play a major role in organizing activities in everyday life, particularly in both ecclesiastic and then civil government. Because of the continuity of aural memory, many documents produced by the laity for use in civil government and business echoed the oral tradition [6, pp. 12–18]. Impetus for business texts led English historians such as Galbraith to believe that "all through the ninth century there is a steady increase in the number of surviving business documents—wills and charters—written in English. . . . lay society must have been more developed in England than abroad" [13, p. 15]. Thus, lists absorbed much of the oral tradition and began to evolve into various types of practical documents, many of which exemplify instructions.

Third, while use of Latin had traditionally determined literacy in the Middle Ages [14, p. v], the ability of the merchant class to read and to write persisted during this period [14, p. 9]. Literacy for administrative and practical uses likely developed in schools that existed outside the Church, and instruction was conducted not in Latin but by the "low," living tongue [14, p. 10]. As Clanchy [15] and Houston [16] have argued, laymen became more literate to cope with written business, initially in England, with writs from the royal government demanding information or money. As commerce developed, so did the need for literacy in business transactions [16, p. 98]. Thus, lay literacy grew out of bureaucracy and commerce rather than from any abstract desire for education or literature [16, p. 19]. From the demands of bureaucracy emerged instructions needed for a variety of practical purposes. A basic ability to read and then to write enabled people and minimally educated clerks to prepare documents needed for government and business. The form and style they learned by hearing embedded itself in their written texts. Thus the plain "low" language used by the laity persisted from the 11th or 12th century onward. Literacy, as it developed, absorbed characteristics of orality.

The speed at which text developed drew from the embedding of orality in text and the drive for literacy. Development of every kind of estate record occurred between 1250 and 1330 [12, p. 107]. These records, prepared by clerks, whose literacy, although uneven, allowed them to follow model texts and sets of instructions for composing these records and documents. If the 12th century had been a period of making documents, the 13th century became the century of keeping them. The Norman invasion and the making of the Domesday Book nurtured the development of bureaucracy in England, an event that spawned the birth of utilitarian documents. Demands of the Exchequer and courts for documents further compelled literacy of the laity for business and legal purposes [6, p. 329].

Orality and the Character of Practical Instructions

Within this increasing, but uneven, oral-based literacy, the power of formularies and cartularies remains critical to development of instructions, both of which incorporate the oral tradition.

Cartularies, themselves lists of decrees rather than the decrees themselves, were used in both church and government. Listeners heard the decree read aloud, but the decree itself existed only as an item in the list of decrees. And, as Stock [6] and Ong [4] both note, the existence of the list was as powerful as the oral version of the document as these became part of the collective memory of people affected by the cartulary. "What gave a work its identity consisted very little in what it looked like. The work was what it ***said*** when someone was reading it, converting it into sound in the imagination or, more likely, aloud" [5, p. 2]. The lengthy titles of the cartularies continued in the lengthy titles of 17th century proposals and reports (see Chapters 7 and 8, this volume).

Formularies, templates, or example documents written to help clerks prepare legal documents were developed to capture not only the oral rendition but also the content of the document. Their influence is critical in the 14th and 15th centuries when both literacy and textuality were emerging. The unevenness of literacy among legal clerks increased the popularity of formularies, or precedents, for legal documents. The formulary writers remain unknown, but formularies helped ensure that legal procedures remained uniform. Formularies and precedents stand as some of the oldest examples of legal instructions in English. Formularies existed for deeds, manorial accounting, inventories, wills, assessments of a deceased person's movable assets; all with the "aim at imparting the rudiments of the law of testament to the untutored layman. Superficial as the treatment of the subject may be, the same guidance, largely word for word, was still being offered to testators nearly two centuries later" [17, pp. xiii, xiv].

The following example illustrates legal instruction embedded in formulary letters:

> The stewards' letter to the bailiff:
> —A. of B. steward of Sir N. of C. to the bailiffs of Burton greeting. For that on Friday next after the Purification we purpose to visit your parts, we command you that for the said day ye do cause to be summoned the hundred, or the court, according to the custom of the manor, and that all other things that pertain thereto be ready and prepared to the best of your power. Farewell. Given this etc. [18, p. 70]
>
> Another form:
> For that we purpose to hold the general court of the lord bishop of Winchester on Saturday next after the feast of St. Valentine at N., we command you that all who owe suit to the said court be there at daybreak before us to do what of right ought to be done, for we shall go thence to our dinner at NJ. Betimes. Farewell. Given this etc. [18, p. 70]

Similar to other formularies and precedents, the clerk had only to enter names and events as appropriate. The truth of the letter was duly tested when it was read aloud. The reading of formularies aloud preserved them in the minds of those present, while the written formularies allowed both the literate and illiterate to participate in the judicial process. Ceremony and rites added to the seriousness and legality of the legal proceeding. For example, formularies explain the actions or instructions required when a person is giving an oath:

> As to oaths. When one is received into frankpledge he shall swear on the book with bent knee thus. Hear this sir steward N. etc. that I, N. will not be a thief nor the fellow of a thief, nor will I conceal a theft nor a thief but will reveal it to those to whom it should be revealed, and I will bear faith to the lord Henry King of England, and more especially to my lord N., and will be obedient to the precepts of his bailiffs. And having kissed the book he shall give a penny and shall be hidebound to be obedient to his chief pledge. [18, p. 77]

Of the oath of witnesses. Anyone brought as a witness in any lawsuit shall swear thus:

> Hear this sir N. that I, N. will not for love, nor hate, nor fear, nor anything in the world, nor for price, nor reward tell any falsehood or conceal any truth touching the matter about which I am brought as a witness between A. and B. before thee. So help me God and his holy gospels etc. [18, p. 77]

The veracity of the content inserted into the formulary often was achieved by ceremony that occurred alongside the spoken rite, not just the written content itself.

A final category—forms—mirror formularies in their purpose: They provided uniformity in administration and business, and they also aided readers with limited reading skill. Forms existed as verbatim copies of original records, complete with fictitious names, dates, and even figures, like formularies. As Oschinsky also notes [19, pp. 4–5], some specimen records were so detailed that they appear to have been drawn up on purpose to serve as formularies.

13th Century Instructions

With the growth of estates and commerce in the 13th century (both are suggested in the previous example formularies) came the need for written instructions and documentation. Medieval society became increasingly oriented toward the scribe, the written word, the literary/religious text, and the practical document. Commercial occasions drove the need for writing beyond the Church.

In addition to accounts and legal formularies, perhaps the first written instructions in England occurred as letters, which show the day-to-day running of a large estate. Letters stand alone among estate records in giving some direct insight into the way decisions were made and implemented. Harvey classifies these early letters as either personal or administrative. The early business letters had a distinct purpose: to give instructions from the owner or senior officer of an estate to lesser or local officials. Among the earliest administrative letters are those written to and by Ralph de Nevill, Bishop of Chichester (1222–1224) and reveal the use of letters as instructions [12, pp. 112–113]. Similar to published Renaissance instructions, these contain background material that contextualizes the instructions for the reader.

As noted by the English historian Blaauw, who edited this collection, de Nevill used a "plain, business-like manner," a style that eliminates any loss of meaning by translation from Latin to English (1850) [20, p. 37]. For example, letter 386, addressed to his steward, shows the Bishop's anxiety about finding sufficient quantities of winter food, often a major concern in the early 13th century. We can visualize de Nevill, wrapped in his "grey cloak," monitoring how iron and steel were added to his plows [20, p. 41]. See Figure 5-1.

The collection also includes letters from Salvage, his servant and business contact, giving instructions to de Nevill. Salvage, in letter 667, also contextualizes his instructions to de Nevill and uses a similar crisp, direct style [20, p. 45]. See Figure 5-2.

"*Ralph de Nevill, Dean of Lichefield,* to his faithful Geoffry Salvage, greeting.—That you have sold my wheat from Thorp for 22 marcs (£14 13*s*. 4*d*.), as you have informed me, because it could not be sold for a higher price, I am content that it should be so sold. About your purchases also, concerning which you wished to inform me, I commend your prudence, requesting that you so manage my affairs, that I may thank you; know that I have spoken with Sir Richard Duket, that he shall let me have 5000 herrings and 200 wax candles, and a grey cloak, and iron and steel for my ploughs; and therefore I order you, that, as soon as you can, you go to him, and agree with him about all these things. Be mindful of the herring which the prior of Nuwic gave me, namely, 5000, in order to receive which it behoves you to be at Nuwic either the third day before the feast of St. Martin, or the third day after the feast of St. Martin; about the other herring which you know of, I leave the whole to your discretion. Farewell. I have quite lost the letters of Abraham de Cruezford, of the tenor of which I am entirely ignorant. Farewell."[3]

Figure 5-1. Example 1300 century instructions letter.

"*To his Reverend and excellent Lord Ralph, by the grace of God Bishop of Chichester*, his faithful servant, Simon de Senliz, greeting, and faithful service.—I send you now 19 pigs, from your manor at Aldingeburn, and, as soon as the pigs of your other manors shall be fat, I will send them to you. Signify to me, if you please, if Thomas of Cirencester (Cirecestrie) has sent to you any message about procuring oxen, and if he shall not have done so, let me know; also whether you wish that I should buy any oxen in those parts, and how many you wish I should buy, since intelligence has been given me of a certain fair in those parts, in which good oxen are often sold at a reasonable rate. Let me know, if you please, about these and other matters, your good will and pleasure, by the bearer of this.—May my lord fare well."

Figure 5-2. Another instructions letter.

13th Century Estate Manuals

Because of the importance of farming, at least four separate instructional treatises emerged for use in early English estate management and farming. These manuals represent a major leap in the development of instructions, although they targeted those able to read either French or Latin. Each manual, written for a distinct audience and/or purpose, illustrates writers who sought to meet the information needs of their readers. As noted by Oschinsky, the most recent editor of these documents, "these texts were written for a rising profession of estate administrators who, together with lawyers practicing in common law, received their theoretical training outside the university by the study of practical manuals" [19, p. 3]. And, as Richardson wrote, business instruction occurred both within Oxford and tangential to the university education [21, p. 25]. These works were prepared during the heyday of high scholasticism and high farming on the estates of the great English landlords. *Rules* and *Husbandry* were prepared for the use of officers of a particular estate, while *Seneschaucy* and *Walter of Henley* were intended as teaching tools for estate managers who learned estate management outside the university setting. Ultimately, *Walter of Henley* provided the greatest historical influence of the four in terms of content that surfaced 200 years later in the works of John Fitzherbert, Thomas Tusser, and Gervase Markham, principle agrarian writers of the 16th and early 17th centuries. The works of these early writers illustrate the interplay of orality and textuality in meeting the literacy levels of a new class of readers eager for knowledge.

Walter of Henley, who became a Friar Preacher in midlife, presents his text as a sermon, the advice of a father to his son. The structure of the treatise—prologue, text, and epilog—reflects sermonic form. Oschinsky suggests that this text may have been delivered as a lecture, likely at Oxford. Prescribed courses in business training are known to have existed there, which provided instruction in the art of letter writing, the formulation of writs, deed, and accounts. Within such a course, instruction in estate management would merely have been an extension of the syllabus. The close relation between the *Seneschaucy* and *Walter* suggests that the author gave his lecture to an audience already familiar with the details of estate administration, as set out in the *Seneschaucy*. The writer may have proceeded from there, elucidating and glossing the chapters of that text [19, p. 148]. As Oschinsky also states, in the evolution of instructional text, *Seneschaucy* and *Walter* both serve as useful examples of a text prepared for hearing, memorization, and for reading. In short, both *Seneschaucy* and *Walter* shared a common purpose, oral *and* textual instruction. This shared purpose explains the rationale for the inclusion of both texts in many extant compilations.

Seneschaucy

Within the sections of *Seneschaucy*, the only one of the four that has a table of contents to suggest the reading purpose of the document, the instructions follow a rigid pattern: the steward should do this, the steward should not do that, and so on. Passages are often rounded off with a brief recapitulation of the main thought, at times expressed in the form of a wise saying—here underlined in the final item of the excerpt below. *Seneschaucy* incorporates constant use of "doubling," closely related terms for emphasis (presented in italics in the example below) to aid understanding. Doubling gives the listener or reader time to absorb the argument when the work is read aloud. Like many medieval works prepared for oral delivery, sentences are additive and aggregative rather than analytic. (The c. stands for paragraph.) For example,

> These are the chapters on how to *manage and improve* manors which are in the care of stewards and bailiffs.
>
> Ch. I Here begins the office of the steward
>
> c. The estates steward out to be *knowledgeable and loyal and capable* of administering the lands profitably; he ought to know the law of the country so that he can defend actions outside the lord's estate, can give confidence to the bailiffs who are under him, and can instruct them.
> The steward ought to make his circuit twice o three times yearly and visit the manors in his charge. On these occasions he ought to inquire into rents, services, and customary payments which have been concealed and are being withheld; and he ought to inquire also into franchises of courts, lands, woods, meadows and pastures, waters and mills, and of any

other estates which belong to the manor and have been disposed of without warrant; by whom and why.

c. And if it lies within his powers to amend the aforesaid things, by legal means, without doing wrong to anybody, he may do it, and if he cannot he ought to bring such matters to the notice of his lord that he may intervene should he wish to uphold his rights.

c. The steward should not have power to sell wardship, or marriage, or escheat, nor to dower any lady or woman, nor to take *homage or suit*, nor to sell or make free a vilein without special warrant from his lord. And the steward ought not to be chief accountant for the things of his office, for he ought on the account of each manor to answer for his *doings and commands and improvements*, and for *fines and amerciaments* of the courts where he has held pleas as another, because no man can or ought to be judge or justice of his own doings. [19, p. 26]

Its simple presentation, lacking the density of legal text, suggests that *Seneschaucy* was also intended for use by a wider, less qualified audience than the lawyers for which it had been principally designed. Additional readers seem to have been the lords of estates interested in the management of their lands and to anyone interested in estate accounting [19, p. 98].

Walter of Henley

Although seen as complementary texts, *Seneschaucy* and *Walter of Henley* differed in their purposes. While *Seneschaucy* provided instructions for large baronial estates, *Walter of Henley* sought to teach students of estate management the efficient running of a *single* manorial unit and some aspects of farming [19, p. 15]. *Walter of Henley* went beyond *Seneschaucy* by providing readers detailed instructions. Note his instructions for plowing:

> First knowe that a quarantine ought to have 4 roddes in breadthe and 40 roddes in lengthe and the kings rodde (or perche) is sixteene foote and a halfe; and then hathe the acre in breadthe 66 foote. When you have gone up and down 33 tymes, with a forrowe of a fote broade, then is an acre plughed.
>
> But to have your forrowe narrower, make youre tourney 36 tymes up an downe, and then be yow at thende of 72 coteries, which doe amount to 6 leagues; for 40 perches in lengthe make one coterie and 12 coteries make one league. Now shall the steere or oxe by verie poore (weake) which cannot goe faire and softely/ three leugues of way and retourne againe from morning tylle noone (nyghte). [19, p. 315]

The lack of chapter divisions of the original manuscripts, many of which contain ¶ "markers" designed to help the lecturer keep his place, soon gave way to better textual organization: an introduction, chapter divisions, and chapter headings, but where these occur in various versions lacks consistency but may illustrate the difference between a sermon and a teaching document. Because *Walter of Henley* gave his instructions in sermonic formula, the treatise begins:

c. The father was in his olde age and sayed to his sonne: Good sonne, live wisely according to God and according to the worlde.

c. Towardes God, think often of the passion and of the death that Jhesus suffered for youe and love him above alle things (and honour him) and his commaundements keepe youe.

c. As to the world thinke of the wheele of fortune howe a man mounteth by little and little to richnesse, and when he is at the toppe of the wheele, how by mischaunce he falleth by little and little into provertie and after that into miserie. Wherupon I pray you that according to that youre landes by woorthe by yeare by exente youe doe order your life and no higher at alle.

c. If youe may your landes amende, either by tllage (thrifte) or by stock of cattaile or by any other provision above the yearly extente putte (turne) that overpluis into money, for if corne sayle [or stock die] or fier doe happen or any other mischaunce then wille that be somewhat worth to you which/you have in coyne. [19, p. 309]

Walter of Henley, following the lecture/sermon format, states his case, as he does in this excerpt, before he gives reasons for it the command he makes.

Husbandry

The third of the four medieval treatises, *Husbandry*, unlike *Walter of Henley* and *Seneschaucy*, deals with different aspects of accounting. It reads as a compilation of memoranda, tables, and advice on auditing. This text echoes the use of lists, described by Goody [2] and Ong [5] (1983), but shows how text with embedded lists could be developed for mnemonic purposes. *Husbandry* exemplifies the least organized of this early group of estate management instructions and suggests that the writer prepared the work, at least initially, to serve as a reference for himself or another person on the estate to use when information was called for during the audit. Rather than an introduction, common to texts prepared for reading, *Husbandry* begins with the oath taken by the accountant and the accountant's clerk before the audit began. The topics—instructions—that follow suggest the order of the audit and explain the requirements and the reasons: cc. 3–8.

At the point in the audit, where the audit of the money account began, the compiler included lists enumerating money sources to support memory so that no item was forgotten, to help the auditor when "hearing" the account. Note the detail of the embedded lists:

c. One owes first for one's arrears if there are any.

And then for assized rents, farms of mills, dovecots, fairs, and markets, for profits derived from the lord's privileges concerning drink, ale, flax, and all kinds of common rights, for stallage, toll, wardsilver, and Peter's pence.

> And then for pleas, for the issues of the manorial court, pledges, reliefs, escheats, marriage fines; for casual receipts from scutage, tallages, payments from heirs, leyrwite; for bargains, acres put out to rent, or for any other property if any rent has freshly accured since the last account.
>
> You ought also to hear audit concerning the fisheries in rivers in open country, in small steams, in ditches, marshes, or fish-ponds; and of all kinds of produce of gardens and curtilages; or herbage of the yard, orchards, and gardens and of herbage in forest wards and in smaller division, in woods and parks, low shrubberies and underwood; and of the agistment of stock and of rents from pastures and herbage sold in gross. And bee it known that he who renders account ought to name all the places which he keeps back for the grazing of the manorial stock—and their value—just as carefully as the grazing which he sells, so that the value of the manor may be more fully ascertained. [19, pp. 433–435]

14th Century Instructions

Additional examples of business letters with accounts can be found in extant letter collections, such as the Stoner letters and papers, 1290–1483. Many of the early letters appear in Latin, but more appear in English by the close of the 15th century. However, a combination of medieval Latin account format and English phrasing characterize these documents. Letters and accounts prepared by businessmen, such as "A Weaver's Bill for Cloth to Mistress Stoner, 21 December 1468," emphasize that the commercial classes were becoming skilled in English. In a sense, the bill, a list of items owed, becomes an instruction to pay for the cloth used. "Item" comes from Latin, but the other words are English. See Figure 5-3.

Other examples include a bill for "Wax and Tapers" supplied to Dame Elizabeth Stoner, September 1478, and from the shoemaker [22, Vol. 30, p. 74]. These bills are not just item lists but sentences. Other bills include one for cloth and two mercer's accounts. These documents support Laquer [23], Clanchy [15], Bennett [24], and Richardson's [25] positions that English people found reading and writing both important and useful for a variety of reasons, not the least of which was the need to conduct business and receive payment for services as documented by text.

16th-Century Agriculture and Estate Management Manuals

John Fitzherbert

The next step in the evolution of instructions in agriculture and estate management can be best exemplified in the early printed instruction manuals of the English Renaissance, which began with John Fitzherbert, an Englishman writing for English men and English women.

95. A WEAVER'S BILL FOR CLOTH TO MISTRESS STONER
21 Dec., 1468

Item to my Mastres Stonor xvj erdys of brode clothe, wyte, ij.d.

Item also vj erdys of fine clothe, j. d. ob.

Item also xvj. Erdys of roset of fine brode, ij. D. ob.

Item also vij. Erdys of ikersey that was made of the same roset, j. d.

Item also xiiij erdys of cors roset brode, ij. D

Item also xij. Erdys of fine . . . re blew, v. d.

Item also xij ellis of roset kersey, j. d.

Summe xvij. S. x. d.

Althyng rekekynnyd between Willm. Demnyst wever of Walyngton for wevyng of all maner of clothe, saving a white pece the wyche is at fuller. And this rekenygge unto Seynt Thomas nexte afore crystemas in the viiij yere of Kyng Edward the iiijth payyd. [22, Vol. 29, p. 67]

Figure 5-3. Example bill, a list of items owed, implying orders to pay.

Fitzherbert's *Booke of Husbandrie*, recorded 12 editions between 1523 and 1573 and another 6 before 1600 [27, p. 136]. Fitzherbert's debt to both *Seneschaucy* and *Walter of Henley* in terms of content has been carefully drawn by agriculture historians. However, Fitzherbert, writing nearly 300 years after *Seneschaucy* and *Walter of Henley*, recognized that he had different readers than those anticipated by the treatises prepared for seigniorial estates.

The 1530 husbandry book exemplifies the earliest printed English husbandry manual written for the English country farmer to help him improve his crop yields. The size and character of this work, and others that emerged in the 16th century, differ markedly in language and size from the classically influenced works, such as Xenophon's, a work of over 600 pages, which would have been unaffordable to the rural farmer. Fitzherbert's style exemplifies many qualities of the "low" or plain style but also qualities of the oral tradition—additive syntax, aggregative rather than analytic statement, empathy, and emotive allusions [4, pp. 36–49].

Booke of Husbandrie [27] contains four books. Each deals with a specific group of tasks: Plowing and Sowing; breeding, ordering, and usage of cattell; setting, lopping, cropping, using of tymber, grading and planting; and ordering of a household. Within each section, instructions proceed with minimal logic, an observation that can be seen in the table of contents, located at the end of the book. Fitzherbert provides no specific order of duties, but he does provide self-contained instructions for each topic. We can almost see Fitzherbert talking to and with his readers about each segment of farm management. For example, in the first book, he moves from plowing to planting of specific crops, to soil maintenance, to weed recognition, and finally to harvesting. Each chapter—18 in Book One—suggests how Fitzherbert himself performed these duties.

Like many modern technical writers, he provides technical descriptions before instructions. His general technical description of the plow, which appears in the opening pages of Book One, precedes description of the constituent parts (Chapters 3–5). Chapters 6–8 then give general instructions for plowing. Fitzherbert's technical description of the plow anticipates modern technical descriptions: introduction, listing of parts, and separate sections for each part.

As he moves into active instruction, his language becomes descriptive and precise; he writes as if he is talking with his readers; he uses an additive style, a major characteristic or orality. For example, Chapter 11 of the second book, "How to cure the worme in a Sheepes foote," shows how carefully Fitzherbert uses precise, additive description and metaphor to help readers visualize the process:

> Many time it happeneth among Sheepe, that they have a worm in their foote which maketh them to halt: take that sheepe and look betweene his clawes, and there you shall find a little hole, as much as a great pinnes head, and therein growth five or sixe black haires, about an inch long, or somewhat more, take a sharpe poynted knife, and slit the skinne a quarter of an inche long above the hole, and as much beneath, and put thy one hand in the hollow of the foote under the hinder clea, and set they thombe above, almost at the slitte, and thrust thy finger underneath forward, and with your other hand take the blacke haire by the end, or with the kinifes point take hold thereof, then pull the haire by little and little, and thrust after thy other hand with thy finger and thy thombe, and there will come out a worme like a peece of fleame, as much as a little finger, and when it is out, put a little Tar in the hole, and it will be quickly wel. [27, p. 52]

Fitzherbert's instructions also provide a window into the issues facing the Renaissance English farmer: draining land, treating water meadows to make hay, building sheepfolds, mowing grass, and making a ewe love her lamb, combined with chapters on cattle, horses, and other stock. The final book on the character of the farmer's wife, the diseases she would be required to treat, their remedies, and lopping, cropping, and grafting fruit trees completes the portrait of the rural farm. Yet the oral tradition, with its appeal to memory, sharply appears again as a short chapter in the fourth book: a "rude ryme" to teach a gentleman's servant, who could probably not read, "to say, every time, when he taketh his horse . . . not forget any implement behynde him." The mnemonic verse is composed in abab, cdcd, eff gg:

Pursee, dagger, cloake, nightcap, keircheffe,
  shooinghorne, budget, and shoone [shoes],
Speare, male, hoode, halter, sadelcloth, spurres,
  hatte, with thy horse-combe:
Bowe, arrowes, sworde, bulker, horne, leishe,
Gloves, string, and they bracer.
Penne, paper, incke, parchmente, reed waxe, punisse,
  and bookes doe, thou remember.
Penknife, combe, thymble, needle, threde, poyte,
  least that by chaunce thy garth breake:
Bodkyn, knyfe, rubber, give they horse meat,
  See he be shodde well, make merry, sing if thou can,
  And take heede to thy needments, that thou lose none. [27, p. 146]

His final Chapter 53 is a short poem entitled "The author to his Booke." Fitzherbert's [27] use of orally based expression throughout his work supports Ong's [5] contention that writing was taught to improve speaking also, as speech did not separate from rhetoric until the early 19th century.

Thomas Tusser

Thomas Tusser (1515–1524?) authored the second most significant published book on agricultural instructions in England in the 16th century. Tusser lived an unremarkable life as a chorister and then an unsuccessful farmer. He studied at Trinity College, Cambridge, but was unable to farm land at Suffolk, which was continually flooded by the ocean. Unable to support himself, he died a pauper, even though his only agricultural book enjoyed sustained popularity throughout the 17th century. A total of 23 editions appeared between 1558 and 1638, a testimony to its popularity.

Tusser's initial book, *A Hundreth good Points of Husbandrie* (1557) [28] grew to *Five Hundred Points of Husbandrie* by 1573, the second edition [29]. Both books used verse to instruct farmers on care of their land from January to December. From the beginning, the book was organized about the 12 months of the year. Tusser explains, in rhyme, what farming tasks should be done each month. Similar to a number of versions of the medieval manuscripts just discussed, Tusser used format—centered headings and numbered verses—to reveal the plan of his instructions. The continued popularity of this work, which contained useful information on farming, despite increasing numbers of prose works on farming, supports the possibility that the verse was shared among farmers who could not read but could remember and quote many of the points captured in rhymed and numbered quatrains. The work is clearly organized. And, as noted by agricultural historian McRae [26], many points reflect advice provided in *Walter of Henley*:

ℭ September.

20 Threshe sede and go fanne, for the plough may not lye:
September doth bid, to be sowing of rye.
The redges well harrowde, or euer thou strike:
is one poynt of husbandry, rye land do like.
21 Geue winter corne leaue, for to haue full his lust:
sowe wheate as thou mayst, but sowe rye in the dust.
Be carefull for sede, for such sede as thou sowe:
as true as thou liuest, loke iustly to mowe.
22 The sede being sowne, waterforow thy ground:
that rain when it cummeth, may runne away round.
The diches kept skowred, the hedge clad with thorne:
doth well to drain water, and saueth thy corne.
23 Then furth with thy slinges, and thine arowes & bowes:
till ridges be grene, kepe the corne from the crowes.
A good boye abrode, by the day starre appere:
shall skare good man crowe, that he dare not come nere.
24 At Michelmas, mast would be loked vpon:
and lay to get some, or the mast time be gon.
It sa[u]eth thy corne well, it fatteth thy swyne:
In frost it doth helpe them, where els they should pine.

By 1573, when *Five Hundred Points* [29] was published, Tusser added abstracts placed before the points for each month. *Five Hundred Points* can be described as a technical report in poetry! Tusser's use of the term *abstract* may well be the first occurrence of it in English.

Of all the English agriculture books, *Five Hundred Points* [29] most echoes the oral tradition of medieval and early Renaissance agricultural lore and practice. Nevertheless, much of Tusser's advice surfaced in agriculture books that appeared well after until the close of the 17th century. Probably because Tusser used verse to provide instruction in agricultural practices, he was ignored by C. S. Lewis in his history of 16th century literature, although later historians have alluded to the value of his work for farmers with limited literacy [30, p. 26]. Because he used poetry, his popularity has been diminished by major agricultural historians. Even well-educated Royal Society Fellows of the 17th century, such as John Worlidge and Walter Blith, who sought to place agriculture on a scientific footing, quoted Tusser.

Late 16th and 17th Century Estate and Agriculture Instruction Manuals

By the last half of the 16th century, more estate management works were appearing in print. Derivatives of the medieval estate management works appeared under the authorship of writers such as Robert Mascall (first instructional work on trees [31], Thomas Hill (first gardening book) [32], Reynolde Scot (first book on how to grow and harvest a single crop) [33] and then Gervase Markham [34–38], John Worlidge [39], Walter Blith [40], Leonard Meager [41], along with Samuel Hartlib [42], and other members of the Royal Society. As McDonald notes, more than 90 English instructional works on agriculture had been printed by the close of the 17th century [30].

By the end of the Renaissance, gothic print replaced blackletter. Well-developed tables of contents made access to lengthy works easier. The differences among estate manuals of these writers show development of instructional writing in book format and a continued shift from oral-based technical writing to texts designed to be used apart from the oral context in which the instructions were developed. The format of many of these books shows that they were developed for reading rather than hearing. But the power of the oral tradition persisted in sentences, particularly in works published before 1670.

Thomas Hill

Little is known about Thomas Hill, who wrote the first gardening book published in English in 1563. He lived in London during a time when every house had its garden. On the title page of this small book, a woodcut provides the layout of the small Elizabethan garden with its penthouse entrance, the "herber" and well, the inner enclosure surrounded by a trellis, the knot garden in the center, and the formal arrangement of the beds. The book includes designs for the dwarf shrub mazes laid out with hyssop, thyme, savory, lavender cotton, or marjoram.

A Most Brief and Pleasant Treatyse Teachynge How to Dress, Sowe, and Set a Garden [32] illustrates the move from verse instructions to linear text instructions, but Hill's style heavily incorporates the qualities of the academic oration. This work thus provides a benchmark for assessing the shift to textual instructions. For example, Hill provides specific instructions for planning, planting, and tending gardens of flowers and herbs, but he begins with an exordium discussing garden placement, irrigation sources, and climate. His style is copious and lofty:

> And, last, the wynde be to be considered, for that the south wyndes be hot and moyst, the North wyndes cold an drye, and the East wyndes be bytwene hotenesse and coldness, temperate, but yet dryer than the West wyndes, and the East wyndes generally be always hotter than the west, and yet the West wyndes be sumwhat moyster than the East. [32, p. 11]

While his instructions for gardening proceed chronologically, his style is subjective, emotive, and rhythmic rather than objectively descriptive. Like Fitzherbert [27], he uses additive style. For example, in discussing how to choose the garden location, he states,

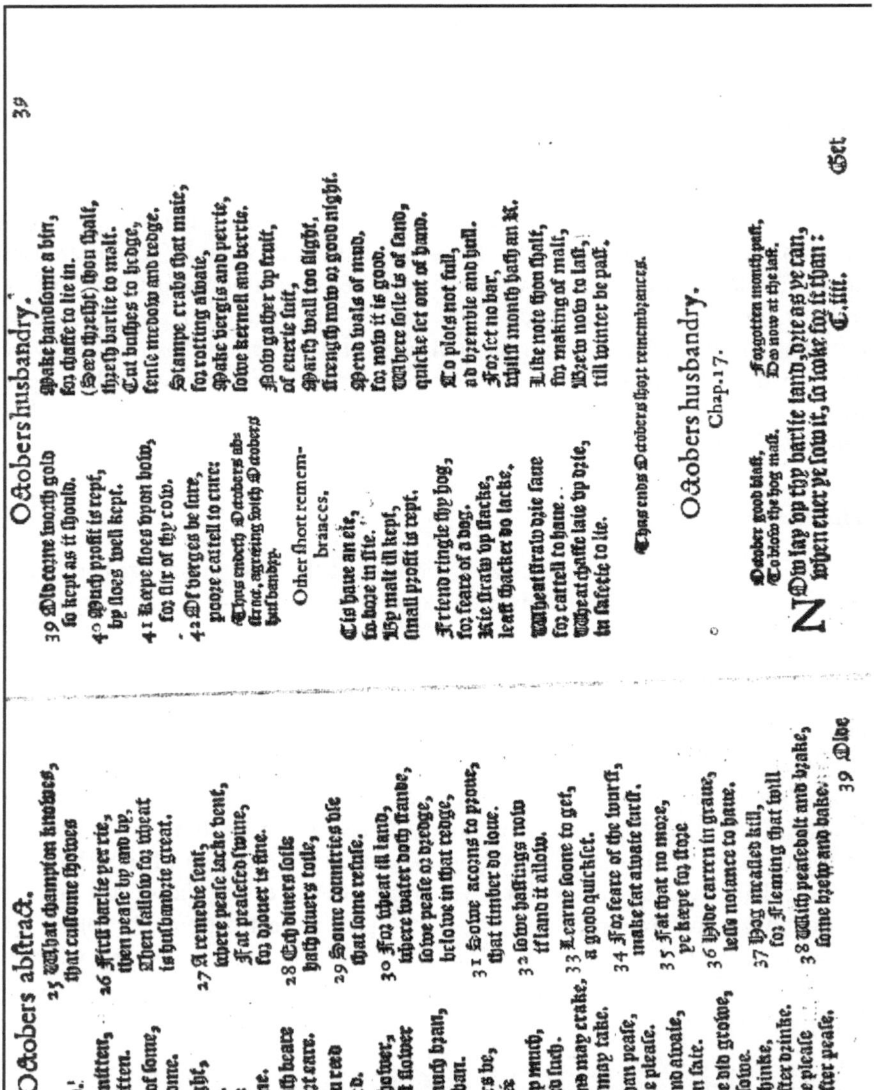

38 Octobers abstract.

A rottenlie mould,
is land worth gold.

12 Why wheat is smitten,
good lesson is written.

13 The iudgement of some,
how thistles do come.

14 A iudgement right,
of land in plight.
land all forlorne
not good for corne.

15 Land barren doth beare
small straw, short eare.

16 Here maist thou reed
for soile what seed.

17 Tis tride erie hower,
best graine most flower

18 Grosse corne much bran,
the baker doth ban.

19 What croppers be,
here learne to see

20 Few after crop much,
but noddies and such.

21 Some woodland may crake,
three crops he may take.

22 First barlie than pease,
then wheat if ye please.

23 Two crops and awaie,
must champion saie.

24 Where barlie did growe,
laie wheat to sowe.
yet better I thinke,
sowe pease, after drinke.
And then if ye please
sowe wheat after pease.

25 What champion knowes,
that custome showes

26 First barlie per rie,
then pease vp and by,
Then fallow for wheat
is husbandrie great.

27 A remedie sent,
where pease lacke vent,
Fat peasefed swine,
for drouer is fine.

28 Ech diuers soile
hath diuers toile,

29 Some countries vse
that some refuse.

30 For wheat ill land,
where water doth stande,
sowe pease or dredge,
belowe in that redge,

31 Sowe acorns to proue,
that timber do loue.

32 sowe hastings now
if land it allow.

33 Learne soone to get,
a good quickset.

34 For feare of the wurst,
make fat awaie furst.

35 Fat that no more,
ye keepe for store

36 Hide carren in graue,
less noiance to haue.

37 Hog measled kill,
for Fleming that will

38 With peasebolt and brake,
some brew and bake.

39 Olde

Octobers husbandry. 39

39 Old corne worth gold
so kept as it should.

40 Much profit is kept,
by floes well kept.

41 Keepe floes vpon how,
for flix of thy cow.

42 Of verges be sure,
poore cattell to cure:

Thus endeth Octobers abstract, agreeing with Octobers husbandry.

Other short remembrances.

Tis haue an eie,
to. bore in stie.
By malt ill kept,
small profit is rept.

Friend ringle thy hog,
for feare of a dog.
Rie straw vp stacke,
least thacker do lacke.

Wheat straw drie saue
for cattell to haue.
Wheat chaffe laie vp drie,
in safetie to lie.

Make handsome a bin,
for chaffe to lie in.
(Seed threshe) thou shalt,
thresh barlie to malt.
Cut bushes to hedge,
fense medow and redge.

Stampe crabs that maie,
for rotting awaie,
Make vergis and perrie,
sowe kernell and berrie.

Now gather vp fruit,
of euerie suit,
Marsh wall too slight,
strength now or good night.

Mend wals of mud,
for now it is good.
Where soile is of sand,
quicke set out of hand.

To plots not full,
ad bremble and hull.
For set no bar,
whilst month hath an R.

Like note thou shalt,
for making of malt,
Brew now to last,
till winter be past.

Thus ends Octobers short remembrances.

Octobers husbandry.
Chap. 17.

October good blast,
To blow the hog mast.

Forgotten month past,
Do now at the last.

Now lay vp thy barlie land, drie as ye can,
when euer ye sow it, so looke for it than:

C.iiii. Get

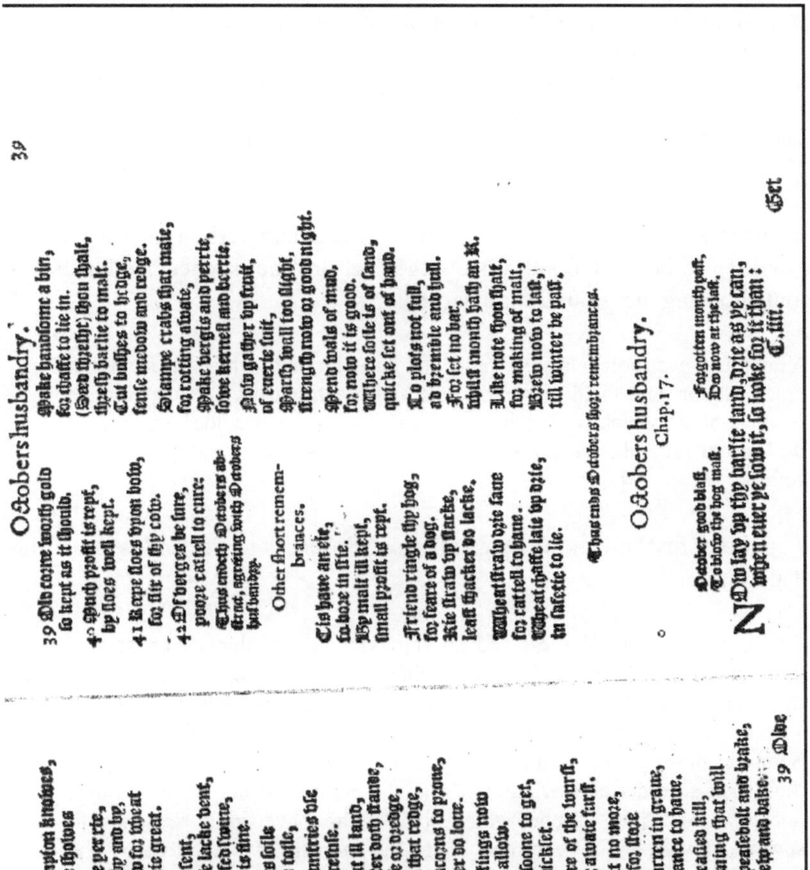

Figure 5-4. Example abstract from *Five Hundred Points of Good Husbandry* (1573).

> This ground also must be swete, which you shall easelye trye by taste therof if that out of a parte of grounde whyche most mislyketh you, you take a clodde puttynge it in an earthen panne, moystning it with freshe and sweate water, and dippying your finger therin, taste on your tonge how it refresheth and savoureth, and by that you shall knowe the swetenesse of the same. [32, p. 15]

Hill consistently uses doublets in adjectives, nouns, and verb, a primary quality of text written to be read aloud to slow the pace for the listener. He has this to say about choosing the ground for a garden:

> You must evermore chose a fatte and lose ground, which needeth but small labour; and yieldest plentifullest and greatest fruit. But that ground is vilest of all, which is drye and thick also, and leane or colde. And among all kind of grounds, the chalky and red ground, be vilest, and moste to be refused. [32, p. 15]

Hills's additive, rhythmic, oral style reflects a closeness to life and is a pleasure to read aloud:

> For asmuche as in Harvest and spryng tyme, many thinges be sowen: therefore those partes of the garden that thou wylt sow in Harvest that they maye serve thee in winter, and those partes also which thou wilt sow in the spryng tyme, that they may bring forth theyr fruyte in the Sommer, those partes thou shalt appyle and turne to the North. And this thou shalt so order and dygge, when as the earth shall not be over moyst nor dry. And that ground which thou wylte have in a readynes to sow in harvest, must be turned vp in the moneth of May, that by the heate of the sonne all the sommer through, yet may breake and lose the cloddes, and burne or wither the rotes of the herbest left in the earth. [32, pp. 25–26]

Reynolde Scot

Reynolde Scot's *A Perfite Platform of a Hoppe Garden* (1574) [33], the first published manual in English on how to produce a single crop [25, p. 151], was directed to poor farmers [33, p. 154]. Unlike Fitzherbert's *Booke of Husbandrie* [27], Scot's *Platform* [33] provides a linear, chronological set of instructions for hop growing and harvesting. His plan, revealed in the table of contents, anticipates a modern plan for writing a set of instructions to enable readers to perform the tasks. Note, in Figure 5-5, that Scot divides his instructions into specific units, each introduced with a ¶.

Scot begins with an overview of the process [33, pp. 1–14], which also includes the equivalent of quality control statements and warnings, a feature used by Marrett in his instructions for glassmaking [33, Ch. 1]. The next four sections

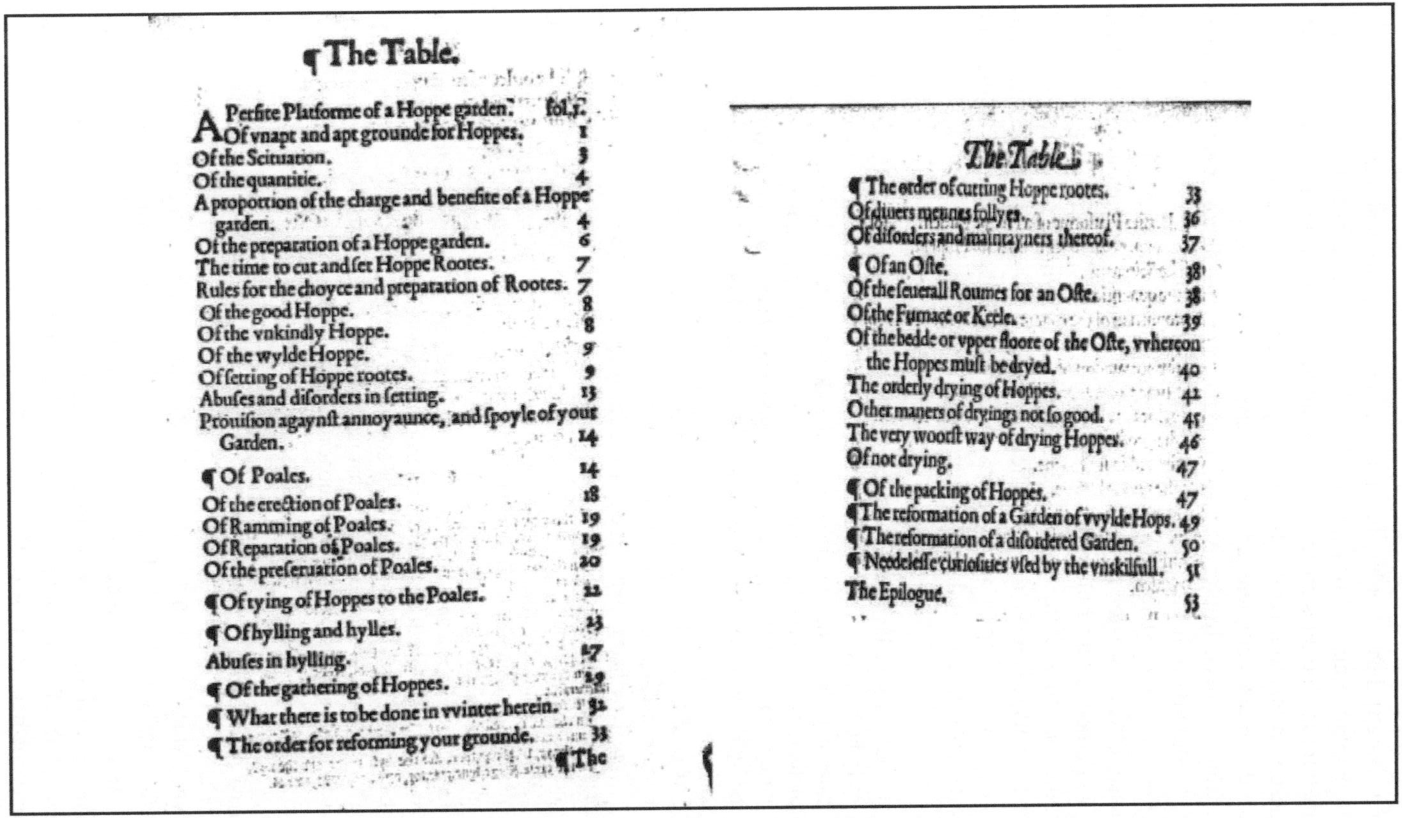

¶ The Table.

A Perfite Platforme of a Hoppe garden. fol.1.
Of vnapt and apt grounde for Hoppes. 1
Of the Scituation. 3
Of the quantitie. 4
A proportion of the charge and benefite of a Hoppe garden. 4
Of the preparation of a Hoppe garden. 6
The time to cut and set Hoppe Rootes. 7
Rules for the choyce and preparation of Rootes. 7
Of the good Hoppe. 8
Of the vnkindly Hoppe. 8
Of the wylde Hoppe. 9
Of setting of Hoppe rootes. 9
Abuses and disorders in setting. 13
Prouision agaynst annoyaunce, and spoyle of your Garden. 14
¶ Of Poales. 14
Of the erection of Poales. 18
Of Ramming of Poales. 19
Of Reparation of Poales. 19
Of the preseruation of Poales. 20
¶ Of tying of Hoppes to the Poales. 22
¶ Of hylling and hylles. 23
Abuses in hylling. 27
¶ Of the gathering of Hoppes. 29
¶ What there is to be done in vvinter herein. 32
¶ The order for reforming your grounde. 33
¶ The

The Table.

¶ The order of cutting Hoppe rootes. 33
Of diuers meanes follyes. 36
Of disorders and maintayners thereof. 37
¶ Of an Oste. 38
Of the seuerall Roumes for an Oste. 38
Of the Furnace or Keele. 39
Of the bedde or vpper floore of the Oste, vvhereon the Hoppes must be dryed. 40
The orderly drying of Hoppes. 41
Other maners of dryings not so good. 45
The very woorst way of drying Hoppes. 46
Of not drying. 47
¶ Of the packing of Hoppes. 47
¶ The reformation of a Garden of vvylde Hops. 49
¶ The reformation of a disordered Garden. 50
¶ Needelesse curiosities vsed by the vnskilfull. 51
The Epilogue. 53

Figure 5-5. Table of Contents. Reynolde Scot, *A Perfite Platforme of a Hoppe Garden* (1574).

[33, pp. 14–28] discuss poles, including a quality control statement [33, p. 27]. The remaining nine sections discuss harvesting, curing, packing, and ground reformation. He includes more warnings and quality control statements throughout these sections too. Of the 16th century instructions I have examined, Scot's appear the most complete and fully anticipate the organization of modern instruction manuals. Ironically, the Epilogue, with its almost epideictic style, contrasts sharply with the style of the instructions and reflects the style of an academic oration. His use of two distinct styles clearly indicates that he selected a concise, direct style for the actual instructions, but he understood that a high style would not suit the bulk of his work.

Another striking point is that Scot provides drawings, integrated into the text, to help his reader see the action he is describing. In Figure 5-6, Scot explains how to plant the hop roots. The instructions excerpted begin with the first full paragraph on p. 10 and conclude at the top of p. 11.

> If your Garden be one acre in bignesse, & lye square, leave betweene every hole, three yardes, or eight foote at the least in space, as well that the hilles may be made the greater, that the Hops of one Poale reache not to another, as also that the Sunne may the more freely and universally passé through your Garden, which by tis meaness may yearely be plowed betwixt the hylles, whereas otherwise it must be digged, a more tedious and costly businesse.
>
> If your Garden be very little, you may sette the hilles somewhat nearer togither, namely seven foote in sunder.
>
> Your lyne being layde level, you must dygge underneath every threede or pinne placed upon the same, a hole like to a Pitfall, one foote square, and one foote deepe.
>
> When you have made twenty or thirty holes, take up so many rootes, from where you bestowed them, as ought to be set therein, and go to worke on this wise, always watching a tyme (if you maye) that the wynde be in some part of the South or West, but be not so scrupulous herein, that you overslip the Moneth of April, least Salomons sayng be spoken of you: Hee that regardeth the wynde shall not sowe, and hee that hath respect to the Clowdes shall not reape. For he that neglecteth the Moneth of Aprill, shall have a bad season to cut or plant hopes. [33, pp. 10–11]

Gervase Markham

Gervase Markham (1568?–1637) [34–38] stands as one of the most influential agricultural writers of the 16th and 17th centuries and the most prolific technical writer. As author of 8 collections of verse, 2 dramas, 6 literary prose titles, and over 20 instructional works on estate management and animal husbandry, he was one of the first empiricists in animal husbandry, perhaps the founder of veterinary medicine [43, p. 100].

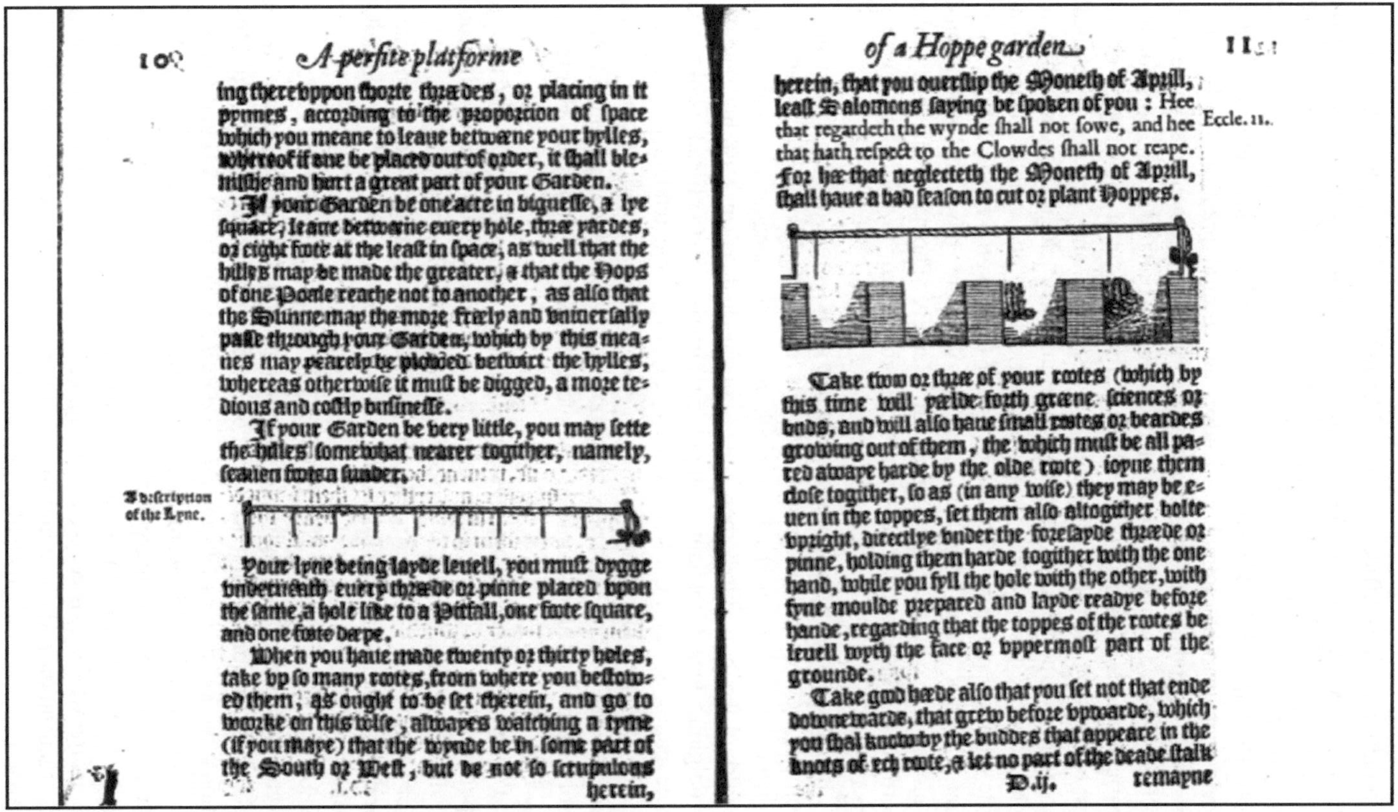

108 *A perfite platforme*

ing thereuppon shorte threedes, or placing in it pynnes, according to the proportion of space which you meane to leaue betweene your hylles, whereof if one be placed out of order, it shall blemishe and hurt a great part of your Garden.

If your Garden be one acre in bignesse, a lyke square, leaue betweene euery hole, three yardes, or eight foote at the least in space, as well that the hills may be made the greater, & that the Hops of one Poale reache not to another, as also that the Sunne may the more freely and uniuersally passe through your Garden, which by this meanes may pearely be plowed betwixt the hylles, whereas otherwise it must be digged, a more tedious and costly businesse.

If your Garden be very little, you may sette the holes somewhat nearer togither, namely, seauen foote asunder.

A discription of the Lyne.

Your lyne being layde leuell, you must dygge underneath euery threede or pinne placed upon the same, a hole like to a Pitfall, one foote square, and one foote deepe.

When you haue made twenty or thirty holes, take up so many rootes, from where you bestowed them, as ought to be set therein, and go to worke on this wise, alwayes watching a tyme (if you maye) that the wynde be in some part of the South or West, but be not so scrupulous herein,

of a Hoppe garden 111

herein, that you ouerslip the Moneth of Aprill, least Salomons saying be spoken of you: Hee that regardeth the wynde shall not sowe, and hee that hath respect to the Clowdes shall not reape. For hee that neglecteth the Moneth of Aprill, shall haue a bad season to cut or plant Hoppes. Eccle. 11.

Take two or three of your rootes (which by this time will yelde forth greene sciences or buds, and will also haue small rootes or beardes growing out of them, the which must be all pared awaye harde by the olde roote) ioyne them close togither, so as (in any wise) they may be euen in the toppes, set them also altogither bolte upright, directlye under the foresayde threede or pinne, holding them harde togither with the one hand, while you fyll the hole with the other, with fyne moulde prepared and layde readye before hande, regarding that the toppes of the rootes be leuell wyth the face or uppermost part of the grounde.

Take good heede also that you set not that ende downewarde, that grew before upwarde, which you shal know by the buddes that appeare in the knots of ech roote, & let no part of the deade stalk remayne

D.ij.

Figure 5-6. Scot, *A Perfite Platforme of a Hoppe Garden* (of the setting of hoppe roots).

Markham, a military officer, possessed knowledge of Latin, French, Spanish, Italian, and possibly Dutch. A prodigious, well-educated writer with broad practical experience, Markham knew farming, husbandry, horsemanship, and military science, topics that surfaced in the 40 separate titles he produced between 1593 and 1637 [44]. His books were packed with instructions valuable to estate management, animal husbandry, agricultural practices, early veterinary medicine, and equine breeding, training, and management. As Mullet [43, pp. 106, 118] noted, Markham was one of the foundational writers of the coming scientific method. Markham's output was prodigious, and the breadth of his works exemplify mature instructional style—clear and detailed—technical description, incorporation of useful visuals, and clear, logical organization of text. Markham's instructional works in agriculture and estate management provide a critical point in the development of these works in the 17th century. Several excerpts from Markham's works illustrate how his instructional text moved beyond that of Walter of Henley, Thomas Hill, and John Fitzherbert. Markham remained the premier contributor to agricultural writing before 1800.

In his instructional works, Markham wrote in a simple, direct style. Unlike Fitzherbert, he avoided literary ornament, even though he had read and drew from humanistic works on agriculture, estate management, and animal husbandry. Like both Fitzherbert and Scot, he had a sense of his readers. So prodigious was Markham's output on agrarian instructional books that example pages of several offer the best way to enable readers to grasp his style. For example, *The Husbandman's Jewel* (1695), a handbook of concise instructions for dealing a variety of problems that a landowner might need to resolve, provides an example of his succinct style, as shown in Figure 5-7 [37, p. 17].

As Poynter, Markham's bibliographer, notes, *Farewell to Husbandry* [36], first published in 1620, represents Markham's highest achievement in instructional discourse [44, p. 100]. *Farewell to Husbandry* illustrates his mastery of land management and his knowledge of how to transform gorse-covered tracts into arable farmland with careful tilling. His style, shown in Figure 5-8, is concise but conversational.

Markham's Master-peece, first published in 1610 [38], has been called "the first veterinary manual in English to enjoy a reputation as the standard work on the subject, a position which it occupied for more than two centuries" [44, p. 116]. Additional works, such as *Cheap and good husbandry* (1614) [34] and *Country contentments* (1615) [35] were directed to rich merchants who, having built for themselves a new country mansion with its carefully designed 7garden, began to look for instructions on how to maintain and manage the estate. While many works existed on these topics, Markham provided works with useful organization and direct and specific instructions for a wide range of tasks that had to be completed on country estates.

To deſtroy all ſorts of Vermin.

TO Kill Rats and Mice, take Wheat Flower and bitter Almonds, and make them into a Paſt, and lay it into their holes, and it will Kill them.

Another, caſt Hemlock ſeed, into their holes, and it will Kill them.

To catch Moles, lay a head of Garlick or Onion before the holes, and they will Immediately come forth.

To Kill Weaſels, Wheat Flower, Sal Armoniack made into a Paſt, with ſome Honey, throw it where the Weaſels uſually come, and they will eat it and it Kills them.

To Kill Piſmires, Origanum beaten to Powder, and ſtrewed before their holes and it Kills them.

To Kill Bugs, or Fleas, take Rue and Wormwood a good quantity, boil them about a quarter of an Hour, then take ſome common Salt, the more the better and uſe it as before.

To Kill Lice and Nits and to cure Scabby Heads; Staveſaker, and Freſh Butter mixed together, and the Head anointed with it, kills Lice.

To Kill Caterpillars, take Lees of Oyl and Ox piſs, and boil them together, then caſt it upon the Trees, or Buſhes and it kills them.

To preſerve Cattle from Flies, take oil wherein Bakeleer hath been boiled, and anoint the Beaſt with it, and they will not come near them.

Figure 5-7. An example of instructions by Markham.

Agricultural and Estate Management Writers of the 17th Century

By the middle of the 17th century, new writers, such as Walter Blith, John Worlidge, and Leonard Meager, produced works that contrast sharply with those of Fitzherbert, Scot, and Markham but exemplify continued development of both knowledge and instructional text to convey the new information with more focus on a prose that lacked oral residue. All these writers owed a debt of gratitude to Samuel Hartlib [42], one of the original Royal Society founders, who sought to improve England through improved agriculture, as point emphasized in Chapter 1.

Walter Blith

The English Improver Improved, third edition, 165 [40], by Walter Blith, a captain in Cromwell's army, vigorously sought to improve agricultural practices

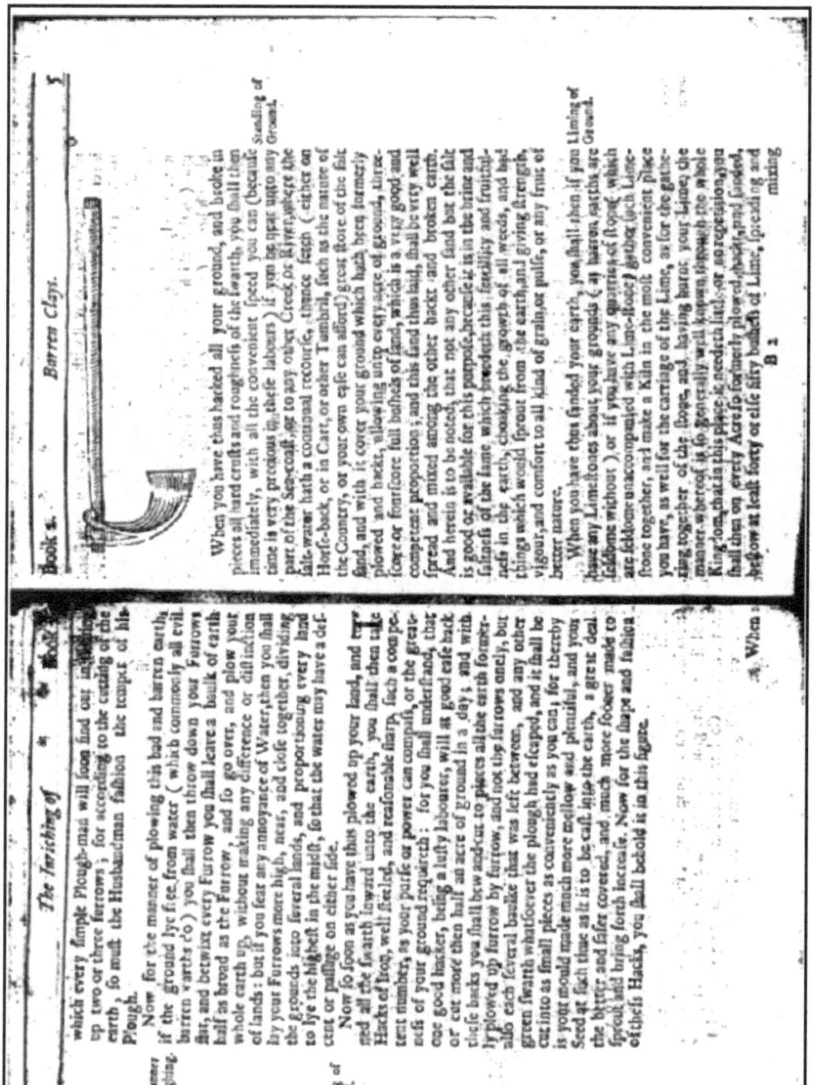

4 *The Inriching of* Book 2.

which every ſimple Plough-man will ſoon find out in turning up two or three furrows; for according to the cutting of the earth, ſo muſt the Husbandman faſhion the temper of his Plough.

The manner of Ploughing. Now for the manner of plowing this bad and barren earth, if the ground lye free from water (which commonly all evil barren earths do) you ſhall then throw down your Furrows flat, and betwixt every Furrow you ſhall leave a baulk of earth half as broad as the Furrow, and ſo go over, and plow your whole earth up, without making any difference or diſtinction of lands: but if you fear any annoyance of Water, then you ſhall lay your Furrows more high, near, and cloſe together, dividing the grounds into ſeveral lands, and proportioning every land to lye the higheſt in the midſt, ſo that the water may have a deſcent or paſſage on either ſide.

Hacking of Ground. Now ſo ſoon as you have thus plowed up your land, and turned all the ſwarth inward unto the earth, you ſhall then take Hacks of Iron, well ſteeled, and reaſonable ſharp, ſuch a competent number, as your purſe or power can compaſs, or the greatneſs of your ground requireth: for you ſhall underſtand, that one good hacker, being a luſty labourer, will at good eaſe hack or cut more then half an acre of ground in a day; and with theſe hacks you ſhall hew and cut to pieces all the earth formerly plowed up furrow by furrow, and not the furrows onely, but alſo each ſeveral baulke that was left between, and any other green ſwarth whatſoever the plough had eſcaped, and it ſhall be cut into as ſmall pieces as conveniently as you can; for thereby is your mould made much more mellow and plentiful, and your Seed at ſuch time as it is to be caſt into the earth, a great deal the better and ſafer covered, and much more ſooner made to ſprout and bring forth increaſe. Now for the ſhape and faſhion of theſe Hacks, you ſhall behold it in this figure.

When

Book 2. *Barren Clays.* 5

When you have thus hacked all your ground, and broke in pieces all hard cruſts and roughneſs of the ſwarth, you ſhall then immediately, with all the convenient ſpeed you can (becauſe time is very precious in theſe labours) if you be near unto any part of the Sea-coaſt, or to any other Creek or River, where the ſalt-water hath a continual recourſe, thence fetch (either on Horſe-back, or in Cart, or other Tumbril, ſuch as the nature of the Country, or your own eaſe can afford) great ſtore of the ſalt ſand, and with it cover your ground which hath been formerly plowed and hackt, allowing unto every acre of ground, threeſcore or fourſcore full buſhels of ſand, which is a very good and competent proportion; and this ſand thus laid, ſhall be very well ſpread and mixed among the other hackt and broken earth. And herein is to be noted, that not any other ſand but the ſalt is good or available for this purpoſe, becauſe it is in the brine and ſaltneſs of the ſame which breedeth this fertility and fruitfulneſs in the earth, choaking the growth of all weeds, and bad things which would ſprout from the earth, and giving ſtrength, vigour, and comfort to all kind of grain, or pulſe, or any fruit of better nature. Standing of Ground.

Liming of Ground. When you have thus ſanded your earth, you ſhall then if you have any Limeſtones about your grounds (as barren earths are ſeldome without), or if you have any quarries of ſtone (which are ſeldome unaccompanied with Lime-ſtone) gather ſuch Limeſtone together, and make a Kiln in the moſt convenient place you have, as well for the carriage of the Lime, as for the gathering together of the ſtone, and having burnt your Lime, the manner whereof is ſo generally well known through the whole Kingdom, that in this place it needeth little or no repetition, you ſhall then on every Acre ſo formerly plowed, hackt, and ſanded, beſtow at leaſt forty or elſe fifty buſhels of Lime, ſpreading and

B 2 mixing

Figure 5-8. Example page from Markham, *Farewell to Husbandry* (1620).

in England during the Commonwealth. His comprehensive treatment of estate management includes the first systematic plan for "alternate husbandry." He recommends breaking up all inferior grasslands and shows the production loss from constant pasturage. He discusses both overplowing and renovating pasturage. He describes and shows sketches of the sowing and wheel plows. He recommends many green fodder plants, which, along with root crops, effected a complete revolution in the cultivation of the soil in his day [26, p. 97].

The work opens with nine different prose and poetic exordia to the wide range of readers he believed he had to convince about the solid value of his proposed instructional methods. Unlike Markham's, his instructions are argumentative, as he believed he had to appeal to a wide range of stakeholders, from Cromwell, to members of the gentry, the university, husbandmen, farmer, tenants, cottagers, laborers, and commoners. His change of approach and style for appealing to each level of reader is also worth close examination. Although Blith [40] quotes Tusser in Chapter 13 on the importance of enclosure farming, Blith looks ahead and includes instructions for toolmaking, including water engines (windmills for draining bogs and fens,) and plows. Figure 5-9 shows an excerpt from his instructions for land improvement.

John Worlidge

In 1669, John Worlidge, a member of the Royal Society, produced a book that became known as one of the enlightened works on agriculture and is perhaps the most complete work on agriculture and estate management produced in the 17th century. Worlidge addressed the work to the gentry and yeomanry of England. According to McDonald, Worlidge took a scientific approach to agriculture. As a result, biographers credit him as "compiler of the first systematic treatise on husbandry, having in his experience gathered into a focus the scattered information published during the period of the Commonwealth [30, p. 119] Yet Worlidge also alluded to Tusser in discussing enclosure farming [39, Ch. 2].

Figure 5-10, excerpted from the third edition of *Systema Agriculture* [39], shows a drawing of the grain drill, one of several technologies Worlidge recommended, along with instructions on building the drill. Worlidge explained that the drill would sow seed fast or slow and that no harrowing was necessary.

Worlidge likely knew of this device, introduced by Evelyn in one of the Royal Society papers discussed in Chapter 8 (see Figure 5-7) possibly because of the length of the book, it contains a 10-page table of contents, which he calls "the Analysis," which begins with a one-page summary page listing the contents of each of the 14 chapters. Organization and format suggest effort to make the work accessible.

Worlidge's work signifies the advances in agriculture knowledge that moved ahead along with the sciences ordained by the Royal Society during the last half of the 17th century. As McDonald noted [30, p. 116], Worlidge's book, first

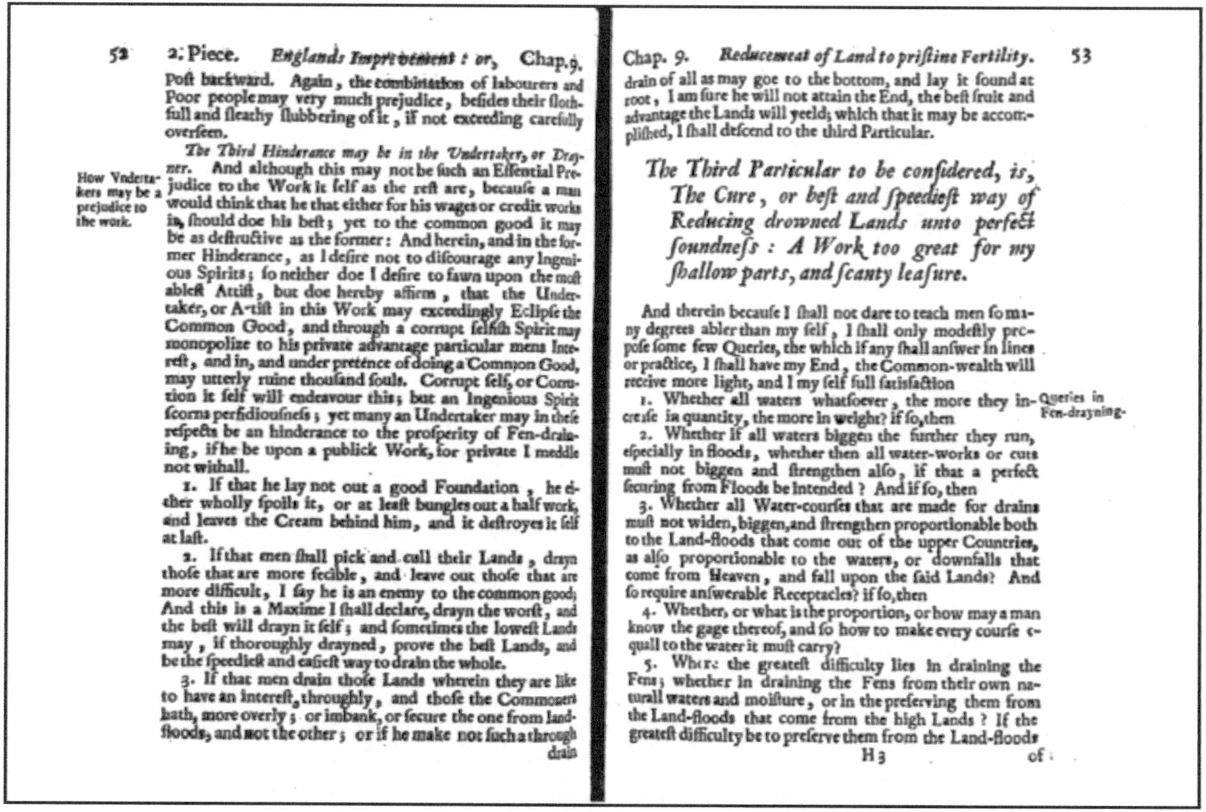

52 2. Piece. *Englands Improvement : or,* Chap. 9.

Poſt backward. Again, the combination of labourers and Poor people may very much prejudice, beſides their ſlothfull and ſleathy ſlubbering of it, if not exceeding carefully overſeen.

How Undertakers may be a prejudice to the work.

The Third Hinderance may be in the Undertaker, or Drayner. And although this may not be ſuch an Eſſential Prejudice to the Work it ſelf as the reſt are, becauſe a man would think that he that either for his wages or credit works in, ſhould doe his beſt; yet to the common good it may be as deſtructive as the former: And herein, and in the former Hinderance, as I deſire not to diſcourage any Ingenious Spirits; ſo neither doe I deſire to fawn upon the moſt ableſt Artiſt, but doe hereby affirm, that the Undertaker, or Artiſt in this Work may exceedingly Eclipſe the Common Good, and through a corrupt ſelfiſh Spirit may monopolize to his private advantage particular mens Intereſt, and in, and under pretence of doing a Common Good, may utterly ruine thouſand ſouls. Corrupt ſelf, or Corruption it ſelf will endeavour this; but an Ingenious Spirit ſcorns perfidiouſneſs; yet many an Undertaker may in theſe reſpects be an hinderance to the proſperity of Fen-draining, if he be upon a publick Work, for private I meddle not withall.

1. If that he lay not out a good Foundation, he either wholly ſpoils it, or at leaſt bungles out a half work, and leaves the Cream behind him, and it deſtroyes it ſelf at laſt.

2. If that men ſhall pick and cull their Lands, drayn thoſe that are more fecible, and leave out thoſe that are more difficult, I ſay he is an enemy to the common good; And this is a Maxime I ſhall declare, drayn the worſt, and the beſt will drayn it ſelf; and ſometimes the loweſt Lands may, if thoroughly drayned, prove the beſt Lands, and be the ſpeedieſt and eaſieſt way to drain the whole.

3. If that men drain thoſe Lands wherein they are like to have an intereſt, throughly, and thoſe the Commoners hath, more overly; or imbank, or ſecure the one from land-floods, and not the other; or if he make not ſuch a through drain

Chap. 9. *Reducement of Land to priſtine Fertility.* 53

drain of all as may goe to the bottom, and lay it ſound at root, I am ſure he will not attain the End, the beſt fruit and advantage the Lands will yeeld; which that it may be accompliſhed, I ſhall deſcend to the third Particular.

The Third Particular to be conſidered, is, The Cure, or beſt and ſpeedieſt way of Reducing drowned Lands unto perfect ſoundneſs : A Work too great for my ſhallow parts, and ſcanty leaſure.

And therein becauſe I ſhall not dare to teach men ſo many degrees abler than my ſelf, I ſhall only modeſtly prepoſe ſome few Queries, the which if any ſhall anſwer in lines or practice, I ſhall have my End, the Common-wealth will receive more light, and I my ſelf full ſatisfaction

1. Whether all waters whatſoever, the more they increaſe in quantity, the more in weight? if ſo, then — Queries in Fen-drayning.

2. Whether if all waters biggen the further they run, eſpecially in floods, whether then all water-works or cuts muſt not biggen and ſtrengthen alſo, if that a perfect ſecuring from Floods be intended? And if ſo, then

3. Whether all Water-courſes that are made for drains muſt not widen, biggen, and ſtrengthen proportionable both to the Land-floods that come out of the upper Countries, as alſo proportionable to the waters, or downfalls that come from Heaven, and fall upon the ſaid Lands? And ſo require anſwerable Receptacles? if ſo, then

4. Whether, or what is the proportion, or how may a man know the gage thereof, and ſo how to make every courſe equall to the water it muſt carry?

5. Where the greateſt difficulty lies in draining the Fens; whether in draining the Fens from their own naturall waters and moiſture, or in the preſerving them from the Land-floods that come from the high Lands? If the greateſt difficulty be to preſerve them from the Land-floods

H3 of

Figure 5-9. Example page showing format and organization from Blith, *The English Improver, Improved* (1653).

Of Arrable Land and Tillage. 47

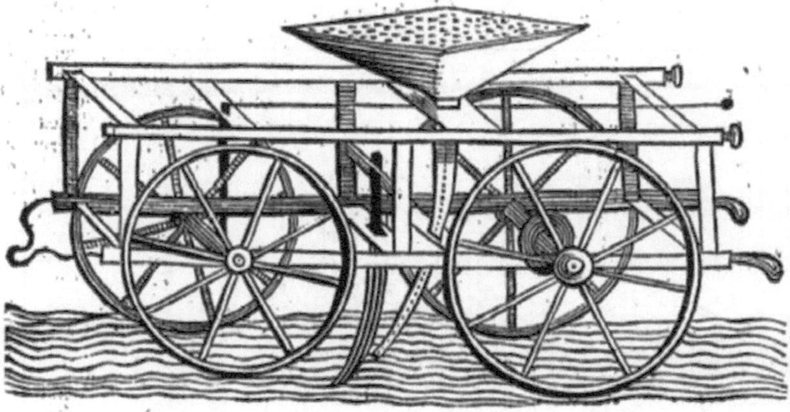

Firſt make a Frame of ſmall ſquare pieces of timber of about two or three inches ſquare, the breadth of the Frame about two foot, the height about eighteen inches, the length about four foot, more or leſs as you pleaſe; place this Frame on two pair of ordinary Pough-wheels. The Axeltree of the two foremoſt wheels is to lock to either ſide as doth the fore Axeltree of a Waggon, for reaſons hereafter ſhewn, the hindermoſt Axeltre being of Iron and ſquare in the middle, muſt be fixed to the Center of the wheels, that the Axes and wheels may move together; Then about the middle of the Frame in the bottom let there be fixed an Iron Inſtrument, or of Wood, pointed with Iron, like unto a Coulter, made a little ſpreading at the bottom in the nature of a ſhare, made to paſs through two Mortoiſes on the top for its greater ſtrength, and made alſo to be wedged higher or lower according as you will have your furrow in depth; the uſe whereof is onely to make the furrow ſo that you muſt make the point thereof of breadth onely to move the Earth, and caſt it or force it on either ſide, that the *Corn* may fall to the bottom of the furrow, then over this *Share* or *Coulter*, a little behinde it may a Wooden-pipe be made to come from the top of the Frame to the lower end of the *Share*, tapering downwards, and as neer as you can to the *Share* to deliver the *Corn* immediately as the ground is opened, and before any Earth falls in, that what Earth doth afterwards fall in, it may fall on the *Corn*. This pipe is to proceed out of a large Hopper fixed on the top of the Frame that may contain about a buſhel more or leſs, as you think fit, but that the *Corn* may gradually deſcend, according to the quantity you intend to beſtow on an Acre, at the very neck of the Hopper underneath in the ſquare hollows thereof, muſt be fitted in the edge of a wheel of wood, about half an inch thick and proportionable to the cavity of the neck, the wheel need not

be

Figure 5-10. Drawing and instructions: Building a Grain Drill, Worlidge, *Systema Agriculturae* (1681).

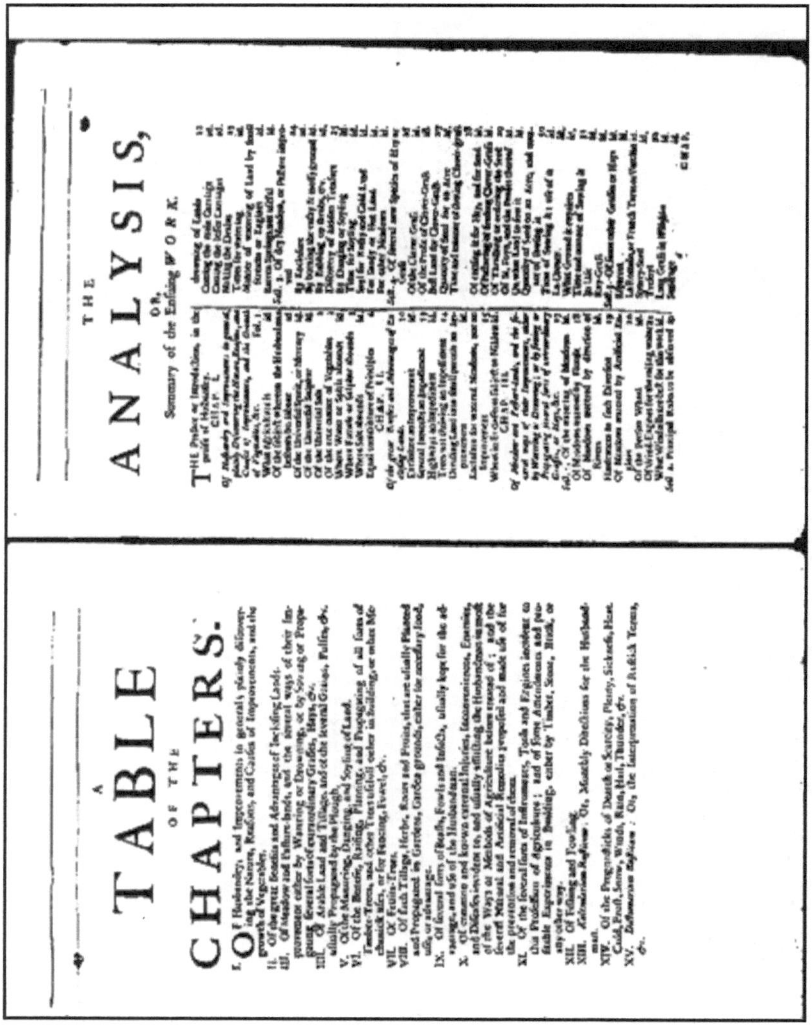

Figure 5-11. Opening Organizers, Worlidge, *Systema Agriculture* (1681).

published in 1659 [39], stands as the most useful and enlightened collection of observations that had been published in England. Yet even in this work, the oral tradition emerges. The book opens with a "Frontispiece," directly across from a poetic "Explanation of the Frontispiece" containing 20 rhymed couplets The drawing illustrates the topics he would cover in *Systema Agriculturae*, and the rhymed explanation serve to remind us that the oral tradition still found a place in the heart of the most determined scientist.

Leonard Meager

Agricultural instructional works by Leonard Meager illustrate the change in instructional writing style by the closing years of the 17th century. Meager also authored *The English Gardner: Or, the Sure Guide to Your Planters & Gardners* (1688) [41], which mirrors the approach he used in his larger work, *The Mystery of Husbandry* and also allows us to see the shift in instructional style from that used by Hill in 1563 [32]. His book begins with three dozen designs for mazes, with empty pages supposedly left for the reader to make notes. Unlike Blith, Meager focuses on precise instructions. Figure 5-12 shows his instructions for planting fruit:

As shown in the above passage and Figure 5-12, Meager's instructional style is thoroughly modern. His instructions could be understood by today's readers. While page design is not the books' strength, Meager does have a half-dozen pages of tables showing types of fruits he discusses and recommends. This book illustrates that by the close of the 17th century, information on gardening as well as estate management had moved beyond the capabilities of poetry. And, as shown in works published after Fitzherbert, what came to be called "plain style" characterized instructional materials designed for readers who would use these works.

By Meager's time, writers and printers had developed layout strategies for producing clear tables. Meager's knowledge of how to use tables for accessibility stands in stark contrast to navigation works discussed in Chapter 4. See Figure 5-13, also from *The English Gardner*, which presents detailed instructions for gardening and superb readability [42].

FURTHER PERSPECTIVES IN THE HISTORY OF INSTRUCTIONS

As I mentioned at the beginning of this chapter, agricultural and estate management instructions illustrate only one of many subject areas of early English practical writing that illustrate the use of instructions. Other topics worthy of research include military science, navigation, herbals, medicine, household management, and science. Those interested in military science will again find the influence of Gervase Markham, who wrote five books on military science directed at soldiers. Example instructions in English government abound, and

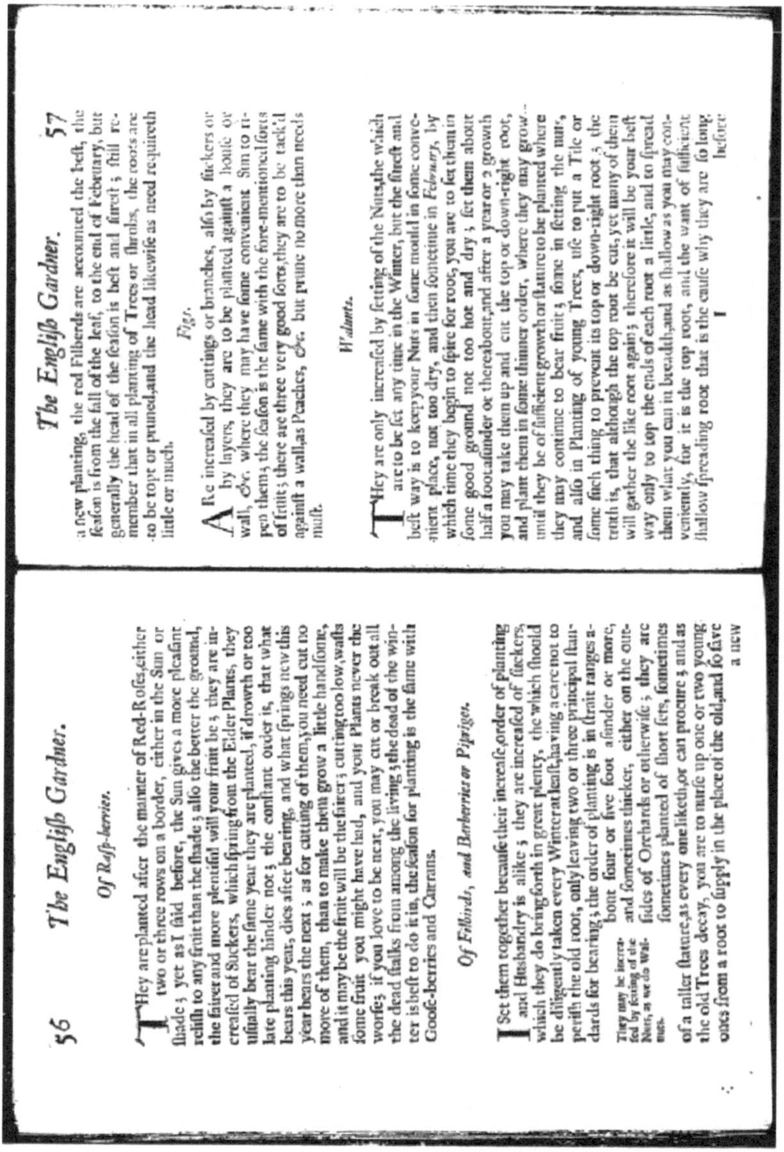

Figure 5-12. Example page, Meager, *The English Gardner* (1683).

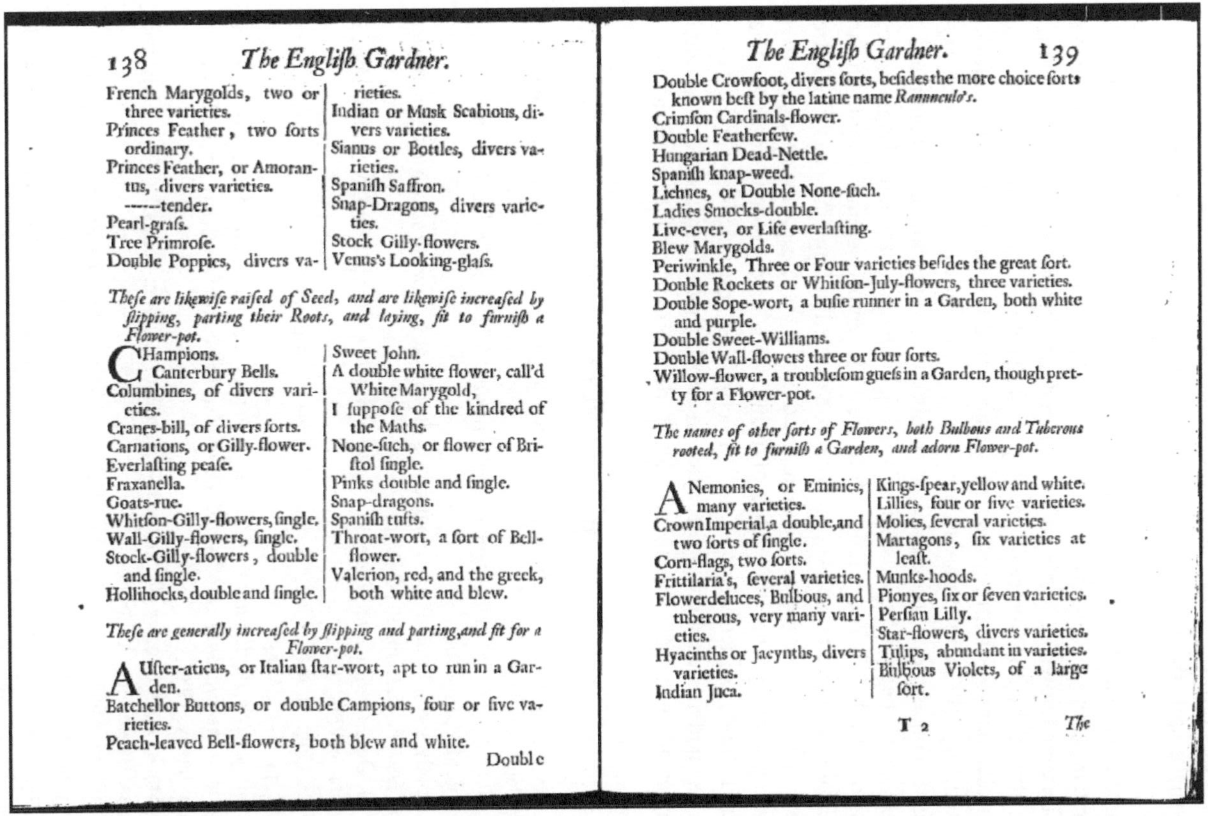

138 *The Engliſh Gardner.*

French Marygolds, two or three varieties.	rieties.
Princes Feather, two ſorts ordinary.	Indian or Musk Scabious, divers varieties.
Princes Feather, or Amorantus, divers varieties.	Sianus or Bottles, divers varieties.
------tender.	Spaniſh Saffron.
Pearl-graſs.	Snap-Dragons, divers varieties.
Tree Primroſe.	Stock Gilly-flowers.
Double Poppies, divers va-	Venus's Looking-glaſs.

Theſe are likewiſe raiſed of Seed, and are likewiſe increaſed by ſlipping, parting their Roots, and laying, fit to furniſh a Flower-pot.

CHampions.	Sweet John.
Canterbury Bells.	A double white flower, call'd White Marygold,
Columbines, of divers varieties.	I ſuppoſe of the kindred of the Maths.
Cranes-bill, of divers ſorts.	None-ſuch, or flower of Briſtol ſingle.
Carnations, or Gilly-flower.	
Everlaſting peaſe.	Pinks double and ſingle.
Fraxanella.	Snap-dragons.
Goats-rue.	Spaniſh tufts.
Whitſon-Gilly-flowers, ſingle.	Throat-wort, a ſort of Bell-flower.
Wall-Gilly-flowers, ſingle.	
Stock-Gilly-flowers, double and ſingle.	Valerion, red, and the greek, both white and blew.
Hollihocks, double and ſingle.	

Theſe are generally increaſed by ſlipping and parting, and fit for a Flower-pot.

AUſter-aticus, or Italian ſtar-wort, apt to run in a Garden.
Batchellor Buttons, or double Campions, four or five varieties.
Peach-leaved Bell-flowers, both blew and white.

Double

The Engliſh Gardner. 139

Double Crowfoot, divers ſorts, beſides the more choice ſorts known beſt by the latine name *Ranunculo's.*
Crimſon Cardinals-flower.
Double Featherfew.
Hungarian Dead-Nettle.
Spaniſh knap-weed.
Lichnes, or Double None-ſuch.
Ladies Smocks-double.
Live-ever, or Life everlaſting.
Blew Marygolds.
Periwinkle, Three or Four varieties beſides the great ſort.
Double Rockets or Whitſon-July-flowers, three varieties.
Double Sope-wort, a buſie runner in a Garden, both white and purple.
Double Sweet-Williams.
Double Wall-flowers three or four ſorts.
Willow-flower, a troubleſom gueſs in a Garden, though pretty for a Flower-pot.

The names of other ſorts of Flowers, both Bulbous and Tuberous rooted, fit to furniſh a Garden, and adorn Flower-pot.

ANemonies, or Eminics, many varieties.	Kings-ſpear, yellow and white.
	Lillies, four or five varieties.
Crown Imperial, a double, and two ſorts of ſingle.	Molies, ſeveral varieties.
	Martagons, ſix varieties at leaſt.
Corn-flags, two ſorts.	
Frittilaria's, ſeveral varieties.	Munks-hoods.
Flowerdeluces, Bulbous, and tuberous, very many varieties.	Pionyes, ſix or ſeven varieties.
	Perſian Lilly.
	Star-flowers, divers varieties.
Hyacinths or Jacynths, divers varieties.	Tulips, abundant in varieties.
Indian Juca.	Bulbous Violets, of a large ſort.

T 2 The

Figure 5-13. Example page with text, tables, and lists from Meager, *The English Gardner*.

material from early administrative documents will lead researchers into fully developed examples of policies.

If this first effort to track the emergence and development of a major genre has been successful, those of us in technical communication can see how history influences what we do in technical communication today; for example, orality declined but never departed technical writing. How text "sounds" often helps us understand if the text, interiorized, can be understood. We also see, again, that what became known as "plain style" emerged long before Bacon. Early technical writers knew the importance of developing instructions for the needs of intended audiences, many of them rural farmers who would not have been able to follow instructions written in humanistic prose. As the characteristics of oral style declined, conciseness and subject-verb-object sentence patterns continued to harden. As has been shown elsewhere [9, 45], development of plain English prose emerged from the Chancery long before the 17th century scientists argued for the virtues of unadorned prose. And direct prose existed in English as early as Alfred [46, pp. 137–174]. In addition to unadorned prose, these early instruction writers increasingly recognized the value of using visuals to enforce content. They prepared their instructions, technical descriptions, introductions, and chapters much as modern technical writers do. In short, modern instructions maintain principles of style, organization, and design developed over four centuries ago.

REFERENCES

1. J. Goody, *Literacy in Traditional Societies*, Cambridge University Press, Cambridge, England, 1968.
2. J. Goody, *The Domestication of the Savage Mind*, Cambridge University Press, Cambridge, England, 1977.
3. J. Goody and I. Watt, The Consequences of Illiteracy, *Comparative Studies in Society and History, 5*:3, pp. 304-345, 1963.
4. W. J. Ong, *Orality and Literacy: The technologizing of the Word,* Methuen, London and New York, 1982.
5. W. J. Ong, Orality, Literacy, and Medieval Textualization, *New Literary History, 14*:1, pp. 1-12, 1984.
6. B. Stock, *The Implications of Literacy: Written Language and Models of Interpretation in the Eleventh and Twelfth Centuries*, Princeton University Press, Princeton, New Jersey, 1983.
7. B. Stock, *Listening for the Text*, Baltimore and London: Johns Hopkins University Press, Baltimore and London, 1990.
8. C. Glenn, Medieval Literacy Outside the Academy: Popular Practice and Individual Technique, *College Composition and Communication, 4*:4, pp. 497-508, 1993.
9. J. H. Fisher, *The Emergence of Standard English*, Lexington: University Press of Kentucky, Lexington, Kentucky, 1996.
10. R. Adolph, *The Rise of English Prose Style*, MIT Press, Cambridge, Massachusetts, 1968.
11. R. Crosby, Oral Delivery in the Middle Ages, *Speculum, 11*:1, pp. 88-110, 1936.

12. P. D. A. Harvey, English Estate Records, 1250-1330, in *Pragmatic Literacy East and West, 1200-1300*, Richard Britnell (ed.), Boydell Press, Woodridge, Connecticut, pp. 107-118, 1997.
13. V. H. Galbraith, *The Raleigh Lecture on History*, British Academy, July 1935.
14. J. W. Thompson, *The Literacy of the Laity in the Middle Ages,* Burt Franklin, New York, 1960.
15. M. T. Clanchy, *From Memory to Written Record: England 1066-1307* (2nd Edition), Blackwell, Oxford, UK and Cambridge, USA, 1993.
16. R. A Houston, Literacy in Early Modern Europe, *Culture & Education 1500-1800*, Longman: London and New York, 1988.
17. *Legal and Manorial Formularies: Edited from Originals at the British Museum and the Public Records Office*, In Memory of Julius Parnell Gilson, Keeper of Manuscripts and Egerton Librarian in the British Museum, Oxford University Press, Oxford, England, 1933.
18. *The Court Baron, Being Precedents for Use in Seignorial and Other Local Courts, Together with Select Pleas from the Bishop of Ely's Court of Littleport*, Frederic William Maitland and William Paley Baildon (Eds.), *The Publications of the Selden Society* (Vol. IV), London: Bernard Quaritch, 15 Piccadilly, London, 1890.
19. D. Oschinsky, *Walter of Henley and Other Treatises on Estate Management and Accounting*, Clarendon Press, Oxford, England, 1971.
20. W. H. Blaauw, Letters to Ralph de Nevill, Bishop Chichester (1222-1224), *Sussex Archaeological Collections, iii,* pp. 35-71, 1850.
21. M. G. Richardson, Business Training in Medieval Oxford, *The American Historical Review, 46*:2, pp. 59-280, 1943.
22. C. L. Kingsford (ed.), *The Stoner Letters and Papers, 1290-1483* (2 Vols.), Royal Historical Society London, 1919.
23. T. Laquer, The Cultural Origins of Popular Literacy in England 1500-1850, *Oxford Review of Education, 2*:3, pp. 255-275, 1976.
24. H. S. Bennett, *English Books & Readers 1603-1640*, At the University Press, Cambridge, England, 1970.
25. M. R. Richardson, *Middle Class Writing in Late Medieval London*, Pickering & Chatto, London, 2011.
26. J. McRae, *God Speed the Plough: The Representation of Agrarian England, 1500-1660*, Cambridge University Press, Cambridge, England, 1996.
27. J. Fitzherbert, *Booke of Husbandrie*, London, 1530.
28. T. Tusser, *A Hundredth Good Points of Husbandrie*, London, 1557.
29. T. Tusser, *Five Hundred Points of Husbandrie*, London, 1573.
30. D. McDonald, *Agricultural Writers: From Sir Walter of Henley to Arthur Young, 1200-1800*, Horace Cox, London, 1908.
31. L. Mascall, *A Book of the Arte and Maner, Howe to Plant All Sorts of Trees*, London, 1572.
32. T. Hill, *A Most Briefe and Pleasaunt Treatyse, Teaching Howe to dress, Sowe, and Set a Garden,* Violet and Hal W. Trovillion (eds.), Trovillion Private Press, 1946, Herri, IL, 1543.
33. R. Scot, *A Perfite Platform of a Hoppe Garden, and Necessary Instructions for the Making and Maynteneance Thereof*, London, 1574.

34. G. Markham, *Cheap and Good Husbandry*, London, 1614.
35. G. Markham, *Country Contentments*, London, 1615.
36. G. Markham, *Farewell to Husbandry*, London, 1620.
37. G. Markham, *The Husbandman's Jewel*, 1695.
38. G. Markham, *Maister-peece*, London, 1610.
39. J. Worlidge, *Systema Agriculture Being the Mystery of Husbandry Discovered and Laid,* London, 1681.
40. W. Blith, *The English Improver, Improved; or, the Survey of Husbandry Surveyed*, London, 1653.
41. L. Meager, *The English Gardner: or, a Sure Guide to Your Planters & Gardners. In Three Parts*, London, 1688.
42. S. Hartlib, *The Complete Husbandman; or, a Discourse of Husbandry, Both Foreign and Domestic*, London, 1659.
43. C. F. Mullet, Gervase Markham: Scientific Amateur, *Isis, 35*:2, pp. 106-118, 1944.
44. F. N. L. Poynter, *A Bibliography of Gervase Markham*, Oxford University Press, Oxford, England, 1962.
45. M. Richardson, The Dictamen and the Influence on Fifteenth Century English Prose, *Rhetorica, 2,* pp. 207-226, 1983.
46. E. Tebeaux, *Emergence of a Tradition: Technical Writing in the English Renaissance, 1475-1640*, Baywood, Amityville, New York, 1997.

CHAPTER 6

Technical Writing in Renaissance and 17th Century Shipwrightery: From Orality to Textuality

INTRODUCTION

As Chapter 1 suggests, research to improve navigation via applied mathematics became a central focus among the first faculty of Gresham College and played a major role in the work of members of the Royal Society. Development of England's commerce and trade depended on improved navigation and then shipbuilding. While the Greshamites focused on mathematics for navigation purposes, and numerous books appeared on navigation, as suggested in Chapter 4 with its focus on format, page design, and visuals, the absence of technical writing about shipbuilding and ship design is noticeable. In this chapter, I explore the absence of technical writing in early English shipbuilding and suggest the reason for its slow development during the English Renaissance and then the 17th century. As this chapter will also show, shipbuilding instructions emerged in a manner very different from agriculture and estate management instructions primarily because of differences in historical contexts.

THE CONFLUENCE OF NAVIGATION AND SHIPWRIGHTERY

Shipbuilding, as a trade, improved because navigation improved. As navigation enabled sailors to travel in open seas and pursue more direct routes between ports, ships had to be built to withstand the rigors of open seas. Examined chronologically, books on navigation written during the 16th and 17th centuries reveal the development of navigation principles, instruments, instructions, and cartography. Some of the most impressive books published in the 16th and 17th centuries in England were navigation texts. These often contained designs and patterns for making navigation instruments. Many works, published in large folios, incorporate carefully crafted graphics—maps, charts, tables, and drawings.

Eden's *The Arte of Navigation* (10 editions, 1561–1630) [1] and Johnson's *The Light of Navigation* [2] exemplify some of the best English Renaissance navigation texts and illustrate format and visual design to enhance usability. In addition, navigation, revealed through its technical writing, stands as a monument to humanistic knowledge as it exemplified the importance of mathematics. Without humanism's influence, the power of mathematics to develop reliable navigation would have delayed improved ship design and growth of shipping and sea commerce in England, Europe, and the Mediterranean.

But while books on navigation increased in quantity and in technical and typographic sophistication after 1550, works on *shipbuilding*, particularly those written in English, were few. The first written treatises on shipbuilding and design were Venetian works published in the 15th century [3, p. 162]. These Venetian manuals described a design system based on geometrical and proportional relationships and a small number of fundamental measurements. The manuals focused on visuals—visual relationships—rather than on linear text [3, p. 163]. Late 15th and early 16th century Venetian books, among the best in the world at that time, show the culmination of Venetian ship technology that existed before the 15th century.

The sophistication of extant Venetian and Spanish books on ship design published in the 16th century reminds us that the Mediterranean world was more technologically advanced than England at the same point in history. The first published book on shipbuilding, *Instrucción Náutica* [4], by Diego García de Palacio, was written in 1586 in Spanish and published in Mexico City in 1587 [3, p. 166]. This work testifies to Spain's naval dominance during the 16th century and the early development of navigation and fundamental maritime theory, which, in many ways, has remained unchanged. J. Bankson, who translated the work into English, states in his Preface that *Instrucción Náutica*

> was the first such work published in the New World and has the further distinction of apparently being the first of its type to have been published in any European language. Its historical value is further enhanced by the author having included information on almost every aspect of contemporary maritime technology within his text, the text is so complete that a twentieth-century shipwright should have little, if any difficulty in using it as a guide for the construction, rigging, and equipping of any of the vessels described therein. Also, the navigating instruments and methods described would enable any mariner with a minimum of mathematical ability and a small knowledge of the heavens to make reasonable navigations in the middle latitudes of the earth's surface. The book enabled seamen to gain nautical knowledge and practice. [5, p. v]

Divided into four books, *Instrucción Náutica* [4] defines the character of many shipwrightery books until the late 17th century: Book One, Book Two, and Book Three discuss navigation principles, while only three chapters in Book

Four describe ship design. The majority of Book Four describes the responsibilities of ship personnel, battle strategy, and extensive definitions of terms used by sailors. With little help from the Preface as to the purpose of the book, the content suggests that it may have been written for wealthy investors interested in becoming involved in shipping. These investors needed knowledge of maritime nomenclature to talk with shipbuilders and sailors. The *Nautical Instruction* thus summarizes what an inexperienced investor would need to know about ships, their basic construction and operations, to enable him to converse with working shipwrights and work crews [3, p. 166]. The prevalence of navigation concepts and definitions presented in *Instrucción Náutica* illustrates how much more was known about navigation than ship design in the 16th century.

This work, and other books on shipbuilding published in the 16th and 17th centuries, were not how-to texts such as those found in medicine, surveying, estate management, and animal husbandry. Shipbuilding "texts," manuscript works prepared for use in shipyards, if they existed at all, were notebooks roughly penned and often highly technical. Venetian treatises prepared by shipwrights recorded detailed elements of ship design for use by foreign workgangs unfamiliar with the designs. As nautical archaeologist Dotson states, these "treatises served as *aides mémoires* and helped conceptual clarification for the masters themselves while their drawings may have facilitated communication of design ideas to the work force. Written treatises diffused through the shipyards of Europe, based upon the Italian example" [3, p. 163].

Matthew Baker wrote the first extant English work on shipbuilding. Baker's "text," dated 1570, is composed of a series of 165 annotated drawings [6] (see Figure 6-1, one of the examples). This large notebook suggests that the drawings served as mnemonics for the shipbuilder and perhaps for shipyard workers, many of whom might not have had more than rudimentary reading skill. Shipwrights intended these notebooks to be cryptic to those who did not work in specific shipyards. These folio-sized drawings include annotations added in handwriting. We can assume that Baker's method mirrors the form of earlier Venetian notebook-manuals developed for use in Venetian shipyards [3, p. 163]. No other texts, manuals, or treatises on ship design seem to have existed in England in the 16th century, the middle period of the English Renaissance.

English printed books on shipwrightery finally began to appear after 1650, nearly a century after Baker's notebook. However, the continued use of texts as *aides mémoires* can be found in books on ship rigging, such as Hayward's *A Full and Perfect Account of the Sizes and Lengths of Riggins for All His Majesties Ships and Frigates* (1666) [7], predominantly a book of tables showing lengths and numbers of ropes and sails, determined by time, tradition, and historical design (see Figure 6-2). Henry Bond's two books, *The Boate-Swains Art*, (16 versions, 1642–1699) [8] and *The art of appareling and fitting of any ship with masts, yards, and cordage* [9] give tables of the sizes of rigging required for a ship of 300 tons and explains a simple method for working out other sizes

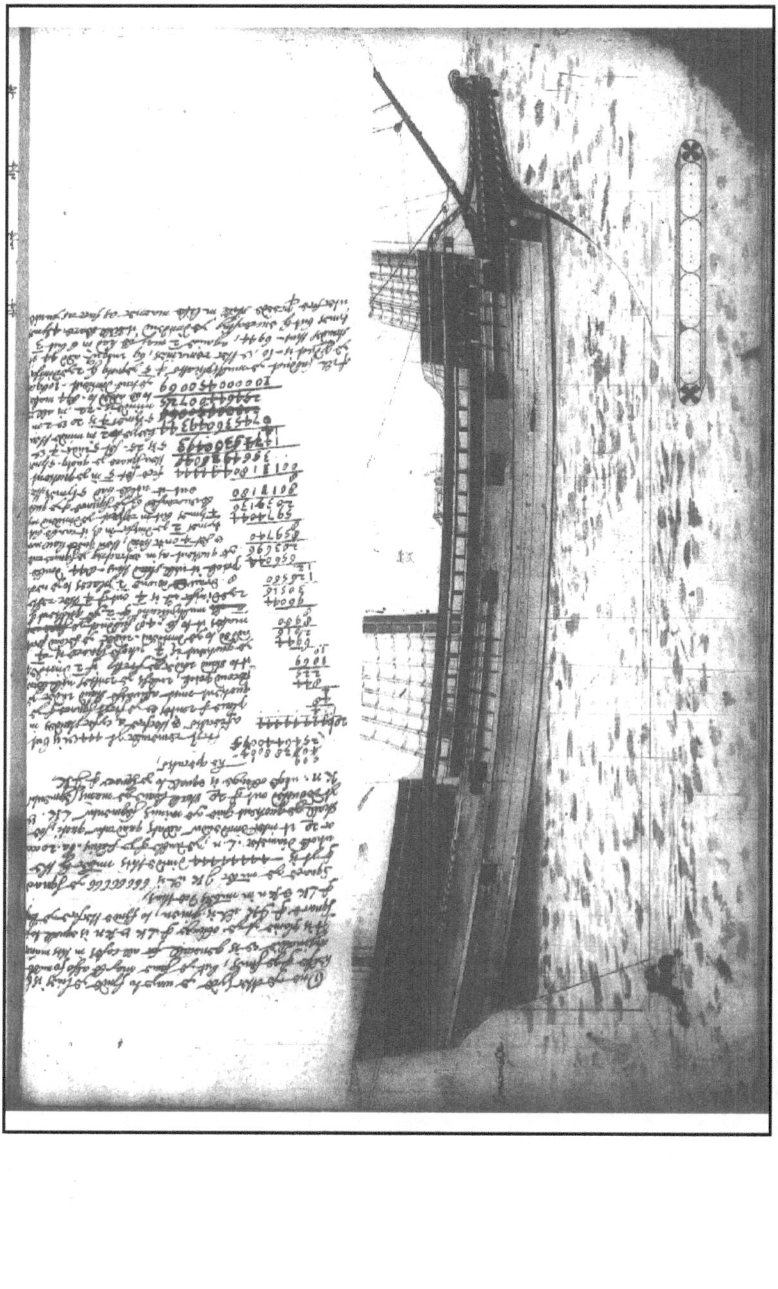

Figure 6-1. Notebook page 122 from *Fragments of Ancient English Shipwrightery* by Matthew Baker. Photo from Magdalene College Library, Cambridge.

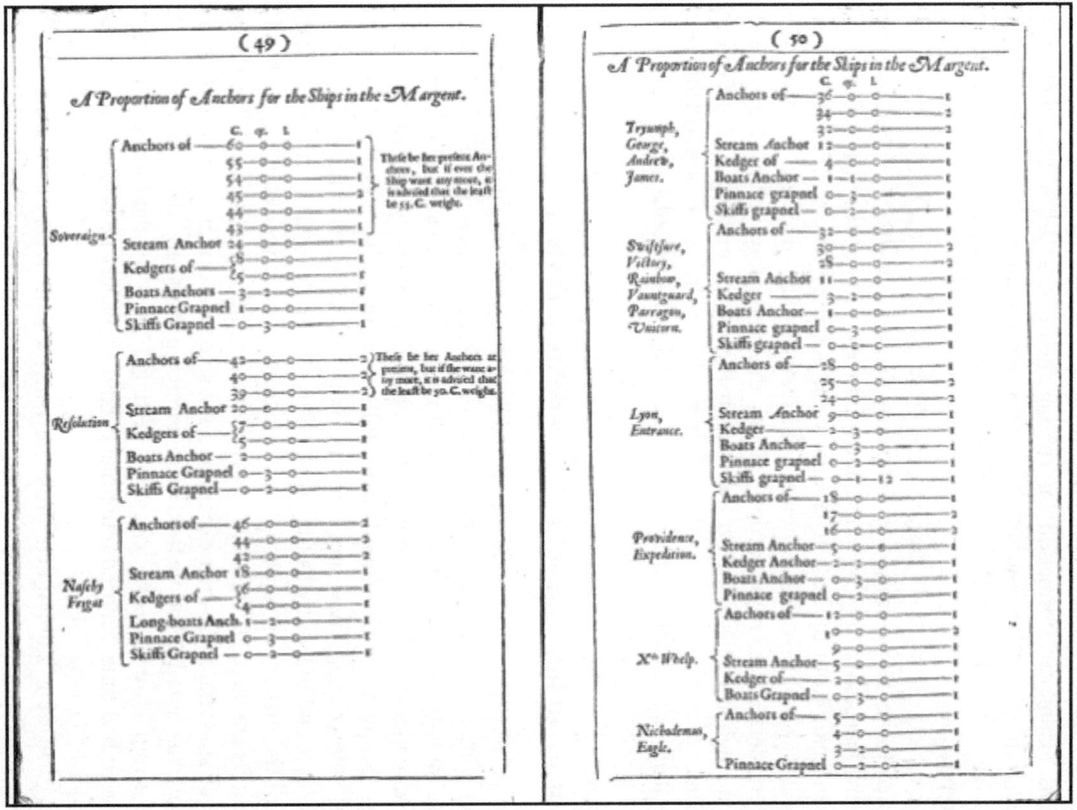

(49)

A Proportion of Anchors for the Ships in the Margent.

		C.	q.	l.	
Soveraign	Anchors of	60	0	0	1
		55	0	0	1
		54	0	0	1
		45	0	0	2
		44	0	0	1
		43	0	0	1
	Stream Anchor	24	0	0	1
	Kedgers of	8	0	0	1
		5	0	0	1
	Boats Anchors	3	2	0	1
	Pinnace Grapnel	1	0	0	1
	Skiffs Grapnel	0	3	0	1

These be her present Anchors, but if ever the Ship want any more, it is advised that the least be 55. C. weight.

		C.	q.	l.	
Resolution	Anchors of	42	0	0	2
		40	0	0	2
		39	0	0	2
	Stream Anchor	20	0	0	1
	Kedgers of	7	0	0	1
		5	0	0	1
	Boats Anchor	2	0	0	1
	Pinnace Grapnel	0	3	0	1
	Skiffs Grapnel	0	2	0	1

These be her Anchors at present, but if she want any more, it is advised that the least be 50. C. weight.

		C.	q.	l.	
Naseby Frigat	Anchors of	46	0	0	2
		44	0	0	3
		42	0	0	2
	Stream Anchor	18	0	0	1
	Kedgers of	6	0	0	1
		4	0	0	1
	Long-boats Anch.	1	2	0	1
	Pinnace Grapnel	0	3	0	1
	Skiffs Grapnel	0	2	0	1

(50)

A Proportion of Anchors for the Ships in the Margent.

		C.	q.	l.	
Tryumph, George, Andrew, James.	Anchors of	36	0	0	1
		34	0	0	2
		32	0	0	2
	Stream Anchor	12	0	0	1
	Kedger of	4	0	0	1
	Boats Anchor	1	1	0	1
	Pinnace grapnel	0	3	0	1
	Skiffs grapnel	0	2	0	1
Swiftsure, Victory, Rainbow, Vauntguard, Parragon, Unicorn.	Anchors of	32	0	0	1
		30	0	0	2
		28	0	0	2
	Stream Anchor	11	0	0	1
	Kedger	3	2	0	1
	Boats Anchor	1	0	0	1
	Pinnace grapnel	0	3	0	1
	Skiffs grapnel	0	2	0	1
Lyon, Entrance.	Anchors of	28	0	0	1
		25	0	0	2
		24	0	0	2
	Stream Anchor	9	0	0	1
	Kedger	2	3	0	1
	Boats Anchor	0	3	0	1
	Pinnace grapnel	0	2	0	1
	Skiffs grapnel	0	1	12	1
Providence, Expedition.	Anchors of	18	0	0	1
		17	0	0	2
		16	0	0	2
	Stream Anchor	5	0	0	1
	Kedger Anchor	2	2	0	1
	Boats Anchor	0	3	0	1
	Pinnace grapnel	0	2	0	1
X^th Whelp.	Anchors of	11	0	0	1
		10	0	0	2
		9	0	0	1
	Stream Anchor	5	0	0	1
	Kedger of	2	0	0	1
	Boats Grapnel	0	3	0	1
Nichodemus, Eagle.	Anchors of	5	0	0	1
		4	0	0	1
		3	2	0	1
	Pinnace Grapnel	0	2	0	1

Figure 6-2. Page from Hayward's book of tables, *The Sizes and Lengths of Riggins for all the States Ships and Frigates*.

via proportions (see Figure 4-8 in Chapter 4) based on numbers of editions recorded by the *Short-Title Catalogs* [10, 11] (the work was popular for well over a century). In contrast, Edward Hayward's 1655 book *The Sizes and Lengths of Riggins for all the States Ships and Frigates* [12] went beyond simple proportions. This small book, composed of 60 pages, mostly double-page tables, contains extensive information for rigging men-of-war. However, the format is predominantly tables, not text. The *aides mémoires* quality of many of these books (see also Chapter 4) shows how audience and purpose drove the design of shipbuilding books.

The variety of methods described in these popular little books—apparently designed to be easily carried by seamen—suggests the range of expertise and mathematical knowledge of the seamen drawn to handbooks. The tables could be used apart from the text if readers were not fully able to read linear text, and tables were likely deemed more visually assessable than text for presenting numbers derived from proportions. Thus, the mnemonic value of these early printed English books suggests the uneven literacy of readers. The prevalence of visuals over text suggests that many readers were more comfortable with tables than text. Despite emergence of practical how-to books for seamen, ship design and construction books did not appear.

Orality as an Impediment to Shipwrightery Texts

Instrucción Náutica [4], with its limited information about ship design, makes a critical point that characterizes the first books about shipbuilding: Shipwrightery as a technology did not readily lend itself to textual rendition. Master shipwrights and apprentices worked in shipyards composed of highly specialized working groups. Shipwrights developed their skills as apprentices and finally became master shipwrights. They succeeded without theoretical training and relied on knack, eye, and ingenuity as they worked from a model with input from sailors. Most of the instruction in design and construction occurred orally, as suggested by Baker's annotated sketches [6]. Master shipwrights, until the late 17th century, designed ships based on their experience, not science. As archaeological historian Hassloff has noted, "The method of determining the shape of a vessel with the aid of a miniature scale-model and so getting a basis for the layout of the ribs must have had time-honored traditions even if its exact age is as yet impossible to establish" [13, p. 171].

Fear also kept shipbuilders from moving beyond techniques contained in memory or cryptic mnemonic notebooks. Ship carpenters and boatbuilders tenaciously guarded their knowledge to ensure their economic survival. So jealous were they of their design knowledge that they often destroyed any drawings, along with models, to prevent their falling into the hands of competitors [13, p. 172]. Protecting the shipwright's expertise may be the most significant reason that shipwrightery texts focusing on design were almost nonexistent in

England until the late 17th century. At that point, increasing literacy, the need for advanced ship design that required more than memory, the emergence of shipbuilding as a major industry, and royal government commitment made continued reliance on oral transmission of shipwrightery impossible.

Because of shipwrightery's stubborn reliance on oral tradition until the closing decades of the 17th century, naval archaeologists believe that knowledge about shipbuilding changed little from the years 600 to 1600. Shipbuilders and merchants continued to make extensive voyages to observe new constructions of wooden ships, but rudimentary equipment and tools also hindered design change and limited innovation as much as shipwrights' reliance on traditional, proven methods. Design change increased costs, which shipwrights had to justify to investors. Safety was also a concern. Without mathematical theory and testing to guide redesign, the reliability of existing designs on ships riding unpredictable seas made sailors and investors fear change. Safety for sailors and cargo remained paramount [14, pp. 21–27].

With limited textual information before the close of the 17th century, knowledge of early ship design has come not from texts on ship design but from naval archaeology, excavations of boats as old as the 12th century, contemporary pictures and drawings of seagoing vessels, painted glass and seals, sketches in notebooks, artists' renditions, and business records. *Essays in the Evolution of the Sailing Ship 1250–1580* [15] provides examples of pictorial sources that have helped historians track changes in ship design. Development of shipping as an industry can only be indirectly determined from extant but incomplete business records.

From Orality to Lists in the Development of Shipwrightery Texts

Aside from notebooks with mnemonic sketches, how did the technical writing of shipbuilding emerge? How did this persistent oral tradition in a major English industry lead to the first shipbuilding texts that appeared near the end of the 17th century?

In addition to the few extant notebooks and treatises, clues to the design of early ships can be found in three additional sources: commercial contracts, ship inventories, and ship accounts [16, pp. 41–43]. From these technical documents we find the earliest evidence of the nomenclature of shipwrightery, which remained unchanged during the height of its oral tradition and shifted from words, then phrases, to full linear text by the latter decades of the 17th century.

Commercial contracts said nothing about ship design but focused on minimum ship capacity. Some charters specified hold dimensions and equipment for ships to be built for special purposes. These contracts, with their measurements, form the conceptual roots of later treatises on shipbuilding. Initially, written contracts—like Renaissance shipbuilding notebooks—were nothing more than *aides mémoire* for witnesses if litigation arose. These bare documents, many from as early as the 13th century, launched the idea of communicating and documenting design by means of writing, which became more popular with the growth of literacy throughout the Renaissance. By the 13th century, the Mediterranean culture was far ahead of England in intellectual development and implemented the contract as proof of financial agreement, a use of writing that followed in England by the 14th century. Increasing literacy and improvements in hull design, although gradual, led to use of written descriptions of designs. The more complex the design, the greater the need for both graphics and supporting text to preserve and transmit design features [3, p. 161].

Inventory lists recorded materials used in ship construction and documented the possessions of an owner. Surviving lists and inventories come from records of governments and wealthy aristocrats rather than private ship owners who probably kept no records. Inventories and contracts assumed that both writer and reader possessed the same knowledge of a vessel, another characteristic of *aides mémoire* prepared for use by members of a closed discourse community. These early inventories provide insight into ship nomenclature and appear to be the first attempt to "write" terms transmitted orally from master to apprentice. Lists often were prepared by minimally literate recordkeepers.

Inventories combined with accounts show ship equipment and structural components, such as the masts, spars, rigging, and other gear [16, p. 43]. As nautical archaeologist Friel notes,

> A difficulty with medieval work accounts, whether for shipbuilding or other activities, lies in deducing how much information has been left out. The aim of the official or officials in charge was to provide a record and justification of the expenditure for which they had been responsible, not to provide a technical record. [17, pp. 16–17]

Fortunately, many of these clerks found that the best way to record expenditure on particular items was to use the terminology of the artisans that they employed and sometimes to state exactly the uses of those items [17, p. 17]. From the earliest English shipbuilding records, technical terms are almost invariably rendered in English, with a few French terms thrown in. Accounts provide the first textualized proof of the existence of maritime nomenclature [17, p. 18].

These inventory lists were not organized about specific ship parts but occurred randomly. Nevertheless, lists from the late 14th-early 15th centuries show the emergence of a model for items included in the inventory. Ship accounts often correspond to inventories [16, p. 50], construction purchases of materials, and the payment of wages. Many accounts, organized by construction stages, showed the items acquired at various points in the building stage. The accounts probably forced the organization of materials into specific sections (timber, boards, ironwork, etc.). Thus, both accounting and the need to track gear define the nature of medieval ship lists, which became more informative with the shift from single words to phrases to show what was purchased [16, pp. 40–54].

As a result, "the best written sources for the technical history of shipbuilding in medieval England are inventories and financial accounts" [17, p. 14]. Friel's examples show some of the earliest shipbuilding accounts [16, pp. 46–48; 17, pp. 202–204]. In examining these lists, we assume that their spelling attempted to capture the sound of the word.

For example, the words *luf* and *bowline* or *luf* and *bowsprit* appeared as early as 1296 in extant English ship inventories. *Bowsprit* was found in the summaries of the inventories of 21 Dutch and German vessels at the ports of Scarborough and Ravenser in Yorkshire in 1294 [16, p. 43]. This list predates the Oxford English Dictionary (OED)[1] note that *bowsprit* was found in 1330 in Robert Brunne's Chronicles. *Luf* appeared as early as 1205 and was certainly in use, but in various spellings, before 1296, the date reported by the OED. *Sailyard* appears in this word form in the OED in 1295. *Bowlines*, in the inventory, predates the OED's date, 1325. *Wyndynghauncer* (winding hawser) predates OED entries for windless. *Tricehokes* ("trice" or windlass) did not appear in the OED until 1357–1358. For researchers, lexicography of inventories provides an intriguing glimpse into the persistence of nomenclature before its textualized form became common some 300 years later.

Humanism and the Emergence of Shipwrightery Text

As I have already mentioned, humanistically inspired technology would ultimately transform shipbuilding into a major manufacturing business and wrestle the carefully guarded knowledge of shipwrightery from master shipbuilders seeking to preserve their craft. Humanism nurtured scientific approaches to navigational problems long before science was even thought to be applicable to ship design. For example, Bede's 8th century treatise on the tides and the early 13th century tide table for London Bridge, compiled by the monks of St. Albans [18, p. 136], reveal attempts to understand the theoretical principles underlying empirical knowledge and the importance of writing these principles [18, p. 164].

Despite development of navigation science that attracted a wide spectrum of educated investors, literacy levels of shipwrights was likely low. Reading skills of apprentices and work crews were likely less than that of shipwrights. The highly specialized nomenclature, learned as apprentices worked in shipyards with master shipwrights, may account for the entrenchment of the nomenclature, much of which persists in the jargon of modern sailors, into the first full-length shipwrightery texts. The importance of nomenclature, apparent in the extensive glossary of maritime terms in *Instrucción Náutica* [4], similarly appeared in nearly all of the early English shipwrightery texts published in England during the

[1] All references to the Oxford English Dictionary are taken from the online OED.

17th and early 18th centuries. The first printed English texts dealing with shipwrightery focused on terminology, with some instructions intermixed, a continued testimony to the pervasiveness of maritime nomenclature. An examination of some standard words illustrates the control, oral and textual, that the discourse community had on the nomenclature of shipbuilding [19, pp. 182–190]. For example, common words such as the following have long histories, as reflected in the Oxford English Dictionary, and meanings remain unchanged:

> **Bulkhead**: According to the OED, was first used in 1496 in the Naval Accounts of Henry VI.
>
> **Cog**: A 14th century Tutonic word. Refers to a broadly built ship with roundish prow and stern. The cog served as a ship of burden or transport and as a ship of war.
>
> **Keel**: First used in 1352 in the Higden Rolls.
>
> **Rudder**: First used in 725.
>
> **Buoy**: Middle English *boie* (1466), from Old French *boue*, probably of Germanic origin.

Other common terms have a similar ancient history and offer interesting lexicography.

Emergence of English Shipwrightery Books (1640–1700)

Only after 1640 in England did principles of ship design and construction begin to appear in print; and most of these, like *Instrucción Náutica* [4], combined navigation principles with concepts of shipbuilding and rigging. The momentous shift from orality to textuality resulted from a concatenation of influences in ship technology and the emergence of literacy [20, pp. 116–120].

First, text replaced oral instruction as navigation science developed and nurtured the need for improved ship design to meet demands of commerce. Ship design was becoming too complex to operate only within the memories of individual ship designers.

Second, by the mid-17th century, shipwrightery was developing into a major commercial operation in all European countries. With industrial development emerged a broad range of readers who now needed books—from an increasing contingent of new investors to new seamen, many of whom still had minimal literacy skills. Growth of sea commerce depended on educating sailors and carpenters.

Third, these preceding events required both mathematicians and astronomers to consider sailors and shipwrights' educational needs. Vernacular texts had to be written on elementary mathematics as well as navigation. Improved tables and charts had to be developed—and written—to be accessible to seamen and to improve safety at sea. Development of navigation instruments required books describing their use. In the early years of Elizabeth's reign, members of the queen's advisory council, such as Dr. William Cunningham, argued for the need for English books that were easy to read so that the new knowledge and information about navigation could be shared with newly literate readers [18, p. 199]. By 1567, popular, concise books on navigation, such as Bourne's *A Regiment for the Sea,* [21] More's *The Carpenters rule . . . very necessary for masons, shipwrights* [22], and Smith's *An accidence or The Path-way to Experience. Necessary for all Young Sea-men* [23] provided basic, unadorned instruction to meet the needs of an increasing number of readers interested in all facets of the maritime industry.

However, more sophisticated books also emerged, such as Leonard Euler's translation of Henry Watson's *Théorie complete de al construction et de la manoeuvre des vaisseaux* (*A Complete Theory of the construction and properties of vessels and practical conclusion for the management of ships made easy to navigators*) [24]. Other more informative popular books, such as Love's *The Mariner's jewel or pocket companion, containing an alphabetical dictionary of . . . naval terms . . . from a manuscript of Sir John Narborough's* (7 editions, 1695–1735) [25], exemplify detailed embedding of concepts within text, which marked the rapid shift from oral instruction to textual instruction by the end of the 17th century.

Comparison of these later books to Smith's *An Accidence or the Path-way to Experience* [23] suggests the increase of literacy and knowledge among the targeted maritime audience. Writers consistently used an oral style to explain terms and then launched into descriptions of how various objects were used. One of the best examples of the definition books is Sir Henry Mainwarings' *The Sea-Man's Dictionary* [26], which provides extended definitions of sea terms as well as derivative words. The book is accessible via a detailed index. For example, Mainwaring spends two pages discussing Anchor, Anchoring or Anchorage, Anchor stock:

> *Anchoring, or Anchorage*. Is when we let fall an *Anchor*, or more, into the Sea, with Cabells to them, so that the Ship may ride-fast by them; we say, there is good Anchoring, where there is showle water, for in deep waters, the Sea hath more force against the Ship, and the *Achor*s are very long a waying upon any occasion: Ground that is not to soft ozee, in which the *Anchors* can have no fast hold: not too hard or knotty, so that it may cut the Cabells: The best ground to ride in, is a stiffe clay, or hard sand: Also, where they may ride out of the way of the tyde: And lastly, where they may ride Land-locked, so as that the sea-gate, can have no power against them: To which may be added, that the Lee-shore on every side is so soft, that if a ship come a-ground, she can catch no hurt, or die that they may

> have sea-roome to let saile, if their Cabells breake, or their *Anchors* come home. That place which hath all these commodities, is good to ride in, and here we say is good Anchoring, or good Anchorage; bad Anchoring, or bad Anchorage, is in a place where all or many of the contrary conditions are to be found.
>
> *Anchor Stock.* This is peece of Timber fitly wrought and flatned at the Nutts, crossing the hookes; The use thereof is so to guide the *Anchor* upon the ground, that one of the hookes may be sure to fasten in the group, without which, the *Anchor* would lye flat upon both the hookes, and take no hold: The proportion, whereof is usually taken from the length of the shancke. [26, p. 4]

ENGLISH SHIPWRIGHTERY TEXTS IN THE LATE 17TH CENTURY: LITERACY, KNOWLEDGE, TECHNOLOGY, AND TECHNICAL WRITING

By the close of the 17th century, a full-fledged synergy had emerged among literacy, knowledge, technology, and the technical writing of shipwrightery. Knowledge and literacy nurtured each other: (1) rapidly improving design emerged from applications of mathematics to shipbuilding problems; (2) technical writing enabled more readers to use technology and decreased reliance on memory as ship design became more complex.; (3) growth of literacy led to the habit of writing down details in works ranging from merchants' handbooks to architectural treatises; and (4) the art of shipbuilding began to be recorded in a growing number of notebooks, handbooks, and finally printed treatises, each focusing on different audiences. Shipwrightery as an industry, rather than a closed craft, could no longer resist change. The need for financial support and the growing demand for more and better ships took shipwrightery to a higher economic level.

One of the best English examples published before 1700 is *The Compleat Ship-Wright*, by Edmund Bushnell, published in London in 1664 (10 editions, 1664–1748) [27]. This slim book contained 72 pages and was written, according to the Preface, to help shipbuilders understand and learn the mathematical basis of shipbuilding. The book marks the transition from shipwrightery as knowledge gained by doing basic actions to the mathematical basis of design. It was the first printed English work that deals purely with naval architecture [28, p. 50] and remained one of the most popular book in shipwrightery until the close of the 18th century. In the Preface, Bushnell describes the problem his book aims to solve:

> I propose to teach those long experienced Ship-Wrights, whose actions hath declared their Abilities to the whole World . . . yet their knowledge they desire to keep to themselves, or at least among so small a number as they can . . . for although some of them have many Servants, and by Indenture . . . bound to Teach them all alike the same Art and Mystery that he himself useth, Yet it may be he may Teach some one, and the rest must be kept ignorant.

Bushnell then states the problem his book attempts to resolve: shipwrights, because they have limited knowledge, can orally instruct their servants to "Hew, or Dub, to Fay a Pesce when it is moulded to his place assigned" [27], but servants cannot understand what they are doing. They remain ignorant because they lack the mathematics needed to advance in the art of shipbuilding. Bushnell says that his book will be "understood by the meanest capacities, if they can but read English, and have the benefit of a little Arithmetick, as Adition, Subtraction, Multiplication, Division" [27 p. A2]. Bushnell's little book provides instruction in the basics of geometry and then uses these principles—drawing perpendicular and parallel lines, for example—to explain how to draw a ship design by applying basic geometric principles. The fundamental arithmetical focus of the book suggests that Bushnell was fully conscious of the knowledge needs of his intended readers. However, Bushnell assumes that his readers have worked in shipyards and are comfortable with the nomenclature of shipwrightery. What they need to know is the mathematical basis for ship design.

Figure 6-3, a foldout drawing of a ship design for which Bushnell provides instructions, is placed next to page 8.

Bushnell's design instructions occur in plain English, as can be seen in the excerpt, Figure 6-4.

Another popular book that marks the shift from orality to textuality in shipwrightery is Miller's *The Compleat Modellist: Shewing the true and exact way of Raising the Model of any Ship* [29], discussed in Chapter 4. See Figure 4-10. The book textualizes the ancient practice of building small ship models and then building full-size ships from models. This small, 24-page pamphlet explains how to draw a rigging plan for use in determining lengths of various ropes. It has two plates and tables of sizes and proportions [29, p. 59]. The work exemplifies those texts that bridged the gaps between oral instruction, *aides mémoires*, and written instructions that complemented the *aides mémoires*.

Miller [29], like Mainwaring [26] and nearly all the maritime writers, used a decidedly conversational writing style, but his extensive tables and drawings helped readers understand the proportions of rigging. Miller included a foldout drawing that shows the rigging of a ship—the names of each rope (see Figure 4-10, in Chapter 4)—and then other tables that relate rope sizes to masts. The book assumes that readers will be thoroughly comfortable with the concept of shipbuilding from models. The chief value of the book is its tables, designed to be used as references by shipwrights and shipyard workers who might able to read only the tables. Miller's style can be seen in his opening segment: Instructions for the raising of the Model of any Ship or Vessel, small or great:

> When you go to raise the Model of any Ship or Vessel, you must in the first place know the Length of her Keel, and the Depth in her Hold, and the Breadth of her Beam.

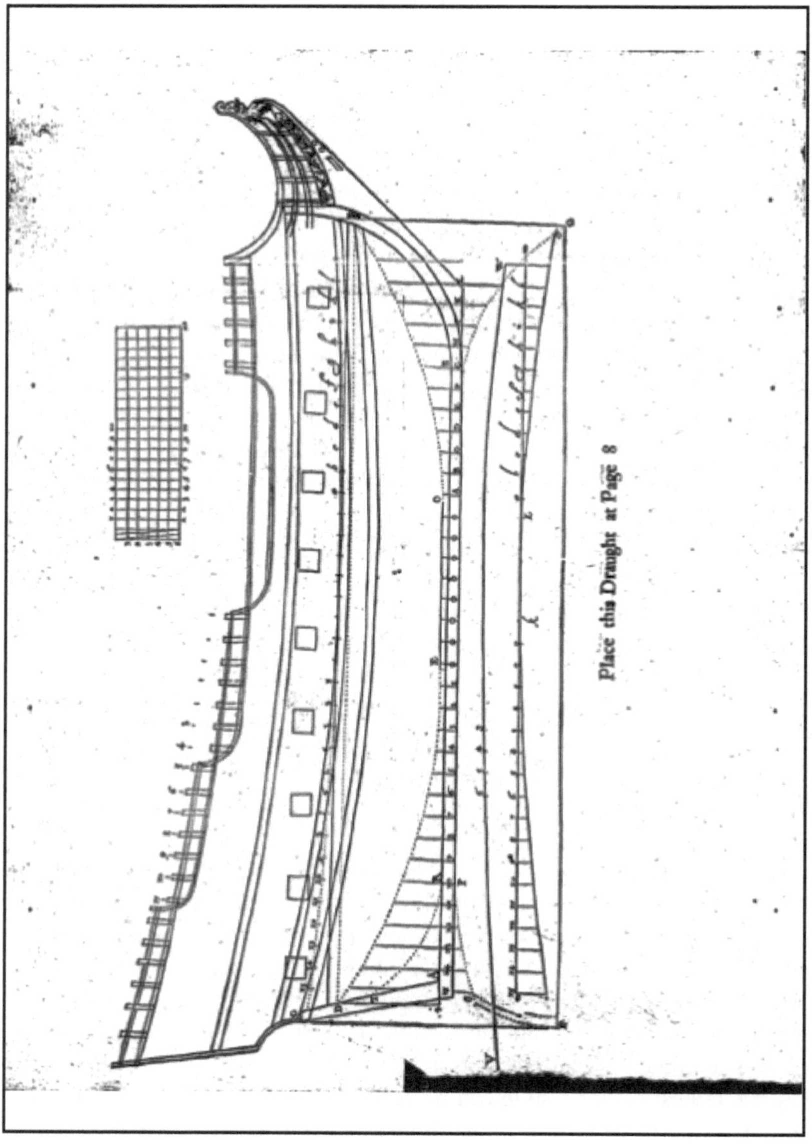

Figure 6-3. Foldout drawing from Bushnell's, *The Compleat Ship-Wright* (1664).

8 *The Compleat* Ship-wright.

ſo then letting the lower ſell of the Ports be two foot from the Gun-decks, the lower edge of the Ports wil be three foot from the water abaſt, and two foot and an halfe afore, in the middle of the Gun-deck 2 foot 9 inches, ſufficient for ſo ſmall a Veſſel, a greater Veſſel would require to have the Guns ſomething farther from the water, then if another Waale be required, firſt ſet off the Ports in their places, that the Waale may ly above the Ports, or elſe he would be cut with the ports in pieces, the upper Deck with height reſpecting the bigneſſe of the Ship, having reſpect to not over building ſmall Ships, to damage their bearing of Sail.

Then for the Head, the length of the Knee would be two thirds of the breadth, ſo then the Knee of the Head in this Draught will be 12 foot 8 inches long, and for his place, as low as conveniently he can, provided that the Rails of the Head, fall not fowl of the halſhols, becauſe that in placing of the Knee low, giveth room to round the Head, and ſteeve it to content: The place of the Knee will be at, or very neer, the upper Waal, the upper edge of the Knee againſt the upper edge of the uper harping, which will be very well for the lower Cheeks of the head to be faced againſt, for by that means they wil be clear of any Seame to avoid Leakings, and will very well bolt the end of the harping, if a Breſt-hook be faſtned alſo within board againſt them, will very well faſten all together.

Then for the ſteeving of him, and rounding the Knee, a regard muſt be had to the lying of the Boltſpreet, leaving room enough for the Lyon and Scrowl under the Boltſprit. Then for the rounding of the Rails, round them moſt at the after ends.

For the heights between Decks and Steeridge, Cabine, Fore-Caſtle, thoſe heights are commonly mentioned in contract by the Maſter or Owners building.

CHAP.

Figure 6-4. Text to accompany Figure 6-3, drawing of a ship to scale in Bushnell's *The Compleat Ship-Wright*.

> First knowing the Length of the Keel, take the Length of the Keel off your scale, and place it on your Paper that you intend to raise your Model on, making two pricks, one with the point of the Compass the other with the others; then draw a Line with your Pen or Pencil of Black-lead, as from A to B. [29, p. 2]

Perhaps the most complete book on shipbuilding that appeared in the 17th century was *Dean's Doctrine of Naval Architecture* [30], published in 1670 by Sir Anthony Deane, a shipbuilder who built 25 ships [30, p. 21]. The book, one of the most important texts in the history of naval architecture, provides more information than Bushnell's book [27] and includes the earliest complete plan of a ship, along with numerous design illustrations. Dean's work marks the point that naval architecture became a modern science. Dean's work, and ancillary works compiled by Brian Lavery, who edited the work, illustrates the full shift to the textual tradition in shipwrightery by the closing decades of the 17th century. Like Bushnell, Dean includes arithmetic in his description of ship design, but he focuses on geometric calculations and their use in indicating the likely performance of the ship. Sophisticated mathematics separates Dean's design instructions from that of Bushnell. Dean's command of proportions enabled him to calculate the displacement of a ship to determine how high gunports should be above the waterline [30, p. 22]. Dean's work suggests a high literacy and educational level of readers involved in the design of English naval ships.

The closing decades of the 17th century also produced ship specifications, such as a contract specification written in 1666–1667 directed to merchant builders, descriptions of bilge pumps, instructions for the care of sails, and business letter reports. The following one, directed to the Navy Board, described changes in construction after Charles II had ordered the shipbuilders to increase the breadth of one of the warships, the *Warspite*. Because the ship had to be redesigned—and could no longer follow the agreed-upon specifications written by Christopher Pett—the status report explains the needed changes. The letter report, excerpted here, lists the needed changes:

> Right Honourable,
>
> According to your warrant to us directed, we have made our repair on board His Majesty's Ship [the *Warspite*], and have strictly compared the performance of the said ship with the builder's contract, and do find all things performed according to the tenor thereof, both in respect of materials and workmanship. We also find several more works to be done by them more than contract, the particulars of which are as followeth:
>
> - One strake of middlebands of six inches thick and 16 inches broad, they being obliged to but 4 inch plank.
> - One transom and a pair of knees, with all bolts to the same, being 13 in number.

- One strake of four inch plank on each side the hatchway of the lower deck, which is bolted into every beam, they being to make a three inch without bolts.

- They have fitted the cisterns and cases about the pumps with the dales for the passage of the water through the side more than contract.

- For a strake of three inch plank wrought about the bows on the strake of spirketting from the stem to the after part of the bulkhead, being an inch thicker than contract.

. . . .

All of which we leave unto your honour's command and remaining

Woolwich, 30th of June 1666 (Signed) Christopher Pett
Jonas Shish.
[30, p. 120]

Despite the increase in the number and type of ship design texts, the concept of the handbook as mnemonic continued until well into the 18th century. William Sutherland's *The Ship-builder's Assistant*, first published in 1711 (7 editions, 1711–1794) [31], is a small book, 165 pages of text, crammed with practical information and a dozen plates that illustrate how to construct specific parts of a ship. This little book also includes two rigging plans. The book appeared in numerous editions, a testimony to its popularity with readers. Its size likely made it easy for sailors and shipwrights to carry and use. Handbooks that focus on definitions also continued, such as Blanckley's *A Naval expository shewing and explaining the words and terms of art belonging to the parts qualities, and proportions of building, rigging, furnishing, & fitting a ship for sea* [32].

Transactions of the Royal Society published before 1700 show the first interest in applying science to shipbuilding: Letters [see 33, published in 1673/1674] and papers on treatments to prevent wood worms [34], effectiveness of lead sheathing on ship hulls [35], propulsion theory, design of oars [36], ventilation systems for cargo holds [37], care of new sails, and design of bilge pumps [38] illustrate the change of focus toward preserving ship design in written text.

The *science* of shipbuilding emerged by the early decades of the 18th century. After *Dean's Doctrine of Naval Architecture* (1670) [30], came Sutherland's *Britain's glory: or, shipbuilding unvail'd. Being a general director, for building and completing the said machines* (7 editions 1717–1729) [39] and Murray's *A Treatise on ship-building and navigation* (1754) [40]. Nevertheless, earlier works, such as Bonds' *The Boate-swains art* [8] and his books on masts and rigging [9] remained popular, as did Bushnell's *The Compleat Ship-Wright* [27] and Miller's *The Compleat Modellist* [29]. The sustained popularity of these first books on ship design suggest the continued range of literacy and educational levels among many shipbuilders.

Shipwrightery and the Emergence of English Technical Writing

Understanding the halting emergence of English shipwrightery texts during the Renaissance and then the 17th century adds a reinforcing segment to the story of the rise of English technical writing. As I have shown in previous studies of English Renaissance technical writing I have published since 1991, modern technical writing echoes much medieval and Renaissance technical writing in genre, format, and presentation [41, 42]. The answer to the fundamental questions—Did technical writing always exist? What did it look like? and How and why did English technical writing emerge and develop?—are questions that I continue to pursue. The answers echo throughout the story of the technical writing of shipwrightery presented here. Based on historical research I have pursued since 1988, I offer four observations explaining the emergence of English technical writing.

First, a substantial quantity of what we now call technical writing, or writing that explained how to perform work, existed long before printing in manuscript or notebook form [43, 44]. And in the 1300–1700 period, for example, more pragmatic writing than humanistic writing existed in England [43]. Analyzed chronologically, these pragmatic texts, more so than other kinds of texts, reveal the importance of tables, visuals, formatting strategies, and page design to organize content and a pervasive plain style that existed long before Bacon and the Royal Society [41, pp. 175–236; 44, 45, 46]. The style of shipwrightery texts, despite an indigenous, nonchanging nomenclature far older than writing, reveals in its earliest textual expression a plain style similar to that used in other how-to discourse [41, pp. 137–174; 43]. For reasons yet to be determined, writers of many pragmatic texts during these early periods saw the value of format, which became evident in all forms of published technical books as orality gave way to textuality [42, 44, 45, 46]. Particularly shipwrightery texts, from their earliest appearance, similarly echo the inherent need for visual communication to capture the intricacies of ship design.

Second, analysis of historical technical writing constantly reveals the emergence of literacy among those groups for whom these works were written and published [42, 46, 47, 48, 49]. As I have found throughout my 25-year examination of early English how-to texts, the emergence of technical writing is the story of the shift from orality to textuality; and technical writing echoes the literacy of both the writer and the intended readers [41, pp. 35–90; 51].

English Renaissance printers were not unlike modern printers: they wanted a market for the books they sold. The numbers of editions of books published from 1475 to 1700, for example, indicates the market size for the books printed during that time frame. The popularity of many technical books attests to the importance of a wide range of pragmatic topics of interest to a public becoming increasingly literate, a reading public that wanted self-help and knowledge beyond the boundaries of oral instruction [49, 51]. In examining these texts, we see a perspective on English culture omitted in many studies of the history of English humanism [50].

Third, what we now call technical or pragmatic writing existed before rhetoric began to influence discourse, particularly how-to texts that emerged outside the pale of humanistic-inspired rhetoric and education [52]. Not until the late English Renaissance did rhetoric, the auspicious progeny of humanism, clearly influence the character of technical writing. My sense is that some kinds of "technical" or "pragmatic" writing influenced nonpragmatic texts. Examination of many Renaissance writers reveals a cross-fertilization of text types—humanistic and technical—in terms of style, organization, arrangement, format, page design, and graphics, as we clearly do in works by two renowned literary writers, Chaucer and Sir Thomas Eliot, and then Gervase Markham, one of the most prolific technical writers of the English Renaissance [41, pp. 91–136; 42, 51]. As seen in writers like these, both literary style and plain style existed side by side throughout the evolution of modern English prose style.

Fourth, historical analysis of early technical writing suggests that technical writing parallels the emergence of text in general. While humanistic writing portrays the literacy of the upper classes, technical writing illuminates the emergence of literacy among the commercial classes, the target audience of many of these how-to books. While religious writing changed little during the 14th–17th centuries in content and style, technical writing changed dramatically. These changes in technical writing show the synergistic growth of both literacy and technology. Medieval and Renaissance technical writing—business letters, descriptions, specifications, leases, contracts, instructions, proposals, and procedures—began to shift rapidly from oral transmission to textual expression, particularly in the 15th century. By the mid-17th century, the oral qualities of much pragmatic writing had faded. While technical writing emerged in response to literacy and technology, this emergence—in terms of both content and style—reveals the rapid decline of the oral tradition. Prose, not poetry, became the necessary medium to communicate information on how to perform work. Increasing knowledge demanded a prose that moved presentation beyond the bounds of traditional humanistic text. Again, the story of technical writing is a story of the shift from orality to textuality in pragmatic discourse.

Needed Research in Shipwrightery

Shipwrightery, which developed within a tightly closed discourse community, exemplifies the clearest resistance to the shift from oral to textual presentation among early technologies in England. Texts revealing the early maritime world offer rich opportunities for historians of early technical writing because few studies of maritime texts written/published before the 19th century have been examined from the perspective of technical writing. Yet resource materials complete enough to launch studies of maritime topics are available to allow students to explore a variety of topics: (1) evolution of discourse on navigation instruments, particularly instructions for their use and technical descriptions of these instruments; (2) dictionaries, discussed in this essay—over six dozen emerged before 1800; (3) instructions on gunnery, rigging, signaling, tactics, and seamanship; (4) evolution of tables, charts, and manuals in English navigational works. Changes in visuals suggest another approach to the oral textual shift. Works such as Adams and Waters' *English Maritime Books Printed Before 1801* [53] provide invaluable resources.

Opportunities for research await the scholar interested in the evolution of literacy, text, visuals, format, description, reports, proposals and procedures—the essential elements of what we still call "technical" writing. Chapters 7 and 8 explore the development of proposals and suggest their relationship to the emergence of reports. Today, both stand as perhaps the two most important genres in our field.

REFERENCES

1. R. Eden (Trans.), *The arte of nauigation by Martin Cortes*, Richard Iugge, London, 1579.
2. W. Johnson, *The Light of Navigation*, Amsterdam: Printed by VVilliam Iohnson [Blaeu], dvvelling upon the vvater, by the Old Bridge, at the signe of the Golden Son-dyal, Anno 1612.
3. J. E. Dotson, *Treatises On Shipbuilding Before 1650*, Frank Howard, *Sailing Ships Of War*, 1400-1860, Mayflower Books, New York, pp. 160-168, 1979.
4. D. Garcia de Palacio, *Instrucíon Nautica para Navega*, Mexico City, 1987.
5. J. M. Bankson (Trans.), *Nautical instruction*, Bisbee, Arizona, 1986, Diego Garcia De Palacio, *Instruccíon Náutica Para Navegar*), Mexico City, 1587.
6. M. Baker, *Fragments of ancient english shipwrighteyr,* Microfilm, Magdalene College, Cambridge, Nautical Archaeology Program Library, Texas A&M University, 1570.
7. E. Hayward, *A Full and Perfect Account of the Sizes And Lengths of Riggins for All His Majesties Ships and Frigates*, Robert Miller, London, 1666.
8. H. Bond, *The Boat-swains art: or the Compleat Boat-Swain*, Richard Mount, London, 1699.
9. H. Bond, *The art of appareling and fitting of any ship with masts, yards, and cordage*, F. R. Boydell, London, 1655.
10. W. Pollard and G. R. Redgrave, *Short-Title Catalogue of Books Printed in England, Scotland, & Ireland, and of English Books Printed Abroad, 1475-1640* (2nd Edition), W. A. Johnson, F. S. Ferguson, and W. F. Panzer (Comps.) (Two Volumes), Bibliographical Society, London, 1976 and 1986.
11. D. Wing, *Short-Title Catalogue of Books Printed in England, Scotland, Ireland, Wales, and British America, and of English Books Printed in Other Countries 1641-1700* (2nd Edition), Index Committee of the Modern Language Association, New York, 1972.

12. E. Hayward, *The Sizes and Lengths of Riggins for all the States Ships and Frigates*, printed by Peter Cole, living in Leaden-Hall, and at the sign of the Printing-press in Cornhil, neer the Royal Exchange, London, 1655.
13. O. Hasslof, Wrecks, Archives and Living Tradition, Topical Problems in Marine-Historical Research, *The Mariners Mirror, 53*:3, pp. 162-177, 1963.
14. R. W. Unger, *The Ship in the Medieval Economy 600-1600*, Croom Helm, London, 1980.
15. B. Greenhill (ed.), *Essays in the Evolution Of The Sailing Ship 1250-1580*, Conway Maritime Press, London, 1995.
16. I. Friel, Documentary Sources and the Medieval Ship: Some Aspects of the Evidence, *The International Journal of Nautical Archaeology and Underwater Exploration, 12*:1, pp. 41-62, 1983.
17. I. Friel, *The Good Ship: Ships, Shipbuilding and Technology in England 1200-1520*, Johns Hopkins, Baltimore, Maryland, 1995.
18. E. G. R. Taylor, *The Haven-finding art* (2nd Edition), Hollis and Carte, London, 1971.
19. G. Hutchinson, *Medieval Ships and Shipping*, Fairleigh Dickinson, Rutherford, 1994.
20. R. Gardiner, *Cogs, Caravels and Galleons: The Sailing Ship, 1000-1650*, Naval Institute Press, Annapolis, Maryland, 1994.
21. W. Bourne, *A Regiment for the Sea: Conteyning most Profitable Rules, Mathematical Experiences, and Perfect Knowledge of Navigation*, H. Bynneman, London, 1574.
22. R. More, *The Carpenters rule . . . very necessary for masons, shipwrights*, Welby, London, 1616.
23. J. Smith, *An accidence or The Path-way to Experience. Necessary for all Young Sea-men, or those that are desirous to goe to sea*, Jonas Man, London, 1627.
24. L. Euler, *A Complete Theory of the construction and properties of vessels and practical conclusion for the management of ships made easy to navigators*, P. Elmsley, London, 1776.
25. J. Love, *The Mariner's jewel or pocket companion, containing an alphabetical dictionary of . . . naval terms . . . from a manuscript of Sir John Narborough's, edited by James Lightbody*, R. Whitledge & S. Sims, London, 1695.
26. H. Mainwaring, *The Sea-Mans dictionary. Or, An Expsxition And Demonstration of all the Parts and Things belonging to a Shippe: Together with an Explanation of all the Termes and Phrases used in the practique of Navigation*, G.M. for John Bellamy, London, 1644.
27. E. Bushnell, *The Compleat Ship-Wright*, W. Leybourn for George Hurlock, London, 1664.
28. R. C. Anderson, Early Books on Shipbuilding and Rigging, *The Mariner's Mirror, 13*, pp. 53-64, 1924.
29. T. Miller, *The Compleat Modelist*, William Fisher, London, 1676.
30. A. Dean, *Dean's Doctrine Of Naval Architectrur, 1670*, Brian Lavery (ed.), Conway Martin Press, London, 1981.
31. W. Sutherland, *The Ship-builder's Assistant: or, some essays toward compleating the art of marine architecture*, R. Mount, A. Bell, and R. Smith (eds.), Mount, Ball, and Smith, London, 1711.
32. T. Blanckley, *A naval expositor, shewing the explaining the words and terms of art belonging to the parts quantities, and proportions of building, rigging furnishing. & fitting a ship for sea*, E. Owen, London, 1750.

33. A letter to the publisher concerning a new way, by an English manufacture to preserve the hulls of ships from the worm, &c, better for sailing and more cheap and durable than any sheating or graving hitherto used, *Philosophical Transactions, 8*:100, pp. 6192-6194, February 9, 1673/74.
34. Of a letter written from, Holland about preserving of ships from being worm-eaten, *Philosophical Transactions, 1*:11 pp. 190-191, Monday, April 2, 1666.
35. T. Hale, *An account of several new inventions and improvements . . . relating to building our English shipping. . . the proceedings relating to mill'd-lead shething*, J. Astwood, London, 1691.
36. W. Petty, *The discourse before the royal society 26th November 1674, concerning the use of duplicate proportions to sundry particulars, together with a new hypothesis of springing and elastic motions*, J. Martyn, London, 1674.
37. S. Sutton, An account of Mr. Sutton's invention and method of changing the air in the hold, and other close parts of a ship; communicated to the Royal Society by Richard Mead, M.D., Physician to His Majesty, fellow of the Royal society, and Royal College of Physicians, *Philosophical Transactions, 42*:462, pp. 42-45, Jan./Feb. 1741/2.
38. B. Martin, *The principles of pump-work, illustrated, and applied to the construction of a new pump without friction, or loss of time, or water, in working*, the author, London, 1776.
39. W. Sutherland, *Britain's Glory: Or, Ship-Building Unveiled. Being a General Director, for Building and Compleating the Said Machines*, T. Norris, E. Tracy, A. Bettesworth, and B. Pickard, London, 1717.
40. M. Murray. *A Ttreatise on ship-building and navigation in three parts*, D. Henry and R. Cave, London, 1754.
41. E. Tebeaux, *The Emergence of a Tradition: Technical Writing in the English Renaissance, 1475-1640,* Baywood, Amityville, New York, 1997.
42. E. Tebeaux, From Orality to Textuality: The Evolution of Technical Description in English Renaissance Technical Writing, *Issues in Writing, 4,* pp. 59-10, 1991.
43. E. Tebeaux, Pillaging the Tombs of Noncanonical Texts. *Journal of Business and Technical Communication, 18*:2, pp. 165-197, 2004.
44. E. Tebeaux, Format and the Evolution of English Accounting Texts, 1100-1700, *Journal of Technical Writing and Communication, 30,* pp. 307-341, 2000.
45. E. Tebeaux, Visual Language: Format and Page Design in English Renaissance Technical Writing, *Journal of Business and Technical Communication, 5,* pp. 246-274, 1991.
46. E. Tebeaux, From Orality to Textuality in Accounting Books: 1556-1680. *Journal of Business and Technical Communication, 7*:3, pp. 322-359, 1993.
47. E. Tebeaux, Ramus, Visual Rhetoric, and the Emergence of Page Design in Medical Writing of the English Renaissance: Tracking the Evolution of Readable Documents. *Written Communication, 8*:4, pp. 411-444, 1991.
48. E. Tebeaux, Technical Writing for Women of the English Renaissance: Technology, Literacy, and the Emergence of a Genre, *Written Communication, 10*:2 , pp. 164-199, 1993.
49. E. Tebeaux, Technical Writing in Seventeenth-Century England: The Flowering of a Tradition, *Journal of Technical Writing and Communication, 29*:3, pp. 219-262, 1999.

50. E. Tebeaux, Women and Technical Writing, 1475-1700: Technology, Literacy, and Development of a Genre, *Women, Science and Medicine, 1500-1700 and Science. In Honor of the 500th Anniversary of Gresham College*, Lynette Hunter, Margaret Pelling, and Sarah Hutton (eds.), University of Leeds, Sutton, United Kingdom, pp. 29-62, 1997.

51. E. Tebeaux, Books of Secrets—Authors and Their Perception of Audience in Procedure Writing of the English Renaissance, *Issues in Writing, 3,* pp. 41-67, 2000.

52. E. Tebeaux, Renaissance Epistolography and the Origins of Business Communication, 1568-1640: Implications for Modern Pedagogy, *Journal of Business and Technical Communication, 13,* pp. 75-98, 1992.

53. T. R. Adams and D. W. Waters (Comp.), *English Maritime Books Printed Before 1801 Relating To Their Operation At Sea*, John Carter Brown Library, Providence, Rhode Island, 1995.

CHAPTER 7

Emergence of Proposals in England: 13th–17th Centuries

INTRODUCTION—EMERGENCE OF PROPOSALS AND REPORTS

When I first developed the plan for *The Flowering of a Tradition*, I knew that I would need to deal with the development of both reports and proposals, foundational genres of technical writing. As I had recognized since 1999, both existed in abundance in the 17th century. I also knew that the history of both the proposal and the report, as we use those terms in technical communication, had not been addressed. When I began to explore the history of both genres, I came to a major conclusion: Both the proposal and the report likely evolved from the petition, an important document that seems to have begun in England with Edward II in the 1270s and continued throughout the 15th through the 17th centuries and well into the 18th century.

As I studied petitions, led by the work of Ormrod [1], Gwilyn Dodd [2], as well as Fisher, Richardson, and Fisher [3] and Myers [4], I observed that the structure of the medieval petition anticipated the structure of modern proposals and also many proposals written during the 17th century. At the same time, I noted that many 17th century printed reports reflected the structure of both modern reports and proposals. With these observations in mind, I offer a common theory for the development of both genres—both reflect the template or formulary of the petition as it existed in the 13th century. Examples of early reports and proposals will support that claim.

The Emergence of Proposals

As Tom Pearsall wrote in the proposal chapters of numerous editions of *Reporting Technical Information*, all projects have to begin with some document that will elicit action. The proposal has become the required document for launching work of all kinds. It has become a standard topic in technical communication curricula because of its importance in the workplace. However, the history of the proposal has not been written, even though I alluded to specific examples in *Emergence of a Tradition*. Determining its origins requires some flexibility in examining early English documents and searching for documents prepared to initiate action desired by an individual or an entity, such as a community, an individual, or a business. With that simple definition in mind, we can find among the earliest petitions in England examples that suggest the modern proposal in terms of purpose: to launch a project or action. From that point, we can track, via archival search, later examples. By the close of the 17th century in England, documents that clearly anticipate the modern proposal emerged with a single goal: to initiate action to achieve change on a variety of topics. This chapter will begin the history of the proposal in England.

MEDIEVAL ENGLISH PETITIONS

As Gwilyn Dodd states, private petitions requested favor or relief (change of circumstances) "within the context of parliament. These petitions, written in a highly formal manner, were submitted to a higher authority, the King or a Parliamentary committee." Petitions "proposed" or sought "royal grace" [2, p. 1], such as a grant, office, pardon, or redress for a grievance [2, p. 2]. However, these early petitions did not use the term *proposal*, perhaps because citizens or subjects petitioning the King may have believed the term *proposal* too demanding a word for inclusion in a document directed to the King or Parliament.

The petition, which emerged with Edward II in the mid-13th century, served as a clever means by which Edward could know issues and grievances his subjects faced [2, p. 26]. The petition allowed commoners to approach the sovereign. As Dodd remarks,

> The use of parliament as a venue for local complaint promised to be decisive in counteracting this unsatisfactory state of affairs [actions of unscrupulous and unpopular local officials]. . . . Presenting a petition to the king at parliament all at once made the Crown more accessible to people who had hitherto remained relatively isolated from royal governance. In short, large-scale petitioning in parliament represented an important stage in the growth and influence of the late medieval English state. [2, p. 33]

Edward II used the petition for a number of purposes: Before Edward's reign, the Crown had used Parliament to resolve complex cases it received from the royal courts, but Edward II opened up, via petitions, a channel by which the population, from peasants, to groups, to communities, and to nobility, could access the Crown—King, counsellors, and justices—directly and on their own initiative [2, p. 19]. The King's power could be made available to all who sought it [2, p. 25]. Edward also wanted to know about the issues experienced by his subjects, which included complaints about his own officials, such as sheriffs. Parliamentary petitions provided an excellent opportunity for the King's subjects to bring attention to the wrongdoing of all royal officials in the localities [2, p. 32]. The action of unscrupulous and unpopular local officials damaged royal prestige and weakened royal authority within local communities. In short, large-scale petitioning in Parliament represented an important stage in the growth and influence of the late medieval English state.

Petitions to Parliament also became an excellent venue for these complaints. From a larger perspective, accessibility of parliament via petition enabled both the Crown and Parliament to understand needs of the English people and provide remediation via the English State [2, p. 33]. Petitions, taken as a continuous whole, provided the Crown a large, continuing flow of information and enabled the government to decide how to use resources to resolve illegal activities and civil needs [2, p. 35]. The power that the petition gave to the King's subjects may explain its continuity throughout the next 500 years. Petitions led to development of a bureaucracy needed to handle the wide range of petitions that appeared and to respond in a timely fashion. As Dodd observed, the King adopted a passive stance in implementing government, and governance often came as a result of the petitions, not the heavy hand of the King.

Edward II and his successors used the petition as a management tool to determine "what was going on" in the lives of his subjects. Even today, many "open-ended" proposals, such as those requested by the Department of Education, allow the requesting entity an opportunity to determine concerns of educational groups who can describe problems in public education and then submit proposals that embody methods of change. In the early 1300s, extant petitions indicate that they came from a range of groups—commoners, individuals of importance, groups, communities, even abbeys. During the high noon of petition submission, Parliament developed a method of sifting petitions to facilitate response, a method that anticipates modern organizations that solicit and then receive large numbers of proposals that are then transmitted to the proper reviewers based on topic. The following shows the disposition of petitions in 1324 [2, p. 79]:

Dodd compiled the disposition of petitions presented in 1324 [2, p. 79]. Less than 30% were rejected:

Total Number of Petitions	**191**
• Handled by parliament	66
• Deferred to chancery	40
• Deferred to the treasurer and barons of the exchequer	26
• Deferred to the justices	8
• Deferred elsewhere	7
• To be resolved at common law	17
• Rejected	27

At some point between 1316 and 1322, those who received the petitions, perhaps the clerks, began to collate common grievances and placed them on a single schedule presented directly to the King and council for consideration [2, p. 133]. Pollard's description of the receivers of petitions and clerks provides a useful survey of how the process worked [5].

Segments of Petitions

As Baldwin [6, pp. 282–283] and then Fisher, Richardson, and Fisher [3, p. 21] have determined, the petition contained three to five distinct parts, all presented in linear text and using a rather formulaic language:

The address—clearly separate from the petition to announce the intended audience. Initially, this audience was the King. However, as the numbers of petitions increased, the intended readers could have been the Commons or the Lords.

Identification of the petitioner—This part usually began with a verb to describe the supplicatory nature of the document. This line began with a verb, such as "besecheth the support of the Commons, etc. Or, Please hit your wyse discrecions to consider. . . ."

Statement of grievance or difficulty [exposition]—The next two stages included separately, or comingled, the grievance and then the suggested remedy. These parts incorporated the greater part of a petition's prose. In this section, the petitioner explained what was wrong and what was needed from the Crown.

Request for redress—This segment often provided an explicit request (proposal) of what should be done.

Appeal for remedy—The final, closing statement appeared as a formulaic appeal, usually framed in obsequious language to emphasize that the petitioner knew his subservience to the King and/or the Parliament and did not want to seem presumptuous. These final three stages—grievance/complain, request for redress, and the suggested remedy—formed the bulk of the petition and occurred as a narrative. The language of the narrative made clear that subjects who petitioned the King needed to adopt an appropriate supplicative, subservient voice that underscored the power difference between the King and his subjects.

Fisher, Richardson, and Fisher provide this 1424 example:

[Address] To the kyng our souerayn lord & to my lords of the counseil.

[Identification of the Petitioner] Beschet humebely your most obeisant subget Thomas Burton knight

[Exposition] how that the saide Thomas hath accompted in your Escheqer for the kepyng of the Erle of Ewe. Arture of Britayne & Bursegand that was Marchal of ffrance And ouer CCCli released & lv li sxvj s viij d disallowed ther ys clerly founden due to the saide Thomas . . .

> *[Petition]* That hit like to your benign grade be auive of my said lords of your said counseil to consider the long labour and heuy and dreadful charge of the said Thomas & the seide sommes above released and disallowed and in discharge of the soule of the kyng your noble fader that was. Whom god assoil ordaigne swyche paiement or agreement for the said Thomas of the sommes aforesaid as thinketh to your saide benign grade be auis of my said lords of your counseil.
>
> *[Valediction]* for the love of god and dede of charitee. [3, p. 21]

As shown in this example, in petitions from the 13th to the early 17th centuries, the overall tone reinforced a sense of deference and humility on the part of the petitioner and upheld the supreme authority of the King.

While many petitions were short, others were lengthy, and length increased over time to ensure that all the facts were given. The petitioner had one opportunity to make the case, and overelaboration proved preferable to lack of detail. Like modern proposals, petitions were scrutinized by a small group of men (the triers [auditors], members of the king's council, senior judges, or senior Crown officials.) The names of the triers always appear in the rolls of Parliament [7]. In short, the situations that these documents relayed to the King were constructed within what appears to have been a broadly recognized petitionary canon, which acted as a template into which the specific details of the case were inserted. The use of formularies in early writing, particularly legal writing, as shown in Chapter 5, was well established in estate management a century earlier.

By the 14th century, petitions were answered by attaching writs (orders) of the privy seal to them instead of the earlier custom of writing endorsements straight onto the back of the petitions themselves. Therefore, the petitioner and the clerks who processed petitions needed to have the information contained within them transferable into royal commands in the most straightforward and efficient manner possible. Thus, the language of the request mirrors the language of the instruction.

Petitions, in the language they use, represent the political discourse expected of those approaching the King and Parliament in medieval England. The address, which came to be separated from the main body, may have occurred as a way to separate petitions for the commons from those directed to the Lords and the King.

Then, in the 14th century, when the Lords and commons became the target audience of petitions, this group was addressed as "right wise and discreet," emphasizing dignity and qualification, and the adjectives used for the King tended to emphasize power. Yet both groups addressed illustrate the use of obsequious language.

From the late 14th century onward, petitions elaborated the details of the request, particularly those written in English, most likely to ensure that enough information was included to convince the triers of the merits of the argument.

Sufficient information was imperative, as many petitioners were not present when the petition was orated. And petitions did not receive a second chance once the triers determined their answer. The process of hearing the petitions was oral; the majority were scrutinized by a small and highly qualified body of men, the triers, (auditors), members of the King's council, senior judges, or senior Crown officials. Most were resolved with no further input.

The Parliament Rolls of medieval England provide numerous examples to those interested in the petition as it evolved. One petition, submitted to Richard II in 1385, contains a number of terse petitions, followed by Richard's often tart reply:

> *Petition*:
> Also, that the estate of the household of our lord the king be inspected once a year, or more often if necessary by the chancellor, treasure and clerk of the privy seal for the time being, and be amended as need be at their good discretion. And similarly, that the statues made in former times concerning the same household be strictly upheld and kept in all respects, and duly executed according to the tenor of the same.
> *Answer*:
> As to the first article of the petition, the king will do as he chooses. And as to the second, the king wills it. [7, p. 25]

The goal of this petition (actually a proposal) was to emphasize the necessity of the King being guided by his council rather than by informers on the desirability of appointing suitable persons as the King's chief ministers and on the need for reform in both the exchequer and the royal household, with the aim of preserving and increasing the revenues of the Crown rather than dissipating them in the form of grants and pardons. Thus, we find a document, entitled "Advice," to emphasize this necessity. This document merely restates the petition [7, p. 28]. As Given-Wilson states, "These proposals were of a radical nature, especially the attempts to subject both the exchequer and the royal household to external review and to restrict the royal patronage which all kings saw as an essential component of kingship" [7, p. 2]. Thus Given-Wilson recognized that this advice was indeed a proposal.

The Evolution of Petitions to Proposals

As we look at the large number of extant petitions beginning with the reign of Edward II and continuing throughout the 17th century, we see that the "petition" serves two functions:

First, it *reports* a problem, along with sufficient explanation, to allow the auditors or triers (auditors) who passed judgment on petitions that required the King, the Crown, or some branch of Parliament to make a decision, hopefully in favor of the petitioner [5, p. 210]. With this characteristic, the petition form or template, as we shall see in the following chapter, could be used for reports. And petitions often incorporated *memoranda*, information documents prepared by the clerks for Parliament to summarize the issues of petitions and to serve as a point meaning "to remember."

Second, the petition, a persuasive document, must provide convincing information to support the solution to the request given by the petitioner and encourage intended readers to respond favorably. Thus, the obsequious language of the first petitions supported the petitioners' claim by emphasizing their subservience to the King. This style of subservience, continued throughout the 17th century, even though "plain" or direct English made its presence felt by the 1660s. And we can find many *reports* that maintain to some extent, depending on the intended audience, the subservient social status of the writer.

Before examining petitions, we need to define terms. First, bills and petitions were entreaties formally presented to a higher authority. In Parliament Rolls, the two words (rolls and petitions) have the same meaning, but bills led to legislation [2, p. 1]. As Dodd points out, a number of types of petitions existed, based on their point of initiation: communities, individuals, the nobility, and even Parliament [2].

As I will next show, in the 17th century, propositions, which occurred in conjunction with many petitions, began appearing as full-length proposals that anticipate modern proposals. Many of these proposals came from men who nurtured the founding of the Royal Society and/or became some of the first members. This chapter explains the rationale for the development of proposals and discusses several major proposals of the century.

16th CENTURY: THE SHIFT FROM PETITIONS TO PROPOSITIONS AND PROPOSALS

The entrenchment of the petition format can be seen in a 1621 petition to the Commons by the dyers. Note how the pattern, established over 300 years earlier, remains, although the address and the name of the petitioner are separated from the petition, a format change that occurred within a century of the appearance of the petition (see Figure 7-1). The rationale was to help the clerks determine, at a glance, which government body should receive the petition. Note the subservient, obsequious tone and the form [8].

A 1642 "humble petition" submitted from "Both Houses of Parliament" to the King, opens with the standard petition in its conventional format. The style is appropriately differential in its request for redress of wrongs that undermine the King and the people. Parliament is directly petitioning the King and understands the importance of difference in power here. Following the petition, however, appear 19 propositions that develop in detail the basic requests of the petition [9] The petition opens on page 191.

To the moſt Honourable Aſſembly of the Commons houſe of Parliament.

The humble Petition of the Wardens and Comminalty of the Art or myſterie of Dyers in London.

Moſt humbly ſhewing,

THat diuers Lawes haue heretofore beene made, prohibiting the vſe of Logwood *alias* Blackwood, which did not produce that good effect which was intended, by reaſon that no recompence was allowed to ſuch as ſhould take paynes in and about the ſeizure thereof, That his Majeſtie for preuention of the deceits, that might ariſe by the vſe thereof, hath made ſeuerall reſtraints by his Highneſſe Letters Pattents, his Bookes of Rates and otherwiſe, thereby to hinder the importation of the ſame; That of late his Highneſſe being informed, that the vſe thereof was moſt neceſſary and not preiudiciall to any of his Subiects for the dying of certaine courſe Stuffes, Linnen, Caddoes, Hats, Caps, Gloues, Leather, Thread, and many other Haberdaſhery wares, which as was alleaged, could not conueniently be dyed without Logwood, did thereupon by His Letters Pattents licence Sir *Tho. Compton* Knight and his Aſſignes, to import yearely fiftie Tunnes of Logwood and no more, and to take and ſeize into his hands and to his owne vſe the moitie of all Logwood imported by others.

The Petitioners further ſhew, That notwithſtanding all the pretences of art and cunning to bee vſed, in the mixture and compoſition of the ſaid wood to Die withall, yet the ſame is a moſt falſe and deceitfull colour; That vnder pretence of the importation of fiftie Tunne yearely, much more may bee conueyed into this Realme, and vſed to the defrauding of His Majeſties good intention.

That by reaſon that the ſaid Wood is ſo licenced to be imported, many Dyers doe ſecretly vſe the ſame in the dying of Clothes and Stuffes in many parts of this Realme, which cannot conueniently be reformed vnleſſe the importation thereof be wholly prohibited.

That by the vſe thereof not onely the Subiects of this Realme are deceiued, and the Art or myſterie of dying, both at home and abroad much ſcandalized, but alſo the ancient Trade of Clothing in this Kingdome is thereby greatly decayed.

The Petitioners therefore moſt humbly pray, That both the importation and vſe thereof, may bee to all intents and purpoſes, (by ſome ſeuere Law, to bee made in this preſent Seſſion of Parliament,) vtterly prohibited and aboliſhed.

Figure 7-1. 17th century printed petition (1621).

THE
Humble Petition of both Houses
Of
PARLIAMENT.

> Your Majesties must humble and faithful Subjects, the Lords and Commons In Parliament, Having, nothing in their thoughts and desires, more precious and of higher esteem (next to the honour and Service of God) then the just and faithfull performance of their duty t Your Majesty, and this Kingdom; And being very sensible of the great distractions and distempers, and of the Imminent dangers, and calamities which those distractions and distempers are Like to bring upon Your Majesty, and Your Skubjects; All which have proceeded from the substill informations mischievous practices, and evill Councells of men, disaffected to Gods True Religion. Your Majesties Honor, and safetie, and the public peace and prosperity of Your People, after a serious observation of the Causes of those mischiefs; Do in all Humility, and sincerity, present to Your Maujestie, their most dutifull Petition, and Advice, That out of Your Princely Wisedom, for the establishing Your Own Honour, and Safety, and Gratious Tendernesse of the Welfare, and Security of your subjects, and Dominions. You will be pleased To grant and accept these their humble desires and Proposition, as the most necessary effectuall Means through Gods blessing of removing those jealousies, and Differences, which have unhappily faln betwist You, and r People, and procuring both Your majestie, and them a constant course of Honour, Peace, and Happinesse.

In short, the above text, the standard petition, appears as page one. Since Parliament wants to resolve the tensions between the King and ultimately Parliament, which represents the people, the actual petition does not emerge until line 18: "You will be pleased to grant and accept these their humble Desires and Proposition, as the most necessary effectuall Means through Gods blessing of removing those jealousies, and Differences, which have unhappily faln betwixt you, and Your People" and asking for a "constant course of Honour, Peace and happiness.: Then, on the next four pages, we find "XIX Propositions (proposals)," which fully develop the petition. I give only the first two here.

Propositions sent to his Majesty.

> 1. That the Lords and others of your Majesties Councell, and such Officers and Ministers of the State, either at home or beyond the Seas, may be put from your Privy Councell, and from those Officers and imployments, excepting such as shall be approved of by both Houses of Parliament; and that the persons put into the places and imployments of those that are removed; may be approved of by both Houses of Parliament, and that al Privy Counsellours shall take an oath for the due execution of their places, in such form as shall be agreed upon by both Houses of Parliament.

> 2. That the great Affaires of the Kingdome, may not be concluded or transacted by the advice of private men, or by any unknown or unsworn Counsellors, but that such matters as concerne the publike, and are proper for the High Court of Parliament, which is your Majesties great and Supreme Councell, may be debated, resolved, and transacted, onely in Parliament, and not elsewhere. And such as shall perfume to doe any thing to the contrary, shall be reserved to the Censure and judgment of Parliament; and such other matters of State as are proper for your Majesties. Privy Councell, shall be debated, and concluded, y such of the Nobility and others, as shall from time to time be chosen for that place, by approbation of both Houses of Parliament, and that no publike Act concerning the Affaires of the Kingdom, which are proper for your Privy Councel, may be esteemed of any validity as proceeding from the Royall Authority unless it be done by the advice and consent of the Major part of the Councell attested under their hands, and that your Councell may be limited to a certain number not exceeding twenty five, nor under fifteen, and if any Counsellors place happen to be voyd in the intervals of Parliament, it shall not be supplied without the assent of the major part of the Councell, which choyse whall be confirmed at the next sitting of Parliament, or else to be voyd.

Point 1 means that Parliament wants to approve the King's appointees. In short, Parliament is asking for "shared governance." The second (request asks that "Affairs of the Kingdom" be debated in Parliament, not in secret, with no input from Parliament. "The language is elegant and respectful but also specific.

The final paragraph of the petition summarizes the petition and ends with a final appeal and well wishes to the King. This final appeal directly mirrors the final statement of the medieval petition. Parliament wants control of revenue, which they assure will benefit the King "and maintaining the Royall Honour, Greatnesse, and safety of Your Majesty, and Your Prosperity."

> And these our Humble Desires being granted by your Majesty, We shall forthwith apply ourselves to Regulate Your present Revenue In such soft, as may be for Your best advantage, and likewise to settle Such an ordinary, and constant increase of it as shall be sufficient to Support Your Royall Dignity in honour and plenty, beyond the Proportion of any former Grants of the Subjects of this Kingdom to Your Majesties Royall Predcessors. We shall likewise put the Town of Hull into such Hands, as your Majuesty shall appoint, with the Consent and Approbation of Parliament,and deliver up a just accompt of all the Magazine, and cheerfully imploy The uttermost of our Power and endavours in the reall expression and Performance of our most dutifull and Loyall Affections, to the preserving and maintaining The Royall Honou, Greatnesse, and safety of Your Majesty, and Your Posterity.

FINIS.

Whatever the length of the petition or proposal, the document's organization remained basically the same: address, addressee, and then the subject of the petition. This next example combines the addressee with a summary of the problem presented in the proposal: Preventing beer brewed outside England and Wales from being sold in England will increase the profit of the English and Welsh brewers and the Royal Treasury. What follows—Items I through V—lists the positive results of this proposal. The last half of the proposal, specifically the last five paragraphs, explain some of the financial results of beer that has not been brewed in England [10] (see Figure 7-2).

PRINTED PROPOSALS AND THE DEVELOPMENT OF PRINT TECHNOLOGY

As I mentioned in Chapter 1, improvements in print technology and the 17th century English milieu shaped development of practical writing, specifically proposals and procedures. Many fully developed proposals illustrate one of the major social issues of the late English Renaissance and the entire 17th century: the need for improved education. Authors could circulate printed proposals to their intended audience (e.g., members of Parliament) or others who might share their concerns and provide possible financial help. While concerns for modifying humanist education appeared as early as the 1530s, the need for education to prepare students for trades became stronger. Vives, in 1531, recommended the study of cooking, clothing, building, agriculture, and navigation—their history and their use for the profit of society [11, pp. 33–34].

Bacon became the first to argue for the study of trades as a description of industrial processes. Bacon saw the connection between trades, as mechanized work, to natural philosophy. He rejected the notion that experimentation on topics, such as agriculture and the manual arts, was "vulgar" and dishonorable. Bacon, along with many of his followers, rejected scholastic speculations, the target of both Cambridge and Oxford, as useless to improving English society. In the *Advancement of Learning*, he advocated the study of arts such as chemistry, dyeing, class manufacture, enameling, gunpowder, and artificial fires. Understanding these, which he called a History of Trades, would aid understanding and then development of these arts to determine the scientific causes and axioms behind each [11, p. 38]. Trades, once understood, could be improved for the betterment of society. Thus, the fact that many of these topics appeared in how-to works and process descriptions written by members of the Royal Society is no accident.

To briefly reiterate a point made in Chapter 1, after 1640, Bacon's influence led in two directions: First, it initiated the growth of experimental philosophy and the formation of a cooperative group of scientists, which became known as the "invisible college" in 1645 and continued at Oxford in the 1650s, finally evolving into the Royal Society in 1663. Second, Bacon's focus on the improvement of

To the Right Honourable the Knights, Citizens and Burgesses in Parliament Assembled.

The humble Proposal of Lewis Rawlins, *Whether the Restraining of Retailers of Beer and Ale within the Kingdome of* England *and Dominion of* Wales *from Brewing, may not only prove profitable to the Crown, but be expedient; upon due consideration of subsequent Reasons.*

I. BEcause it will supply his Majesties Exchequer in present with three hundred thousand pounds of lawful money or thereabouts; without the least trouble to the King, or Kingdome, to which Sum the present Farmers of the Excise can make no claim.

II. It will advance his Majesties Annual Revenue in the Excise of Beer and Ale for the future, and that considerably.

III. It will save his Majestie near 30000 *per annum*, which hath hitherto been disbursed out of that part of his Revenues in contracting the number of Officers which attend that duty.

IV. It will obviate that general Fraud which is practised by these brewing Retailers, to the detriment of his Majestie some thousands yearly.

V. Hereby the common Brewer will receive encouragement from his encrease of Trade.

VI. No damage can hereby accrue to the Farmers of the Excise, for they still enjoy their contract: The consumption of Beer and Ale will not be lessened by reducing the Brewing only of such Exciseable Liquors to the common Brewer, whilst the Victuallers, Inholders and the like, retail as great quantities as formerly, which in short, is but for Country Retailers to conform to the Example of the City of *London*.

This Proposition may perhaps be excepted against by the Commissioners of the Excise, whose candor in the Kings businesss is hardly justified by the ensuing Narrative.

The Propounder *L.R.* Gager in the year 1573. did exhibit as many Informations in the Excise office against *W.C.* a common Brewer in *London*, as the forfeitures thereupon amounted to above 2000 *l.* some of those were heard by the Commissioners, and *W.C.* the offender thereupon condemned in the sum of 340 *l.* or thereabouts.

The Prosecutor continued solliciting the Commissioners by Petitions, and otherwise during the space of six Moneths, to hear and determine the remaining Propositions, upon which, the said *W.C.* (if Justice had accordingly been done) must as inevitably have been pronounced guilty, as he was upon those which were tryed.

Whilest Justice was thus delayed. *W.C.* makes his Addresss to Sir *William Bucknall* and other Farmers of the Excise, and with the payment of 200*l.* or such like sum to them contracts to purchase his peace, and excuse his crime.

At this Contract *Robert Huntingdon* and *Richard Kingdon* Esquirs, two of the Commissioners connive, and from that day hitherto, the prosecutor could not by any means be admitted to make good the charge against the said *W.C.* a practise infinitely tending to the prejudice of the Kings interest, and contrary to the Act of Parliament in that case made and provided, which gives one moyety of the Fines to the Informer, whereof the Informer *L.R.* hath not as yet received one penny.

Hereupon the Prosecutor being in effect denyed Justice by the Commissioners, humbly acquaints his Majesty by Petition with his greivance, who gratiously referred the examination of the matter to the present Lord High Treasurer, of which so soon as the aforenamed Comissioners had notice, they and the Farmers without any other cause dismissed the said *L.R.* out of his office of Gager, of which injury the Propounder justly complains, and prays upon the whole matter Redress, as the wisedom of Parliament shall judge meet.

Figure 7-2. Petition to keep retailers of beer from brewing beer (1670).

education affected reformers such as Samuel Hartlib, John Dury, William Petty, Abraham Crowley, and John Evelyn. These virtuoso thinkers saw the need for a new kind of education that would mesh with the goals of those who wanted to focus on experimental science.

Abraham Cowley's Proposition for a College

Cowley's *A Proposition for the Advancement of Experimental Philosophy* (1661) contains five distinct sections, each with a heading [12]. His proposition (really a proposal) shows how the proposition developed into the proposal genre. He opens with a preface (actually a summary) that argues for the founding of a philosophical college built on the model he presents in his proposal. The next section, "The Colledge," recommends the location, revenue required, staff, salaries for the staff, and a description of the buildings needed with a justification for each one.

The third section proposes the duties "Of the Professors, Scholars, Chaplain, and other Officers," and their qualifications. The fourth section, "The School," describes the number of boys to be admitted (about 200 for four classes) and their age, beginning with boys of about 13. He also states that students would not be charged to attend. Cowley reflects Bacon's criticism of university education "because it is deplorable to consider the loss which children make of their time at most schools employing, or rather casting away six or seven years in learning of words only" [12 p. 45]. Cowley also describes the curriculum: a carefully crafted classical education, morals and rhetoric, combined with the study of plants, geography, geometry, astronomy, and physical training. The project would be funded by patrons.

In the "Conclusion," Cowley reiterates his goals for the college: "to examine and prove all things of Nature; recover lost inventions; improve all arts we now have; and discover new ones." Like modern proposals, Cowley opens with a summary, then proposes his idea, presents the supporting information, and ends with a reiteration of the value of what he has proposed [12].

William Petty's Proposal for Advancement of Learning

Another initial force in the formation of the Royal Society was Samuel Hartlib, a close friend of Milton, who publicized the theories of Comenius and persuaded Milton to write "Of Education." Petty saw a scientific basis for education as a business that would support both scientists and artificers. Petty's proposed outcomes for this new college reflected the climate of the middle years of the 17th century: overarching position of the new philosophy, rejection of scholasticism, commercial efforts, demands for practical studies of things, applied mathematics, apology for science as the study of God's second book, and a full spirit of utilitarianism [11, p. 45].

In *The Advice of W. P. [William Petty] to Mr.Samuel Hartlib for the Advancement of Some Particular Parts of Learning* [13], Petty targets a new education system. It begins with determining what has been written about experimental learning (a literature review), finding a place for readers to meet and discuss their research. Petty's goal was to establish a literary workshop for children to prepare them with work skills. All children would be taught to read, write in English, do arithmetic, drawing, music (for those so inclined),and a skill. Petty, consistently using direct style, lists the goals of his education proposal. See Figure 7-3 [6, 13].

Petty then argues for a College of Trades-Men, where the history of trades would emerge from experiments executed by students and teachers. The College would provide a rich environment for children, including numerous "hand on" experiences that would prepare students to later "understand all good Bookes" and "smell out the sopperies of bad ones." Unfortunately, Petty provides a linear list of the benefits of his proposed curriculum. But the detail of his plan shows his careful thought and his perceived outcomes for this College:

> Within the walls of this Gymnasium or College, should be a Nosecomium Academicum according to the most exact and perfect Idea thereof a complete Theatrum Botanicum, stalls and Cages for All strange Beastes and Birds, with Ponds and Conservatories for all Exotic Fishes, here all Aimalls capable thereof shou'd be made fit For some kind of labour and imployment, that they may as well be of Use living as dead; here should be a Repositorie of all kinds of Rarities Naturall and Artificiall pieces of Antiquity, Modells of all great and noble Engines, with Designes and Platformes of Gardens and Buildings. The most Artificiall Fountaines and Water-works, a Library of Select Bookes, an Astronomicall Observatory for celestiall Bodies and Meteor, large pieces of Ground for severall Experiments of Agriculture, Galleries of the rarest Paintings and Statues, with the fairest Globes, and Geographicall Maps of the Best descriptions, and so farre as is possible, we would have this place to be The Epitome or Abstract of the whole world. So that a man conversant Within those walls, would certainly prove a greater Schollar, then the Walking Libraries so called, although he could neither write nor read. But if a child, before he learned to read or write, were made acquainted with All Things, and Actions (as he might be in this College) how easily would he Understand all good Bookes afterwards, and smell out the sopperies of bad Ones. As for the Situation, MOdell, Policy Oeconomy, with the number of Officers and Rataines to this Colledge, and athe Priviledges therof, it is as Yet time enough to delineate. Only we with that a Society of Men might be Instituted as carefull to advance Arts as the Jesuites are to Propagate their Religion for the government and managing of it. [13, p. 8]

(6)

Making Mathematicall Inſtruments, Dialls, and how to uſe them in Aſtronomicall Obſervations.
Making Watches and other Trochilick motions.
Limning and Painting on Glaſſe or in Oyle Colours.
Graving, Etching, Carving, Emboſſing and Molding in ſundry matters.
The Lapidaries Art of knowing, cutting and ſetting Iewells.
Grinding of Glaſſes Dioptricall and Catoptricall.
Botanicks and Gardening.
Making Muſicall Inſtruments.
Navarchy and making Modells for buildings and rigging of Ships.
Architecture and making Modells for houſes.
The Confectioners, Perfumers or Diers Arts.
Chymiſtry, refining Metalls and Counterfeiting Iewells.
Anatomy making Sceletons and excarnating bowells.
Making Mariners Compaſſes, Globes, and other Magnetick Devices.

And all, for theſe Reaſons.

1. They ſhall be leſſe ſubject to be couſened by Artificers.

2. They will become more induſtrious in generall.

3. They will certainly bring to paſſe moſt excellent Works, being as Gentlemen, ambitious to excell ordinarie Work-men.

4. They being able to make Experiments themſelves, may doe it with leſſe charge, and more care then others will doe it for them.

5. The *Reſp. Artium* will be much advanced, when ſuch as are rich and able, are alſo willing to make Luciferous Experiments.

6. It may engage them to be Mecænates and Patrons of Arts.

7. It will keepe them from worſe occaſions of ſpending their time and eſtates.

8. As it will be a great Ornament in proſperity, ſo it wil be a great Refuge and ſtay in adverſity and common calamity.

As for what remaines of Education, we cannot but hope

Figure 7-3. Partial list of goals and rationale for Petty's proposed college.

Petty next explains the staff for this college, along with the salary of each person, the required qualifications for each person, their duties. See Figure 7-4.

Those learning and working at the college would contribute to writing the history of trades by assisting those hired. For example, Petty describes how the physician and his students would work together to write the history of medicine. This collaborative project would produce a much-needed work and benefit the student, who would thus learn medicine:

> The Physician must be a Philosopher, skill'd at large in the Phenomena of Nature, must understand the Greek Tongue, be well read in good Authors, And seene in the practice of all the ministrant parts of Physick, willing to Instruct the foreward all that are under him, his work shall be twice every Day deliberately to visite and Examine all the Sick, and after due Copnsideration Of their Condition, to prescribe them convenient Medicines, and shall dictate In Latine to the Vice-Physician attending him, the History of their severall Diseases (excluding impertinencics) he shall see all Patients in outward Griefes (to whom he administreth any inward remedies) opened and Dressed ever now and then, to the end that himself and the Chirurgeon May both have the same intention and scope in their practice. He must take Care that the Chirurgeon and Student keep the History of their Cures likewise, and that the apothecary and Student doe the same in their Pharmacy and Botanicks. Giving the Apothecary directions for the making of new Enquiries and Experiments in his way; . . .
>
> In briefe, he shall have an influence upon all the rest, and all the rest reciprocally pon him, so that he being made acquainted with all the Hstories taken in the Hospitall, Laboratory, Anatomical Chamber,

To the Steward	80	per annum.
To the Phyſician	120	
To the Vice-Phyſician	50	
To the Chirurgeon and Apothec. each	60	
To the Student	25	
To the Chirurg. and Apotheck..mate	20	
To each of the Apprentices	10	
To each ordinary Nurſe	4	
To an Extraordinary by the week	3. ſhil.	

Figure 7-4. Proposed salaries of those working in the college [13, p. 11].

> garden, &c. may give the reason of the most Notable Phaenomena hapning in either of them. All of which he shall commit to writing and Our of them, by the end of the terme of his service, shall collect a Systeme of Physick and The most approved Medicinall Aphorismes; taking notice by the way where those of Hyppocrates are deficient or true, and by how many severall experiments he hath so Found them. He shall either dissect or overlook the dissection of bodies dying of Diseases, and lastliy shall take care that all luciferous experiment whatsoever, may be carefully brought To him, and recorded for the benefit of Posterity. [13, pp. 12-13]

Petty provides a similar description of the jobs of the others employed and the requirements of the students who work with them. Petty concludes his proposal by explaining, in a numbered list, the value of his plan to England and to education. For example he reiterates a point upon which he had Bacon agreed, an idea that underpinted both his and Bacon's issue with the current education. Petty would have none of this approach:

> 14. Boyes in stead of reading hard Herbrew words in the Bible (where They either trample on, or play with Mysteries) or parrat like repeating Heteroclitous nouns, and verbs, might read and hear the History of Faculties expounded, so that before they be bound Apprentices to any Trade, they may foreknow the good and bad of it, what will and Strength they have to do it, and not spend seven years in repenting, And swimming against the stream of their Inclinations. [14, p. 23]

Petty wanted students to learn the Theory of their Trades before they were bound to a master. He also wanted them to be exempt from seven years "of bondage." He sees his plan as producing meaningful knowledge and understanding of tasks involved in a trade. Petty does not explain how this college will be supported, only that an existing school would be used and its curriculum transformed [13].

This 1648 proposal prepared for Hartlib anticipates modern proposals with a letter of transmittal, a statement of purpose, a rationale, goals of the project, a rationale for each goal, and some additional description of the activities of the college. Petty apparently believed that he would continue his discussion of this proposed college with Hartlib at some future date.

Samuel Hartlib: Proposals for Advancement of Husbandry Learning

Samuel Harlib, like John Evelyn and other Royal Society members, wrote proposals that mirrored the plans for the history of trades and the improvement of England. However, unlike proposals by Petty and proposals by Bellers, Cellier, and Haines, whom I will discuss next, Hartlib's proposals exemplify intellectual propositions rather than proposals to initiate action immediately. For example, *An Essay for Advancement of Husbandry Learning* (1651) [14], which proposes a College of Husbandry, could today be called a "concept paper" rather than a proposal. Hartlib begins with 17 skills of husbandry, followed by a proposition to erect a College of Husbandry. The remainder of his proposition deals with how to support the college (cost to students) and the requirements of students.

Mainly, however, Hartlib emphasizes the need for a college that emphasized the teaching of husbandry [14]. His concept of what a college should emphasize is more specific than his concept for establishing the college: "to that most Ancient, and most Noble, and most necessary Trade of all others {viz} good Husbandry, consisting of abundance of Parts, of which these are some. Note his effective use of typography in Figure 7-5.

John Dury's Proposal for Affirming the Protestantism of Chelsea College

One of the Baconians, John Dury, prepared a proposal, which included a supporting letter written by Hartlib, about the need to strengthen the Protestantism of Chelsea College and to prevent any incursions of anti-Christian influences. Dury concludes by advocating written connections with foreign Protestant correspondents, entitled "The Reformed Spirituall Husbandman: With an Humble Memorandum concerning Chelsy College and a Correspondencie with Forreigne Protestants," which includes 10 specific proposals for strengthening the college in its Protestant teachings [15].

John Bellers: Proposals for a Colledge of Industry

Many proposals written by members of the Royal Society exemplify intellectual proposals prepared for discussion of projects the Baconians sought to initiate, except for Petty, who wanted changes quickly. In contrast, other proposals appeared that sought similar goals but developed from a different perspective. One major example of this kind of proposal was written by John Bellers, a Quaker, whose writings passionately advocated his belief that faith cannot be formulated in creeds but must find expression in caring for people [16, p. 3].

Bellers' proposal, the first of his works on the importance of caring for the poor, reflected the problems of England as it began to emerge as a modern industrialized society. By the end of the 17th century, England was becoming a formidable industrial power. Pragmatism controlled national thinking; the State had become far more concerned with creation of wealth than with the welfare of its ordinary citizens; poverty; dispossessed tenant farmers who had no place to go but London; deserted children abandoned to the streets by destitute parents or the death of mothers; resulting crime; homelessness; and disease, all amid England's growing industrial power, drove reformers like Bellers to

1. *Tillage*, or ſetting or ſowing of ſeveral ſorts of Corne and Graine, for the reliefe and ſuſtenance of man and beaſt.

2. The breeding of Cattel,(in which the breeding of *Sheepe* may ſeem particular.)

3. The *feeding* of Cattel.

4. The Uſe of the Dairie.

5. The planting of *Orchards*.

7. The planting of *Gardens*.

o, The breeding and feeding of *Swine*.

. The breeding and feeding of the ſeveral ſorts of *tame Poultry*.

9. The planting of *Hops*.

10. The ſowing of *Hempe*, *Flax*, or *Rape*.

11. The breeding, preſerving and taking of *wilde Beaſts*, as *Conies* &c.

12. The breeding, preſerving or taking of *wilde fowle*, particularly of *Duckes* in and by *a Decoy*.

13. The making and managing of *Rivers*, *Moats*, *Ponds*, &c. for the preſerving and taking *Fiſh* of all ſorts for the uſe and ſuſtenance of man.

14. The planting of *Woad*, and all outlandiſh rare or extraordinary *Roots*, fruits or *plants*.

15. The dreining, fencing, mowing, and making of Graſſe in Meadowes into Hey.

16. The making of *Malt*.

17. And (that now ſo exceeding neceſſary endeavour) the planting all ſorts of *Wood* for Timber or fire.

Beſides

Figure 7-5. Goals of Hartlib's curriculum in animal husbandry [14, p. 2].

propose solutions to these problems that would benefit both the rich and the poor [16, p. 9].

Bellers' *Proposals for Raising a Colledge of Industry* (1696) [17] stands as perhaps the most trenchantly argued proposal of the 17th century. He printed numerous copies for distribution to Parliament and the public. Thus, he wasted no time and opened his proposal with a four-point rationale for his College of Industry. Following this opening segment, a transmittal document, he lists the contents:

1. To the Lords and commons Assembled in Parliament
2. To the Thinking and Publick Spirited
3. The Introduction, with Reasons for providing for the Poor
4. A Specimen, shewing the way of doing it in a College Fellowship.
5. Proposals to the Colledge-Founders
6. Some of the Advantages to the Founders and Rich by it
7. Some of the Advantages to the Poor Collegians
8. Some Rules about Governing the Colledge-Workmen
9. Of the Education of Children
10. An Answer to several objections.
11. A Postscript. [17, p. 2]

Bellers' introduction exemplifies the qualities of a modern proposal introduction: He opens with his argument—that taking care of the poor will benefit the rich. While his introduction is lengthy, he presents his purpose statement at the end of the paragraph 3.

> IT's the Interest of the Rich to take Care of the Poor, and their Education, by which they will take Care of their own Heirs: For As Kingdoms and Nations are subject to Revolutions and Changes Much more (and nothing commoner than) for private Families to Do so; and who knows how soon it may be his own Lot, or his Posterities, to fall poor? Is there any Poor now, that some of their Ancestors have not been Rich? Or any Rich now, that some of Their Ancestors have not been Poor?
>
> View the Cities, Towns, and Counties in this Nation, and See what alterations come in two or three Generations in most Families. Were above one in ten of the Men now House-keepers in London, born There? And but few (in comparison of the Multitude) that have gone out With Estates: And what better is it with Gentlemens younger Children, And the Eldest also, many times.
>
> There is Three Things I aim at: First, Profit for the Rich, (which Will bew Life to the rest.) Secondly, A Plentiful Living for the Poor, without Difficulty. Thirdly A good Education for Youth, that may tend to prepare Their Souls into the Nature of the good Ground. [17, p. 1]

Section 4, "A Specimen shewing how the Rich may gain, the Poor maintain themselves, and Children be Educated, by being incorporated as a Colledge of all Sorts of Useful Trades, that shall work one for another, without other Relief. Suppose Three Hundred in a Colledge, to work the Usual Time or Task as Abroad, and what any doth more to be paid for it, to encourage Industry" lists in two columns the employees needed. See Figure 7-6. His listing arrangement appears designed to help readers fully see the requirements:

After his three-page list, Bellers explains the value of his college to teach industry. He artfully shows throughout his proposal how both rich and poor will benefit. As he presents the main body of the proposal, he numbers and lists each point. Bellers seems to be aware that unless Parliament *reads* his proposal, it will have no chance of consideration. Thus, he begins by showing how the rich may gain, the readers in Parliament, then explains how the poor will gain [17]:

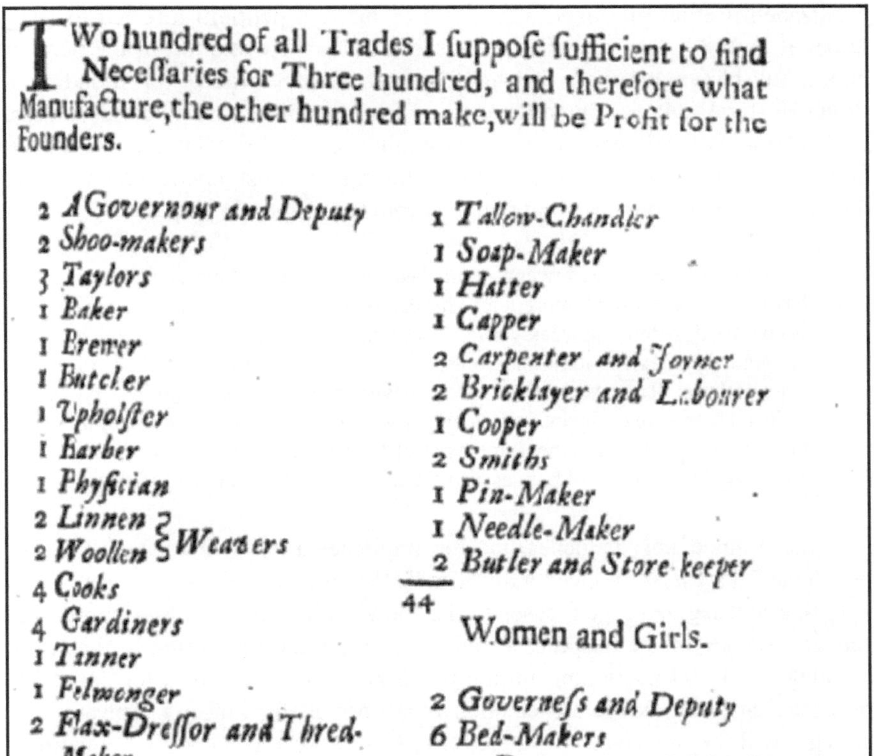

TWo hundred of all Trades I ſuppoſe ſufficient to find Neceſſaries for Three hundred, and therefore what Manufacture, the other hundred make, will be Profit for the Founders.

2 *A Governour and Deputy*
2 *Shoo-makers*
3 *Taylors*
1 *Baker*
1 *Brewer*
1 *Butcher*
1 *Upholſter*
1 *Barber*
1 *Phyſician*
2 *Linnen* } *Weavers*
2 *Woollen* }
4 *Cooks*
4 *Gardiners*
1 *Tanner*
1 *Felmonger*
2 *Flax-Dreſſor and Thred-Maker*

1 *Tallow-Chandler*
1 *Soap-Maker*
1 *Hatter*
1 *Capper*
2 *Carpenter and Joyner*
2 *Bricklayer and Labourer*
1 *Cooper*
2 *Smiths*
1 *Pin-Maker*
1 *Needle-Maker*
2 *Butler and Store keeper*
44

Women and Girls.

2 *Governeſs and Deputy*
6 *Bed-Makers*

B 6 *Nurſes*

Figure 7-6. The teaching staff of Beller's college to prepare children for trades [17, p. 5].

Section 5, "Proposals to the colledge-Founders," itemizes 10 ways the College would be funded.

Sections 6 and 7 list the advantages to the founders and rich and then to the poor students. In each of these sections, he enumerates his arguments.

Section 8 numbers and lists the rules for governance.

In Section 9, he lists and discusses what he believes should be included in the education of these students.

Section 10 again states possible objections and numbered replies to each one. Bellers shows his awareness that for his proposal to have any consideration, he must anticipate his readers' objections. The postscript, a conclusion, again summarizes his proposal.

Elizabeth Cellier—Proposal for A College for Midwives and Foundlings

Elizabeth Cellier presented a proposal (actually a petition) directed to King James II in 1687. This proposal/petition illustrates the evolution of the petition into a full-blown proposal format. Elizabeth Cellier argued for founding a Royal Hospital "and Raising a Revenue of Five or Six-thousand Pounds a Year, by, and for the Maintenance of a Corporation of skilful Midwives, and such Foundlings, or exposed Children" [18]. Cellier, an activist, opens with a strong, direct statement of evidence that will underpin her proposal:

> That, within the Space of twenty Years last past, above Six thousand Women have died in Child-bed, more than Thirteen thousand *chrysome* infants have been buried, within the weekly bills of mortality; above two thirds of which, amounting to sixteen-thousand souls, have in all probability perished, for want of due skill and care, in those women who practice the art of midwifery.
>
> Besides the great number which are overlaid, and willfully murdered, by their wicked and cruel others, for want of fit ways to conceal their shame, and provide for their children, as also the many executions on the offenders.

Cellier immediately launches into her implementation plan for this hospital, which has 15 steps, introduced with "That" [18]. For example, see Figure 7-7.

In this next segment, she focuses on the "how" of her proposal: She emphasizes self-governance of the hospital, a midwives corporation reporting to the King; a college of 1,000 maximum; tuition of £5 per year, to be used for charitable purposes. She also advocates erection of a hospital for housing orphans. The college will be managed by a governess, one secretary, and 12 matrons. Following this implementation plan, she begins a section of rules governing the hospital. She emphasizes internal succession planning to allow promotion from within: Female children shall remain under the direction of the governess until age 21 or marriage that must be approved by the governess [18]. See Figure 7-8.

To remedy which, it is humbly proposed, that your Majesty will be graciously pleased, by your royal authority, to unite the whole number of skilful midwives, now practising within the limits of the weekly bills of mortality, into a corporation, under the government of a certain number of the most able and matron-like women among them, subject to the visitation of such person or persons, as your Majesty shall appoint; and such rules for their good government, instruction, direction, and administration, as are hereunto annexed, or may, upon more mature consideration, by thought fit to be annexed.

That such number, so to be admitted, shall not exceed a thousand at one time; that every woman, so to be admitted as a skilful midwife, may be obliged to pay, for her admittance, the sum of five pounds, and the like sum annually, by quarterly payments; for and towards the pious and charitable uses hereafter mentioned.

That all women, so admitted into the thousand, shall be capable of being chosen matrons, or assistants to the government.

That such midwives as are found capable of the employment, and cannot be admitted into the first thousand, shall be of the second thousand; paying, for their admittance, the sum of fifty shillings, and fifty shillings a year by quarterly payments, towards the pious and charitable uses hereafter mentioned, and out of these the first thousand are to be supplied, as they die out.

That, out of the first sum arising from the admittance-money, one good, large, and convenient house, or hospital, may be erected, for the receiving and taking in of exposed children, to be subject to the care, conduct, and management of one governess, one female secretary, and twelve matron-assistants; subject to the visitation of such persons, as to your Majesty's wisdom shall be thought necessary.

Figure 7-7. Portion of Cellier's proposal for a midwives college.

The final section deals with recommendations for financing the hospital: one treasurer, with all money kept in iron chests with a locking system and audits monitored and conducted by the registrar. See Figure 7-9.

Cellier's proposal begins with the topic (title), followed by a brief rationale, then a concise description of the "colledge," how it will be achieved, and how the plan will be funded. Her proposal fully anticipates modern proposals: what, why, how, and cost [18].

Printed proposals during the 17th century provide us some indication of issues of importance to those who sought change. Among those wanting change were members of the Oxford group that spurred the founding of the Royal Society as well as people like Bellers who, like many members of the Royal Society, saw education as a means of dealing with crime and poverty. Other proposal

Rules for Governing the Hospital of found Children.

That the governess be appointed by his Majesty, as likewise her secretary, and twelve assistants, who are to name twenty-four to be of the government.

That, upon the death of the governess, her place be supplied by her secretary, or such person as shall be chosen by the twelve principal assistants, or the major part of them, and the approbation of his Majesty; that the secretary be chosen by the governess, and approved of by his Majesty, his heirs and successors.

That, upon the vacancy of one of the twelve principal assistants, by death or otherwise, one of the four-and-twenty shall succeed, by election of the governess, secretary, and the other eleven; as also, the number of four-and-twenty shall be supplied, by election of the governess, female secretary, and twelve principal assistants, or the major part of them; and, in all cases, the governess to have three, and the secretary, two voices.

That all rules for governing the children, under five years of age, shall be made by the governess, her secretary, and their assistants; that the government of the whole, under such rules, be in the governess.

That all female children shall continue under the sole government and direction of the governess, until they attain the full age of twenty-one years, or are married by her consent.

Figure 7-8. Governance proposal for Cellier's midwives college.

documents include Edward Chamberlayne's *Academy or colledge: wherein young ladies and gentlewomen may . . . be duly instructed in the Protestant Religion, and in all Vertuous Qualities that may adorn that Sex.* (1671) [19] and over a dozen proposals by Richard Haines, whom I will consider separately, because he exemplifies an understanding of the rhetoric of proposal writing even more so than John Bellers' proposals targeting Parliament show.

Richard Haines—Three Proposals for Work Houses for the Poor

Richard Haines' first proposal, *England's Weal & Prosperity Proposed: Or, Reasons for Erecting Publick Work Houses in Every County* (1681) [20] and Haines' second proposal on a similar topic, *Proposals for Building in Every County a Working Alms House or Hospital: As the Best Expedient to perfect the Trade and Manufactory of Linnen-Cloth* (1677) [21], show his cleverness in adapting his proposal to attempt to convince his readers. Haines' proposals, like Bellers' proposals, recognized that schools that offered education in skills to the poor and dispossed in England's growing industrial economy had to offer something to those who would have to pay for these schools. Haines produced at

That all the money coming to the said hospital, either by annual payments, charity, redemption, or any other ways whatsoever, shall be placed into one common treasury, to be kept in one, or more iron chests; not to be opened, but by the consent of the governess, her secretary, the chief chaplain, or him that shall be governor of the male children, the register, and treasurer; who shall each of them have a key to so many several locks; and the said monies, other than the constant salaries of the officers, and daily maintenance of the children, shall not be applied to any extraordinary use, but such as shall be appointed by the whole government of the hospital; in which number the keepers of those keys, for such purposes, are to be accounted part.

The accounts whereof, and of all monies coming into, or going out from the same, shall be kept by the register; and free access shall be had, at all times, to the same, *gratis*, by the governors, or any of the visitors of the said hospital; and that, once a month, all comings in, and goings out, and all other transactions on that account, shall be, by the register, fairly entered into a book for that purpose, which shall always remain with the governess, and not be taken out upon any pretence whatsoever; and that any person may search the register's book, for the fee of sixpence for one year's search.

Figure 7-9. Proposed rules for financial control in the midwives college.

least a dozen different proposals presenting differing perspectives on his plan for building workhouses and schools for the poor.

Haines adapts different rhetorical strategies as he attempts to have his ideas considered. For example, in a third proposal, *Proposals for Promoting the Woollen-Manufactory* (1679) [22], Haines argues that English wool manufacturing should be preserved and strengthened. Achieving that goal would help the poor and provide new revenue to England. His argument is rigorous and passionate. He wastes no time in justifying his idea, as shown in the excerpt below:

> If Wool be Exported, the same will destroy our Markets abroad, and make Peioke/beggars at home, saith the Clkothier. But if it be kept home, and not Converted, it will do us no good, saith Reason, with whom Experience and Poverty both joyn their Voices; the out-cries of the Latter being heard, not only in every Street, but almost at every Door. The last Parliament hath almost every Session made it a great part of their Business to Hear and Encourage those who had any thing to offer for the Recovery of this Trade; and yet, till the late Bereviat of Proposals published by one Mr. Richard Haines came out, it must be Acknowledged, That neither the Exporter of Wool,

> nor Opposer of the Same, or any others, have offered any certain Expedients for bringing all Idle hands to Industry, whereby the Wool may be Converted as fast As it grows and arises, on such terms, that the Cloth we have to spare May be Exported as fast as 'tis made.
>
> Wherefore, we cannot think it unseasonable, to joyn out Suffrages With him, and stir up all Active Publick Spirits, to promote and encouorage That which will best accomplish this good design: namely
>
> *Work-houses to be Erected, and Stock raised in every County, in Which, poor People, Beggars, &c. shall be Employed in this Manufactory, With such Method of Government as may prevent all such Miscarriages, As have formerly rendred such Houses unsuccessful in our Nations; which He hath, in such his Breviat, declared to make good, and to Answer all Objections in that Case, in his Proposals at large, which undoubtedly may Be done with little Difficulty, because 'tis done and practiced in other Nations, and maiy here as well as there.* [22, p. 2]

Haines, in the final, italicized paragraph, admits that workhouses have had lack of success in every country because they have not been managed properly, a problem he believes his proposal will correct.

In short, Haines' goal in his third proposal is the same as his first two proposals: to present practical ways to help the poor by appealing to those in Parliament who would have to see the wisdom of his ideas. However, in using the woolen industry as a point of departure, he makes his case from a financial perspective.

> In England we may reasonabloy conclude there are of Beggers Petty Felons, Parish-Poor, and oather idle People, which do little or nothing for their Living, 200 Thousand; wherof many who do nothing may earn six pence Per day, and those that earn six Pence might earn twelve pence, when placed together in view of each other, where Correction And Encouragement are duly administered.
>
> It may also be reasonably concluded, that each Person, one with Another, may (with the help of some Clothiers) work as much Wool as will make fifteen pounds worth of Cloth in one Year, which If so, then there will be thirty hundred thousand pounds worth to be Exported more than now is, as the Author of the Proposals does insert. Bkut suppose the quantity fall short, yet we may reasonably conclude, That our Merchants in their course of Trade will near the matter double The Value in the returns of what they Export, so that still the Wealth of the Nation will be more, rather than less, increased, than what is proposed. But if we take no other measures than this, viz. that each Person, one With another, earns six pence per day, and that which is wrought by their Labour shall be exported, 'tis a certain truth, the Nation will gain by their Labour 5000 l. for every day in the year that they work, as by Calculation Doth appear; which, if

> doubled by the Merchants course of Trade, as Aforesaid, it will then amount to no less than 10000 l. per day, clear gain To the Nation. And put case we allow 80 days in the year for Holy days, there Remains 285 days to work in, which, by this modest Calculation, will bring in 2850000 l. in one year clear gain by their Labour, besides the Value of the Wool, and other things which add to the Value of all the Cloth they make. Nor is it to be doubted, but the Work of each Person, one with another, may Amount to 8 d. per day, after some time of practice, so that for the future, By the xpedients proposed, all people, as aforesaid, being brought up to Such habits of Industry, the Wealth of the Nation cannot but increase to the Value proposed. The Case is plain, and an ordinary Capacity, whose Judgment Is not sway'd by corrupt Affection, may be well satisfied. [22, p. 3]

Basically, his proposal provides sixpence/day for the parish poor. Then, if each earns sixpence per day, and their product is exported, England will gain £5,000 for every day they work. Haines seems to assume that the parish poor want to work [22].

His Proposals for establishing a workhouse to manufacture "Linnen-Cloth" [21] takes a similar rhetorical stance. Note, however, that he opens his proposal with an itemized list of reasons for the value of his proposal. The design suggests that he wants to ensure that his readers at least see this page, which contains the summary of the goals. See Figure 7-10.

His introduction ends with a strong rationale, which leads to his proposal statement printed in italics.

> Considering the great Complaints of Poverty, the heavy Burdens most Parishes Lie under to maintain their Poor, which daily increase; the Swarms of Beggars, Vagrants and Idle People in City and Countrey: the great, and 'tis fear'd, Irrecoverable decay of our Ancient Trade for Woollen Cloth; the vast Charge We are yearly at in purchasing Linnen, &c. from other Nations, whereby our Treasure is exhausted, and our Lands fall for want of being improved some Other way, besides planting Corn, breeding for Wool, &c. Which are become of so low a price, as scarce to turn to Account: And understanding, that for remedying thereof, the Improving the Manufactory of Linnen is now under Debate, I have taken the boldness to offer the following P R O P O S A L, Which if thought fit to be put in practice, will (in my opinion) infallibly conduce To all the good Ends desired and intended; vix.
>
> *That there may be Erected in every ounty according to its Extent or Populousnesse,*
> *A greater or lesser Working-AlmsHouse, wherein the Poor may be Continually employed in the Manufactory of Linnen Cloth.* [21, p. 2]

PROPOSALS for Building in every County a WORKING-ALMS-HOUSE *or* HOSPITAL, *as the beſt Expedient to perfect the Trade and Manufactory of Linnen Cloth.*

WHEREBY,

I. All Poor people and their Children from five or ſix years old may be employed and maintained; as alſo all Beggars, Vagrants, *&c.* reſtrained and for ever prevented, and ſo all Pariſhes eaſed of that intolerable Burden.

II. Many Hundred Thouſand Pounds kept at home, which now every year goes out of the Kingdom for *Linnen*, whereby our Wealth becomes a prey to other Nations.

III. Much Land improved in every County to great advantage of Landlord and Tenant.

Humbly Offered to the Conſideration of the great Wiſdom of the whole Nation, now Aſſembled in Parliament.

Figure 7-10. Opening visually-effective summary arguing for the value of linen manufacturing [22, p. 1].

The next section lists the advantages of his proposal.

> The Advantages wherof are evident; For,
>
> I. This Manufactory is an Employment for the weakest people Not capable of stronger Work, viz. Women and Children, and Decrepit or aged people, now that most chargeable; as llikewise For Beggars and Vagrants, who live idly, and by the sweat of other mens Labours, and can no way so effectually be brought to Industry and Order, as when reduced into so narrow a Compass or Confinement under fitly qualified Rulers, Officers, and Regular Government.

> II. These Working-Alms-Houses mayi raise and supply the Nation Yearly with a sufficient Stock of Linnen Cloth (the finest sort Excepted) if true measures be taken, and athe Design effectually Prosecuted; As for example, 1353000 pounds worth of Cloth May yearly be spun in them only, besides what is made in Private Families——Thus demonstrated. [21, p. 2]

He supports Advantage II with three statements justifying his proposed financial benefits, as shown below He uses listing to highlight his argument.

> 1. 'Tis well known by experience, that three quarters of a Pound of Threat worth 12. D. per Pund spinning, will make One Ell of Cloth worth 2s s. per Ell; which Thee quarters of A Pound two Spinners may spin in one day; Hence it follows,
> 2. That 2000 Spinners will spin Thread enough in one day to Make 1000 Ells of Cloth, worth 100 l. And working but 260 days in the year, may spin 26000 Pounds worth of Linnen Cloth in a year.
> 3. Suppose then there be as many public Work-Houses, as There are Counties, which are 52, and in every Work-house, One with another, 2000 Spinners (though in some more, some Less) then according to these reasonable measures, there will Be forementioned Sum of 1352000 l. worth of Cloth spun in One year; which is what we undertook to demonstrate. [21, p. 3]

Next, he lists eight objections he anticipates from members of Parliament. He posits each objection and then follows each one with his rebuttal. His presentation exemplifies his skill in argument and anticipating his readers' objections. The proposal is comprised of proposals and his answer to what he sees as Parliaments' objections. See Figure 7-11, for example.

His final section deals with how his proposal will be managed. He includes a section that argues for the value of hemp, then [22, pp. 9–12] adds a conclusion, which he calls a Postscript, which provides the financial analysis to support his plan.

Haines' proposal for workhouses (1681) [21] opens with a letter of transmittal to the Lord Mayor of London. Following this short letter, he immediately states his proposal, which he follows with reasons, announced with centered italicized subheads. He seems to want to attract his readers' attention with use of layout and typography. See Figure 7-12.

He offers 14 reasons to support his proposal [20, pp. 4–15], each one following the above format. The 14th includes possible objections and answers. Then at the end, he concludes by lapsing into the petition format—the addressee (the House), the (re)statement of the proposal, followed by the request printed in italics: he wants to bring a bill for their consideration. So

Obj. 2. But some will be ready to object, and tell us, *That we talk of brave things, if words would do the Work; but where's the Money for the Building of such great Hospitals? and providing all Tools and Materials will cost many Thousand Pounds.*

I confess here's the Knot, which seems knit by Magick Art; but if it can be untied without cutting or breaking the Thread, then I hope our Proposals will not be rejected.

Answ. To this therefore I humbly Answer, That it may be done by a County-Charge, with as much Satisfaction, Pleasure and Advantage, as to part with 5 *l.* to prevent paying 50 *s. per An.* which I think no wise people will judge to be an hard Bargain; especially if they consider the other vast profit to the Nation, and that thereby they purchase (in the Country) 50 *s. per Annum* more by improvement of their Lands for Hemp or Flax.——As thus;

Figure 7-11. Haines presents then answers the perceived objections [21, p. 5].

May it please this Honourable House,

To onsider, That, as National Sins and Provocations incur The Displeasure of Heaven, so likewise to reflect how much this Present Parliament may concillate to themselves and this poor Nation, the Favour and Blessing of the Almighty; and also how Gloriously the Renown shall to all Posterity be Recorded; when by One Act more than an Hundred Thousand shall be converted, as It were at once, in your day, to their own, and the whole nations Present and future happiness: For which, not only the Poor of this Generation, but their Posterity (yea, every individual in the Kingdom) In all Ages may call you Blessed.

Wherefore if this Honourable House do believe, that what I have With all Humility and Zeal to the Public, offered to your Grave and Pious Consideration, be well pleasing to God, and accordingly be approved of by this Honourable House (as it is by the King's Most Excellent Majesty, who hath been graciously pleas'd to declare, that he would be Ready in his station, to encourage it all he could) then my humble Petition to this Honourable House, is,

England's *Weal and Proſperity Propoſed: Or,* Reaſons *for the Erecting* Publick Work-Houſes *in every County,* &c.

FOraſmuch as our Wool is the main Support of that Trade which maintains and encreaſes the Wealth, Strength and Glory of the *Engliſh* Nation; therefore, of all Commodities of our own Growth, this ought to be moſt carefully maintained and encouraged in that way it is *THUS SERVICEABLE*, which conſiſts not in the Exportation of unwrought Wooll, nor in the Conſumption of our Woollen-Draperies at Home; but in the Exportations of, and quick Markets for our Woollen-Draperies abroad, as faſt as they can be made and ſpared at home; That this may be done, and that for the doing thereof, *Publick Work-Houſes* to be Erected in every County, will be a moſt certain and effectual Expedient. Theſe following Reaſons are prepared for the Satisfaction of all concern'd therein.

Reaſon I.

Becauſe ſuch *Work-Houſes* (are no New Project, but) have already in Fact proved the beſt Expedients to bring all Idle, poor People, Beggars, Vagrants, *&c.* into ſuch a Habit of Induſtry, that there is not a Beggar, *&c.* bred up, or ſuffered in thoſe Countries where ſuch Houſes are Erected and well-governed.

Reaſon II.

Becauſe, without ſuch *Houſes*, in which ſuch poor, and now idle People may be kept to Labour under good Government, it is altogether improbable to convert them to ſuch Habit of Induſtry, and keep them employed in ſuch profitable Manufactures, by which the Trade, Wealth and Safety of the Kingdom may be promoted.

Figure 7-12. Initial proposal for workhouses followed by reasons.

That I may obtain leave to bring in a Bill, which I have prepared, Containing such Clauses, Rules and Methods as may (with Such Corrections and Amendments, as this Honourable House shall think meet) in short time compass the Design in such manner, that all Cheats and Fraunds may be so far prevented, that the Contributers cannot be injured unless they themselves will willfully be regardless of their own Interests, together with such Rules and Methods for prosecuting The Design, as may for ever secure it from failing or ceasing After it is begun. [20, pp. 15-16]

CONCLUSION

This discussion and history of the first proposals—manuscript and then printed—provides technical communicators a sense of the rich history of a major genre we teach and use. In addition, the history of proposals in England owes much to the Baconians and the founders of the Royal Society. The first printed proposals in England, many of which were products of Bacon's History of Trades and his followers' commitment to improved education for the middle class, illustrate the emergence of fully developed proposals. These proposals were based on persuasive arguments and recognition that proposals had to provide logical reasons that would appeal to readers. In short, we see proposals that exhibit the need for rhetoric.

Some proposals were working papers, white papers, or concept papers (to use modern terminology), but others illustrate the major sections of today's proposals: what is proposed, why, how the proposal will be implemented, and in some cases methods to finance the proposal, counterarguments to perceived objections, and explanations of the profit anticipated.

Authors printed copies to disseminate to Parliament and colleagues who shared concern with the author about middle-class education. Printing became the method of spreading problems and possible solutions to problems. And, as Haines' proposals illustrate, some writers wrote with the perspectives of their readers directing the presentation of content.

Obsequious language, by the last half of the 17th century, had given way to language that was clear and direct, a result of many of these writers' connection with the Royal Society. Thus, the appearance of reports during the 17th century (an allied genre, as I suggested at the beginning of this chapter) also illuminated organization and style needed to convey information, recommendations, and supporting analysis with clarity and concision. Chapter 8 will illustrate the commonalities of proposals and reports as they developed from 1640 to 1700.

REFERENCES

1. W. M. Ormrod, Agenda for Legislation, 1322-1340, *English Historical Review, 105,* pp. 1-33, 1990.

2. G. Dodd, *Justice and Grace, Private Petitioning and the English Parliament in the Late Middle Ages,* Oxford University Press, Oxford, United Kingdom, 2007.
3. J. H. Fisher, M Richardson, and J. L. Fisher, *An Anthology of Chancery English*, University of Tennessee Press, Knoxville, Tennessee, 1984.
4. A. R. Myers, Parliamentary Petitions in the Fifteenth Century: Part 1: Petitions from Individuals or Groups, *The English Historical Review, 52*:207, pp. 385-404, 1937.
5. A. F. Pollard, Receivers of Petitions and Clerks of Parliament, *The English Historical Review, 57*:26, pp. 202-226, 1942.
6. J. F. Baldwin, *The King's Council in England During the Middle Ages*, Oxford University Press, Oxford, United Kingdom, 1913.
7. Chris Given-Wilson (ed)., *The Parliament Rolls of Medieval England 1275-1504, Richard II, 1385-1397 (Vol. VII), The National Archives,* The Boydell Press, London, 2005.
8. *To the most honourable assembly of the Commons House of Parliament, the humble petition of the wardens and comminalty of the art or mysterie of Dyers in London* [London : s.n., 1621].
9. *The Humble Petition and Advice of Both Houses of Parliament, with XIX Propositions and the Conclusion Sent unto His Majestie the second of June, 1642*, I. Wright, London.
10. *To the right honourable the knights, citizens and burgesses in Parliament assembled. The humble proposal of Lewis Rawlins, whether the restraining of retailers of beer and ale within the kingdome of England and Dominion of Wales from brewing, may not only prove profitable to the crown, but be expedient; upon due consideration of subsequent reasons* [London: s.n., 1670?]
11. W. E. Houghton, The History of Trades: Its Relation to Seventeenth-Century Thought, *Journal of the History of Ideas, 2*:1, pp. 33-60, 1941.
12. A. Cowley, *A Proposition for the Advancement of Experimental Philosophy*, London, 1661.
13. W. Petty, *The Advice of W. P. to Mr. Samuel Hartlib for The Advancement of Some Particular Parts of Learning,* London, 1648.
14. S. Hartlib, *An Essay for Advancement of Husbandry-learning: Or Propositions for the Erecting College of Husbandry,* London, 1651.
15. John Dury, *The Reformed Spirituall Husbandman: With an Humble Memorandum concerning Chelsy College and a Correspondencie with Forreigne Protestants*, London, 1652.
16. G. Clarke, *John Bellers: His Life, Times and Writings*, Routledge & Kegan Paul, London and New York, 1987.
17. J. Bellers, *Proposals for Raising a College of Industry of all Useful Trades and Husbandry, with Profit for the Rich, A Plentiful Living for the Poor, and A Good Education for Youth*, London, 1696.
18. *A Scheme for the Foundation of a Royal Hospital, and Raising a Revenue of Five or Six Thousand Pounds a Year, by and for the Maintenance of a Corporation of skilful Midwives, and such Foundlings, or exposed Children, as shall be admitted therein: As it was proposed and addressed to his Majesty King James II. By Mrs. Elizabeth Cellier, in the Month of June, 1687.* The Harleian miscellany: or,

A collection of scarce, curious, and entertaining pamphlets and tracts, as well in manuscript as in print, found in the late Earl of Oxford's library. Interspersed with historical, political, and critical notes. With a table of contents, and an alphabetical index, London, printed for T. Osborne, 1744-1746.

19. E. Chamberlayne, *An academy or colledge, wherein young ladies and gentlewomen may at a very moderate expence be duly instructed in the true Protestant religion, and in all vertuous qualities that may adorn that sex also be carefully preserved and secured till the day of their marriage . . .* , In the Savoy: printed by Tho. Newcomb, 1671.
20. R. Haines, *England's Weal & Prosperity Proposed: Or, Reasons for Erecting Publick Work Houses in Every County*, London, 1681.
21. R. Haines, *Proposals for Building in every County a Working Alms House or Hospital; As the Best Expedient to Perfect the Trade and Manufactory of Linnen-Cloth*, London, 1677.
22. R. Haines, *Proposals for Promoting the Woollen-Manufactory*, Promoted, London, 1679.

CHAPTER 8

Emergence of Reports in England, 1475-1700

Reports stand as the heart of what all technical communicators would call "technical writing." As a result, any history of technical writing must involve the origin, definition, and development of reports. In Chapter 7, I offered my theory of the development of the proposal and suggested its link to the report. In this chapter, I extend my theory of the development of the practical report by showing its relationship to the petition and the proposal. This chapter will show how modern concepts of reports—organization, genre, visual displays (drawings and tables), and concise style—all appear by the mid-decades of the 17th century.

The term *report* suggests a transfer of information from one person to another person or group. With this definition, we can infer the writer's awareness of the needs and perspective of the intended readers. As we look at the large number of extant petitions presented from Edward II onward, we can easily define petitions as a form of report: As Dodd states, petitions were documents that allowed Edward's subjects to approach him with their problems, their requests, and recommendations for solutions [1, pp. 19–48]. At the same time, the petitions informed/reported to the King the events occurring in his realm. Petitions asked and proposed solutions while informing, and they clearly exemplify the awareness of the power distance between the petitioner and the King. These short petitions, penned to be read aloud, most likely by the clerks, were directed to the King, parliament, chancery, treasurer, or perhaps the exchequer and served as content guides to the readers. Many, written in Norman French until the 14th century, were likely translated into English by readers (clerks) for the benefit of auditors (triers), who over the years, knew less and less French.

As I discussed in the previous chapter, petitions had an organization, almost a formulary, which anticipates the form of the modern report as well as the proposal.

The address—to the intended audience
Identification of the petitioner
Statement of grievance or difficulty
Request for redress
Appeal for remedy

I believe we can see how the modern informal report format reflects the structure of medieval petitions:

> **To**
> **From**
> **Subject**
> **Background**—material necessary to familiarize readers with the topic.
> **Recommendation/proposed action**—This segment often suggested what the petitioner wanted.

As I examined early examples of "reports," I found that the term appeared in Parliamentary documents as in the early 1200s. Ormrod [2, p. 133] as well as Fisher, Richardson, and Fisher [3, pp. 147–148, 160–161] provide examples of memoranda to Parliament and to the King. These simple memos, many suggesting "reminders," or information to be "remembered," complete with medieval listing devices, item (Also), seem to provide an agenda for Parliament's consideration or points for the King's consideration. For example, this memorandum from Council to the King, dated 1421, illustrates the directness of these early memoranda via a listing of major topics.

> To the kynng. Oure souuereigne Lorde:
> Primerment. That it like yow to graunte letters vndre youre priue seell directed to all youre Capitaines Lieutenantz and bailliez of your towne & marches of Caleis charging thim to supporte youre Tresourier. And his deputees to emprowe. And to sette to fferme all youre lands Rentes Tenementz with all other comoditez of youre towne. And marches afoersaide to the moste prouffit. In confermacon of youre worship and right as they desire to stande in yhoure good lordshippe.
> *Item* that it like you to appoincte what gouuernance that your Castell of Guisnes shall be putte Inne. And what persone shall haue the keypng, and saufegarde therof.
> *Item* that it like you to appoincte what gouuernance that youre Castel of Baunelingham shall be putte Inne. And what persone shall haue the keypng and saufegarde therof.
> *Item* that it like you that youre priue Seell be charged to make warrant soufficeant for youre Tresourier to be allowed in his accompte of all mannere Stuffe and ordonnance. Deliured and pourveied by youre Commandment as it pereth by endenture deliured vn to the clerke of the ordonnance att this type. [3, no. 115, p. 147]

EVOLVING CHARACTERISTICS OF RENAISSANCE REPORTS

Similarly, a 1561 printed report under the name of Elizabeth I, during the third year of her reign, preserves the direct nature of the English report as captured in early (medieval) petitions. Elizabeth's reports, written 150 years later, immediately describe the problem—the subject of the report—followed by description of the issue, and end with an order describing what should be done [4].

Perhaps the directness of the proclamation influenced report design when numerous copies had to be printed. These documents anticipate reports that present and perhaps briefly explain the topic and then provide the expectation or recommended action. Because of the Olde English font used in this report, often called a broadside, I will provide a general explication of the report to show its organization and content: Paragraph 1, summary, Paragraphs 2–3 commands. This exposition explains the content of the public report:

¶ By the Queene,

This public report deals with Queen Elizabeth's intent to stop piracy. The problem: the "colour [flags] of certain of the subjects of the Queene of Scottes, haunting the Seas." She threatens them "with paine of death" and orders these offenders captured. These pirates have captured ships and passengers of the King of Portugal by using "olde" letters of reprisal. Some Englishmen have joined the Scottsmen on their unjust piracy acts. Elizabeth warns that she "will extend her force of armes or justice" against any pirates. She warns them to return immediately or to sail to the nearest port and to explain their purpose, why they were "upon the Seas."

And those who do not return and cease piracy will be subject to attack by English ships to punish them and make them an example to others. She also demands that those captured by the pirates be returned. If the pirates follow her orders, she will provide mercy to them.

Finally, Elizabeth commands that no manner of ship that shall by any colour [by flag] to be suspected of piracy, or that shall not be known to be a notorious merchant, shall not be allowed to pass out of any port until the master, captain, or owner therof shall have put in assurances before the vice admiral and to answer to all manner of things that may be objected against such as have or shall possess, or sail in the same ship.

She also orders that the vice admirals of all coastes, to repress and avoid all pirates and rovers upon the Seas, which her highness wills to be observed by all manner of her subjects. She also charges all Vice admirals, Majors, Sheriffes, Bayliffes, and other of her offers and subjects to have a diligent and careful eye to the execution of this order to avoid uttermost danger for not obeying.

God save the Queen.

¶ Imprintd at London in Powles Churchyard,

by Richard Jugge and John Cawood, Printers to the Queenes Majestie

Some 50 years later, improvements in print technology, particularly sharpness of typeface, combined with an apparent need for conciseness for printing multiple copies of reports for distribution and to control cost, provide a quantity of information reports. While presentation and document design varied, these seventeenth century reports exemplify the following characteristics. These items characterize modern technical reports.

- **Statement of purpose of the report at the beginning of the document.** The report purpose often appeared as a title, centered on the page, which fulsomely states the topic. In many reports, this lengthy topic statement serves as both a summary of the content and the report topic or purpose. Figure 8-1 illustrates this concept in the report on hospital costs. After the report topic/title, we find a listing of costs and descriptions of costs. Note the use of brackets. The concluding paragraph summarizes the meaning of the costs [5].

Figure 8-2 shows Robert Mansell's report on glass [6], which anticipates another modern cost-analysis report. Note that the report begins with title/subject line in larger font, while the purpose statement/summary (centered) follows beneath the subject line. The report content uses italicized headings on the left (exdented headings), with costs and cost analysis on the right. The final paragraph evaluates the costs presented; and the last paragraph, which cannot be included here due to page size, asks for "a speedy hearing and examination of the Preuisses, of costs." Glass was a major trade in England, which may have encouraged Merrett to translate Neri's *Arte of Glass*, discussed in Chapter 1.

- **Conciseness and directness of English Renaissance reports**. Many printed reports appear as one page broadsides, while others may be three pages printed on smaller-size paper, perhaps quarto or duodecimo, for example. Short reports, those of three to four pages, may contain a title page to introduce the report that begins on the next page.
- **Page design and formatting to reveal the content**. Many reports use enumeration of major points.
- **Letters of transmittal to introduce the report**. Transmittal letters appear in many reports, particularly when the report is addressed to person(s) of the nobility or members of Parliament (see Figure 8-3).

The 11th. day of Aprill, 1653.

A true Report of the great Coſts and Charges of the five Hoſpitalls in the City of *London* under the Care of the Lord Mayor, Commonalty and Citizens of *London*, in the maintaining of a very great Number of poore the yeare laſt paſt.

CHildren kept and and maintained this laſt year at the Charge of Chriſts Hoſpitall in the ſaid Houſe, in divers places in this City and Subburbs, and with ſundry Nurſes in the Country ——— 674

The names of all which are Regiſtred in the Books kept in Chriſts Hoſpitall, there to be ſeen from what Pariſhes, and by what meanes they have beene from time to time admitted.

Children put forth Apprentices, diſcharged, and dead this yeare laſt paſt ——— 70

So remaines now in the Houſe, and kept abroad at the ſaid Hoſpitalls Charge — 604

And lately admitted into the ſaid Hoſpitall ——— 200

Which in all is ——— 804

THe ſaid Hoſpital hath divers wayes fallen very ſhort of means formerly Received, *viz.* heretofore many have given Monies privately, others very bountifull at their Deaths: And ſeverall Pariſhes in *London* have ſent in large Contributions, and now but one that ſends any thing at all: Yet notwithſtanding all theſe diſcouragements, by reaſon of theſe hard and miſerable times among poore people, and the great number of diſtreſſed, needy, and comfortleſſe ſuiters: The Governours of the ſaid Hoſpitall, have lately admitted as aboveſaid, into their Care and Charge 200. poor Children, with hopes that divers able and well affected perſons, will liberally Contribute towards the Charges of the keeping of the aforeſaid 804. Children, which yearly amounts unto ſo great a ſum of Money, as the Revenue of the ſaid Hoſpitall doth not diſcharge the Moytie thereof

THere hath been Cured this yeare laſt paſt, at the Charge of St. *Bartholomews* Hoſpitall, of maimed Souldiers and other diſeaſed perſons, to the number of ——— 1096

Many whereof being Souldiers, have been relieved with Money, and other neceſſaries at their departure.

Buried this yeare after much Charge in their ſickneſſe ——— 168

Remaining under Cure, at this preſent, at the Charge of the ſaid Hoſpitall ——— 322

THere have been Cured at the Charge of St. *Thomas* Hoſpitall, this year laſt paſt, of diſeaſed perſons, whereof many being Souldiers, have been relieved with Money and other neceſſaries at their departure. ——— 985

Buried this yeare paſt, after much Charge in the time of their ſickneſſe ——— 152

Remaining under Cure at preſent, upon the Charge of the ſaid Hoſpitall ——— 262

THere hath been brought to the Hoſpitall of *Bridewell*, within the ſpace of one whole year laſt paſt of wandring Souldiers, and other Vagrant people, to the number of ——— 705

Many whereof have been very chargeable to the ſaid Hoſpitall, for Apparel, ſick dyet, and Surgery, beſides the ordinary dyet, and other proviſions and charges expended about them, which could not be avoided by reaſon of their neceſſities; and many of them have been paſſed into their Countries, with Clothing Apparell, and other charges which could not be avoided conſidering their naked and miſerable condition, nor they thence paſſed without ſuch charge. And there are now kept and maintained in Arts and Occupations, and other ſervile works and labours, at the charge of the ſaid Hoſpitall, Apprentices and other perſons, to the number of ——— 177

The Hoſpital of *Bethlem* is of great Antiquity, uſe, and neceſſity, for keeping and curing diſtracted perſons, who are of all others the moſt miſerable, by reaſon of their wants, both for ſoule and body, and have no ſence thereof.

The charge thereof is very great, there being kept and maintained with Phyſick, Dyet, and other reliefe: 43. diſtracted perſons conſtantly at leaſt, beſides the charge of Servants to look to them, they being moſt unruly and not able to help themſelves; and the Rents and Revenues of the ſaid Hoſpitall being very ſmall, not amounting to two third parts of the yearly charge, and therefore this Hoſpitall is a fit object of charity, there having been of late by the bleſſing of God, upon the charge of the ſaid Hoſpital, and the care of thoſe that are intruſted with the ſaid diſtracted people, divers reduced to their former ſenſes.

Figure 8-1. Cost analysis report.

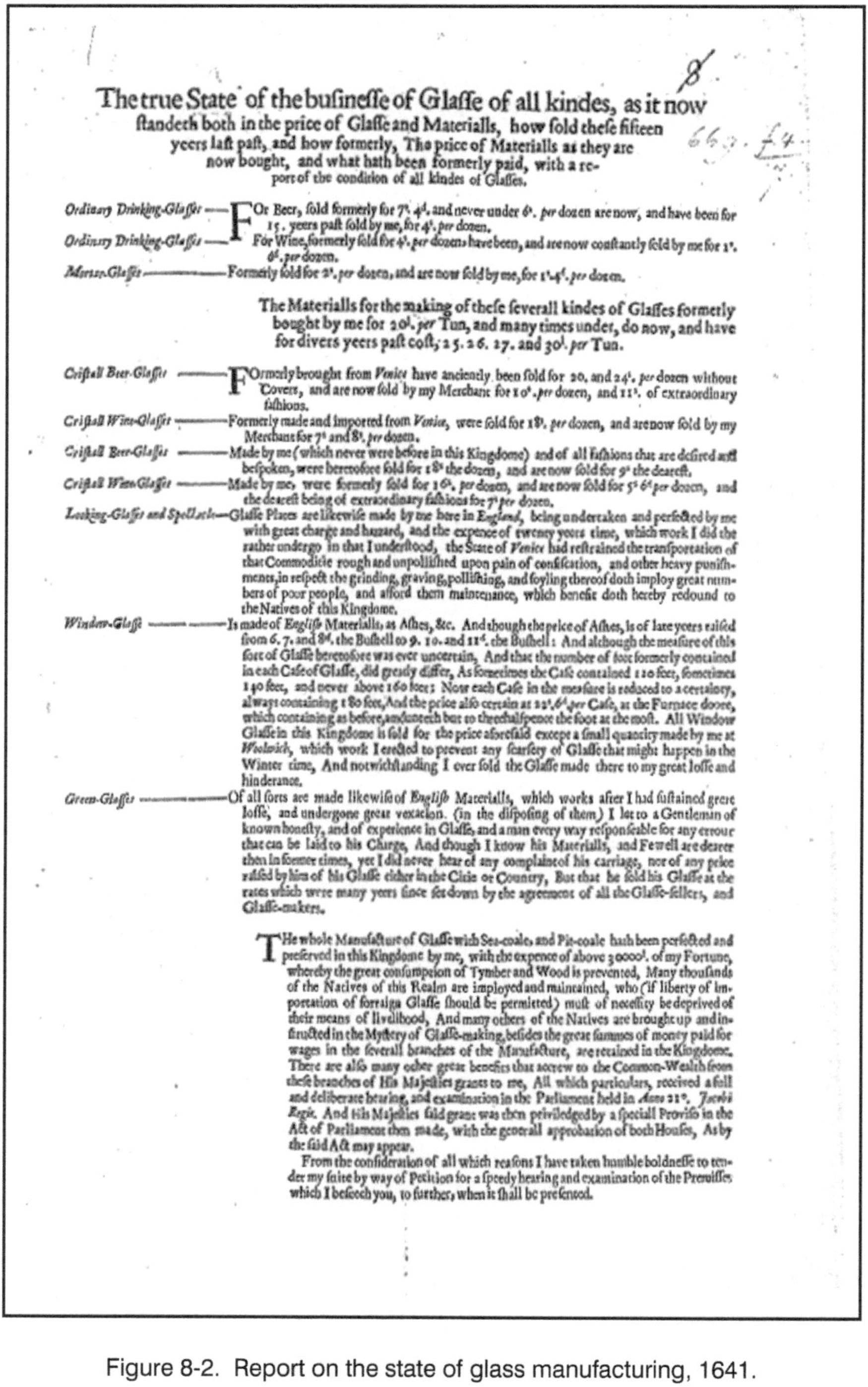

The true State of the buſineſſe of Glaſſe of all kindes, as it now ſtandeth both in the price of Glaſſe and Materialls, how ſold theſe fifteen yeers laſt paſt, and how formerly, The price of Materialls as they are now bought, and what hath been formerly paid, with a report of the condition of all kindes of Glaſſes.

Ordinary Drinking-Glaſſes — FOr Beer, ſold formerly for 7s. 4d. and never under 6s. *per* dozen are now, and have been for 15. yeers paſt ſold by me, for 4s. *per* dozen.

Ordinary Drinking-Glaſſes — For Wine, formerly ſold for 4s. *per* dozen, have been, and are now conſtantly ſold by me for 2s. 6d. *per* dozen.

Mortar-Glaſſes — Formerly ſold for 2s. *per* dozen, and are now ſold by me, for 1s. 4d. *per* dozen.

The Materialls for the making of theſe ſeverall kindes of Glaſſes formerly bought by me for 20l. *per* Tun, and many times under, do now, and have for divers yeers paſt coſt, 25. 26. 27. and 30l. *per* Tun.

Criſtall Beer-Glaſſes — FOrmerly brought from *Venice* have anciently been ſold for 20. and 24s. *per* dozen without Covers, and are now ſold by my Merchant for 10s. *per* dozen, and 11s. of extraordinary faſhions.

Criſtall Wine-Glaſſes — Formerly made and imported from *Venice*, were ſold for 18s. *per* dozen, and are now ſold by my Merchant for 7s and 8s. *per* dozen.

Criſtall Beer-Glaſſes — Made by me (which never were before in this Kingdome) and of all faſhions that are deſired and beſpoken, were heretofore ſold for 18s the dozen, and are now ſold for 9s the deareſt.

Criſtall Wine-Glaſſes — Made by me, were formerly ſold for 16s. *per* dozen, and are now ſold for 5s 6d *per* dozen, and the deareſt being of extraordinary faſhions for 7s *per* dozen.

Looking-Glaſſes and Spectacles — Glaſſe Plates are likewiſe made by me here in *England*, being undertaken and perfected by me with great charge and hazard, and the expence of twenty yeers time, which work I did the rather undergo in that I underſtood, the State of *Venice* had reſtrained the tranſportation of that Commoditie rough and unpolliſhed upon pain of confiſcation, and other heavy puniſhments, in reſpect the grinding, graving, polliſhing, and foyling thereof doth imploy great numbers of poor people, and afford them maintenance, which benefit doth hereby redound to the Natives of this Kingdome.

Window-Glaſſe — Is made of *Engliſh* Materialls, as Aſhes, &c. And though the price of Aſhes, is of late yeers raiſed from 6. 7. and 8d. the Buſhell to 9. 10. and 11d. the Buſhell: And although the meaſure of this ſort of Glaſſe heretofore was ever uncertain, And that the number of feet formerly contained in each Caſe of Glaſſe, did greatly differ, As ſometimes the Caſe contained 120 feet, ſometimes 140 feet, and never above 160 feet: Now each Caſe in the meaſure is reduced to a certainty, always containing 180 feet, And the price alſo certain at 22s. 6d. *per* Caſe, at the Furnace doore, which containing as before, amounteth but to three halfpence the foot at the moſt. All Window Glaſſe in this Kingdome is ſold for the price aforeſaid except a ſmall quantity made by me at *Woolwich*, which work I erected to prevent any ſcarſety of Glaſſe that might happen in the Winter time, And notwithſtanding I ever ſold the Glaſſe made there to my great loſſe and hinderance.

Green-Glaſſes — Of all ſorts are made likewiſe of *Engliſh* Materialls, which works after I had ſuſtained great loſſe, and undergone great vexation. (in the diſpoſing of them) I let to a Gentleman of known honeſty, and of experience in Glaſſe, and a man every way reſponſeable for any errour that can be laid to his Charge, And though I know his Materialls, and Fewell are dearer then in former times, yet I did never hear of any complaint of his carriage, nor of any price raiſed by him of his Glaſſe either in the Citie or Country, But that he ſold his Glaſſe at the rates which were many yeers ſince ſet down by the agreement of all the Glaſſe-ſellers, and Glaſſe-makers.

THe whole Manufacture of Glaſſe with Sea-coale, and Pit-coale hath been perfected and preſerved in this Kingdome by me, with the expence of above 30000l. of my Fortune, whereby the great conſumption of Tymber and Wood is prevented, Many thouſands of the Natives of this Realm are imployed and maintained, who (if liberty of importation of forraign Glaſſe ſhould be permitted) muſt of neceſſity be deprived of their means of livelihood, And many others of the Natives are brought up and inſtructed in the Myſtery of Glaſſe-making, beſides the great ſummes of money paid for wages in the ſeverall branches of the Manufacture, are retained in the Kingdome. There are alſo many other great benefits that accrew to the Common-Wealth from theſe branches of His Majeſties grants to me, All which particulars, received a full and deliberate hearing, and examination in the Parliament held in *Anno* 21°. *Jacobi Regis*. And His Majeſties ſaid grant was then priviledged by a ſpeciall Proviſo in the Act of Parliament then made, with the generall approbation of both Houſes, As by the ſaid Act may appear.

From the conſideration of all which reaſons I have taken humble boldneſſe to tender my ſuite by way of Petition for a ſpeedy hearing and examination of the Premiſſes which I beſeech you, to further, when it ſhall be preſented.

Figure 8-2. Report on the state of glass manufacturing, 1641.

In this one-page tax report, one of the first Royal Society Fellows begins his report with a transmittal letter to Parliament. In this letter, he tells Parliament the meaning of each column in his table of taxes [7]. This transmittal letter anticipates the modern transmittal letter, which tells readers what they can expect in the report that follows and highlights major topics:

> My Lords and Gentlemen,
>
> Designing the Welfare of my Country, I have published Three volumes of Collections for Improvement of husbandry and Trade: In the first whereof, Numb. 13 are some considerations for Subscriptions upon Lives, (now again in Debate) and published by its self in Two sheets: In Other Places of them, are the Matter and Use of Money with several other (I presume) useful Notes. In the Last (now perfected, with an Index) I promised this Table, all the Uses whereof will be hard to enumerate, altho' a great many I shall give in my next Collections: a Table wherof beseeches your Acceptance, as follows.
>
> Of the Proportional Tax in Decimals, I have the whole Computation by me; with it an Arithmetician, in two or three Hours, may proportion each Counties Share of any Number of Thousand Pounds; whatsoever shall be laid. And as it should be laid whooly on Acres or houses, it would prove near, as in the Table; In which I remark, that London, or the Lord Mayor's Jurisdiction only, without the Suburbs in Middlesex and Southward, I can near the Sixteenth Part of the Tax: That Middlesex, abstracted from London, (the Lord Mayor's Jurisdiction) bears near the Two and Twentieth and halst Part of the Tax: And both together, abstracted from Southwark, bears the Ninth and half Part of the TaxL That Cumberland bears but one Penny the Acre towards the Tax; but Middlesex (including London) bears Five Shillings and Eleven Pence the Acre: That Yorkshire has about the Tenth and half parts of the Acres of the whole Kingdom, the Eleventh Part of the houses (much about the same Number with the Bills of Mortality) and bears about the Twentieth Part of the Tax. It seems to me, that the Places Over-charged have about 150 Parliament Men; Those Under-charged, about 130 Men; Those that have no reason to complain, about 220 Men. Whether this Table may shew Reason or Alteration of the Method of Taxing, I submit to proper Judges. The Matter of Fact I here endeavour to Demonstrate; and am,
>
> My Lords and Gentlemen,
> Your most Obedient Servant,
> John Houghton, F. R. S.

The report follows, as shown as Figure 8-3.

Another report, prepared by Cromwell's generals, provided Cromwell and his council with a report of a sea battle. It contains a title page, "The Particulars of all the Late Bloody Fight at Sea . . . A List of the Losse on both sides. . . . And an Account of the whole Fight" [8.] The title page exemplifies a descriptive abstract, cast as an elaborate subtitle, followed by two letters of transmittal, one

An ACCOUNT of the

Acres & Houses, with the Proportional Tax, &c.

Of each COUNTY in

ENGLAND and WALES,

Humbly Presented to the

LORDS and COMMONS.

My Lords and Gentlemen,

Designing the Welfare of my Country, I have Published Three Volumns of Collections for Improvement of Husbandry and Trade: In the first whereof, Numb. 13. are some Considerations for Subscriptions upon Lives, (now again in Debate) and published by it self in Two Sheets; In Other Places of them, are the *Matter* and *Uses* of *Money*, with several other (I presume) useful Notes. In the Last (now perfected, with an Index) I promised this Table, all the *Uses* whereof will be hard to enumerate, altho' a great many I shall give in my next Collections; a Taste whereof beseeches your Acceptance, as follows.

Of the Proportional Tax in Decimals, I have the whole Computation by me; with it an *Arithmetician*, in two or three Hours, may proportion each Counties Share of any Number of Thousand Pounds; whatsoever shall be laid. And if it should be laid wholly on Acres or Houses, it would prove near, as in the Table; In which I remark, That London, or the Lord Mayor's Jurisdiction only, without the Suburbs in *Middlesex* and *Southwark*, bears near the Sixteenth Part of the Tax: That *Middlesex*, abstracted from *London*, (the Lord Mayor's Jurisdiction) bears near the Two and Twentieth and half Part of the Tax: And both together, abstracted from *Southwark*, bears the Ninth and half Part of the Tax: That *Cumberland* bears but One Penny the Acre towards the Tax; but *Middlesex* (including *London*) bears *Five Shillings* and *Eleven Pence* the Acre: That *Yorkshire* has about the Tenth and half part of the Acres of the *whole Kingdom*, the Eleventh Part of the Houses (much about the same Number with the *Bills of Mortality*) and bears about the Twentieth Part of the Tax. It seems to me, that the Places *Over-charged* have about 150 Parliament Men: Those *Under-charged*, about 130 Men: Those that have no reason to complain, about 120 Men. Whether this Table may shew Reason for Alteration of the Method of Taxing, I submit to *proper Judges*. The *Matter of Fact* I here endeavour to Demonstrate; and am,

My Lords and Gentlemen,

Your most Obedient Servant,

John Houghton, F.R.S.

Counties.	Monthly Tax, 1691. l. s. d.	Proportional Tax in Decimals.	Acres.	Houses	Parts of the whole: Tax.	Parts of the whole: Acres.	Parts of the whole: Houses.	Acres per House.	Years Tax per Acre. s. d.	Years Tax per House. s. d.
Bedford.	1793. 15. 6	013032136	260000	14170	a 77 th	a 154 th	a 96 th	21	1 8	35 6
Bucks.	2630. 12. 10	019112114	441000	18370	52	91	63	24	1 5½	34 6
Berks.	2264. 13. 2	016453262	527000	16906	61	75	69	31	1 0½	32 3
Cambridge.	2040. 0. 0	014821067	} 570000	17347	50	70	67	32	1 2	38 0
Ely.	699. 15. 10	005[illegible]84147								
Chester.	1495. 14. 4	010866724	} 720000	21054	86	55	48½	30	0 6½	16 0
Chester-City.	106. 16. 8	000776168								
Cornwall.	3081. 16. 6	022390164	960000	25874	45	42	46	38	0 9½	29 2
Cumberland.	336. 12. 2	002445536	1040000	14825	410	38½	78½	70	0 1	05 6
Derby.	1724. 16. 8	012531308	680000	21155	80	59	55	32	0 7½	19 6
Devon.	6459. 18. 4	046952275	} 1920000	56310	21	21	21	34	0 10	28 8
Exeter.	232. 14. 8	001690860								
Dorset.	2669. 0. 10	019536505	} 771000	21944	51	52	53	35	0 10	29 8
Pool.	21. 19. 4	000159593								
Durham.	647. 13. [illegible]	0047[illegible]5507	610000	15984	211	66	73	38	0 3	[illegible] 9
York.	6933. 10. 4	[illegible]9513	[illegible]	101151	20	10½	11	35	0 5½	15 8
Essex.	6190. 17. 8	045021777	1441000	34819	24	32	33½	36	1 2½	43 0
Glocester.	5017. 0. 6	036470314	} 820000	26764	37½	50	43½	30	1 1½	33 0
Glocester-City.	78. 16. 0	000572500								
Hereford.	2263. 6. 8	016442636	660000	15006	61	61	77½	44	0 10	36 3
Hertford.	2091. 12. 6	015555274	451000	16569	51	89	70	27½	1 5½	39 0
Huntington.	1267. 8. 4	009208073	240000	8217	109	166	142	29	1 3½	37 0
Kent.	6653. 17. 4	048341866	1248000	39242	21	32	29	32	1 3½	41 0
Lancaster.	2013. 7. 0	014627449	1150000	40202	68	35	29	28½	0 5	12 0
Leicester.	2169. 8. 6	015761370	560000	18702	64	71½	62	30½	0 11	28 0
Lincoln.	5150. 4. 0	037417383	1740000	40590	27	23	28½	43	0 8½	30 6
London.	8583. 2. 8	062358430	} 247000	30997	16 } 9½		37½			66 8
Middlesex, Westm.	6081. 0. 0	044179858		69139	22½	162	16½	3½	5 11	21 0
Monmouth.	780. 0. 0	005666878	340000	6490	177	118	180	52	0 6½	28 10
Northampton.	2827. 16. 4	020544736	550000	24808	49	73	47	22½	1 3	27 6
Nottingham.	1746. 16. 0	012690902	560000	17554	79	71½	66½	32	0 9	24 0
Norfolk.	6741. 4. 0	048976362	} 1148000	47180	19½	35	24½	24	1 6	36 2
Norwich.	360. 0. 0	002615481								
Northumberland.	745. 11. 4	005416712	1370000	22741	184	29	51	60	0 1½	07 10
Oxon.	2271. 1. 4	016499819	534000	15007	61	75	61½	28	0 12½	28 8
Rutland.	480. 17. 10	003493785	110000	3263	287	364	357	34	1 0½	35 4
Salop.	2407. 8. 4	017490433	890000	23284	57	45	50	38½	0 7½	24 10
Stafford.	1705. 3. 4	012385416	} 810000	23747	80½	49	49	34	0 6½	17 6
Litchfield.	26. 0. 0	000188896								
Somerset.	5543. 1. 4	040271049	1075000	45686	25	17	26	24	1 4	32 0
Bristol.	398. 16. 8	002897615		5122	345		227			18 8
Southampton.	4378. 17. 4	031813469	1312500	26851	31½	30½	43½	49	0 9½	39 2
Suffolk.	6597. 1. 4	047929200	995000	34422	21	40	34	29	1 7	46 0
Surry.	3194. 0. 4	023205262	592000	34218	41	68	34	17½	1 3½	22 6
Sussex.	3642. 15. 6	026465594	1140000	21537	38	35	54	53	0 9½	40 8
Warwick.	2384. 17. 6	017326663	670000	21973	58	60	53	30	0 10	26 0
Worcester.	2107. 18. 0	015314370	} 540000	20334	62	74	56½	26	0 11½	25 10
Worcester City.	110. 19. 0	0008[illegible]677								
Wilts.	3933. 15. 2	028579655	876000	27093	35	46	43	32	1 1	35 00
Westmoreland.	232. 0. 0	001685533	510000	6501	595	78½	179	79	0 1½	8 6
Anglesey.	251. 7. 4	001826234	200000	1840	550	200	650	109	0 3½	32 4
Brecknock.	565. 0. 11	004105187	620000	5934	241	65	197	104	0 1½	22 10
Cardigan.	211. 11. 7	001537171	520000	3163	650	77	370	164	0 1½	16 00
Caermarthen.	544. 17. 4	003957118	700000	5352	254	57	218	131	0 2½	24 5
Carnarvan.	293. 6. 4	002130286	370000	2765	470	108	411	134	0 2½	25 5
Denbigh.	447. 1. 1	003247981	410000	6398	308	98	183	64	0 3½	16 9
Flint.	237. 14. 8	001727187	160000	3150	580	250	371	51	0 4½	18 1
Glamorgan.	757. 15. 8	005505469	540000	9644	182	74	121	56	0 4	18 10
Merioneth.	201. 12. 2	001464730	560000	2590	685	80	451	193	0 1½	18 8
Montgomery.	553. 4. 4	004019245	560000	5660	249	71½	107	99	0 3	23 5
Pembroke.	653. 0. 0	004744194	420000	4329	201	95	270	97	0 4½	36 2
Haverford West.	30. 6. 10	000220439								
Radnor.	348. 13. 4	002533143	310000	3158	395	129	370	98	0 3½	26 6
Summ Total,	137641. 18. 2	1000000000	39938500	1175951						

LONDON, Printed for *Randal Taylor*, near *Stationers-Hall*, 1693.

Figure 8-3. Tax report with transmittal letter.

from the fleet (a general battle description) and one from the generals who oversaw the battle. This second transmittal letter describes the battle action in greater detail.

Today, we often see in US government reports multiple letters of transmittal directed to different targeted readers of the report. This strategy seems to have begun in the 17th century. Of particular interest is the final page of the report, a two column table, what today we would call a factual summary, that lists numbers killed, ships sunk, ammunition taken (see Figure 8-4).

While numerous reports existed in manuscript, these early printed reports can be seen throughout the 1600–1700 era; and by the late 1700s, the printed report

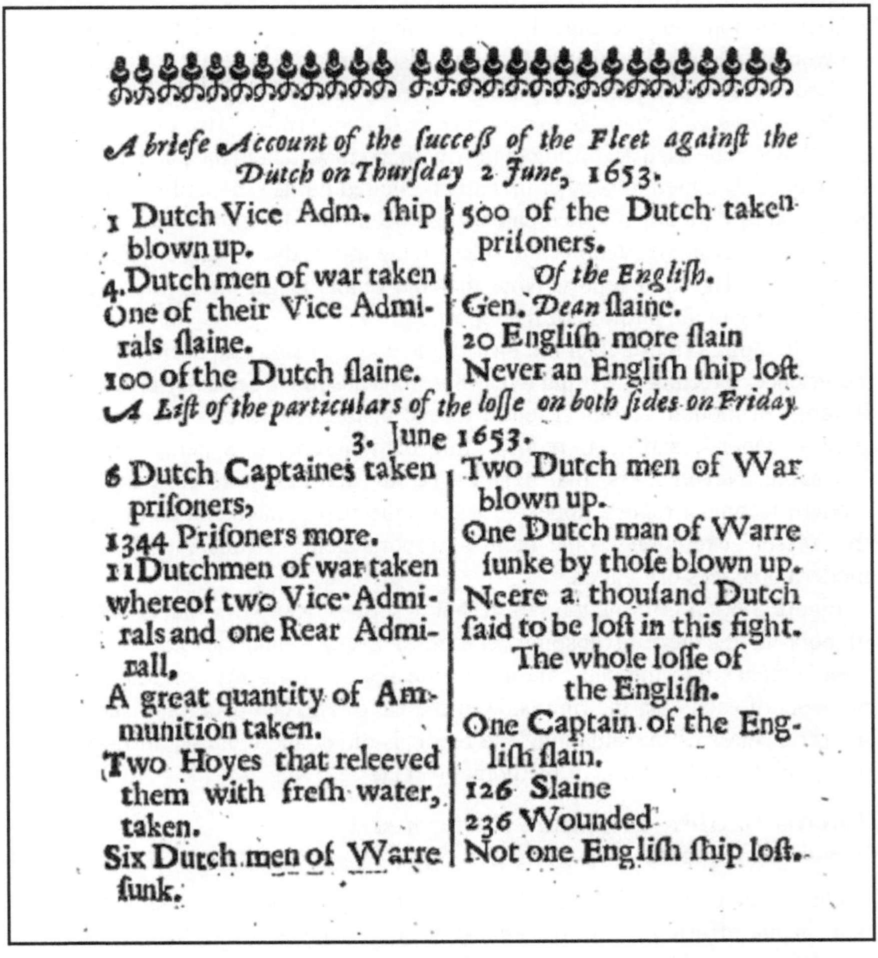

A briefe Account of the ſucceſs of the Fleet againſt the Dutch on Thurſday 2 June, 1653.

1 Dutch Vice Adm. ſhip blown up.	500 of the Dutch taken priſoners.
4. Dutch men of war taken	*Of the Engliſh.*
One of their Vice Admirals ſlaine.	Gen. *Dean* ſlaine.
	20 Engliſh more ſlain
100 of the Dutch ſlaine.	Never an Engliſh ſhip loſt.

A Liſt of the particulars of the loſſe on both ſides on Friday 3. June 1653.

6 Dutch Captaines taken priſoners,	Two Dutch men of War blown up.
1344 Priſoners more.	One Dutch man of Warre ſunke by thoſe blown up.
11 Dutchmen of war taken whereof two Vice-Admirals and one Rear Admirall,	Neere a thouſand Dutch ſaid to be loſt in this fight.
A great quantity of Ammunition taken.	The whole loſſe of the Engliſh.
Two Hoyes that releeved them with freſh water, taken.	One Captain of the Engliſh ſlain.
	126 Slaine
	236 Wounded
Six Dutch men of Warre ſunk.	Not one Engliſh ſhip loſt.

Figure 8-4. Factual summary concluding a sea battle report.

existed in a form that today's technical communicators will quickly recognize. The turning point for the emergence of practical reports came with the increased numbers of printers, who could provide copies of reports targeting Parliament and other influential readers whose financial support may have been sought by the report writer. Information reports to Parliament, for example, required numerous copies. Multiple copies of reports to members of the Royal Society also became necessary. The size of the target audience may have reduced length and produced the need for cost control.

In addition, instruction reports also appear, many of these patents, which provide specific instructions on how an invention or other intellectual property can be used. For example, Nehemiah Grew, the renowned botanist of the Royal Society, explains his method of changing saltwater into drinking water. Following this description and procedure, he presents the patent (really an instructions report) on what someone who wished to use his discovery would be allowed to do [9].

Another example patent, Figure 8-5 *Directions Given by the Patentees, how their Compounded Stuffe is to be Used in Dyeing* (1604) [10], explains exactly how dye is to be used according to the dye-makers who held the patents. The issue of patents, designed to protect inventors, elicited numerous petitions to both the King and to Parliament. Once patents had been issued, reports, such as this one, captured specific instructions and rules on the use of dyes. Figure 8-5 anticipates a modern instruction report. Note that the report has a clear title and moves immediately to the purpose of the report.

In attempting to establish the history of reports, I find that seeing through my experience in technical writing over 38 years leads me to see early printed reports in terms of modern technical report forms: instruction reports, research reports, process reports, statistical reports, information reports, a perspective that is inevitable. Nevertheless, that perspective reveals that these common types of modern technical reports appeared among the first printed English reports. For that reason, examining some of these reports helps us understand where our modern practices originated.

Figure 8-6, an information report that describes, as its title states, the number of poor in the general hospital, combines the title and the introduction and then launches into the data, the main illnesses describes the conditions and the numbers of each—on the right side of the page. The last paragraph summarizes the seriousness of the situation. The report is direct and concise, an fully anticipates modern data reports or memoranda [11].

Samuel Hartlib—Literature Reviews and Research Reports

One of the main contributors to the emergence of reports was Samuel Hartlib, who in his efforts to improve education also believed that change in society depended on the dissemination of information, a point mentioned in Chapter 1. Hartlib contributed to the development of the proposal, but he also contributed to

DIRECTIONS GIVEN BY THE PATENTEES, how their Compounded ſtuffe is to be *vſed in Dying.*

IMPRIMIS all Colours to be made or finished with the said Stuffe, are to be wrought upon good grounds of Woad, according to the lightnesse or sadnes of the colours that are intended to be made.

Item, the Cloth, Kersey, Bayes, or whatsoever else to be dyed, is first to be boyled with Allom and Argall as they use to doe for graine colours in halfe water, and halfe liquor or Float, and then to be landed, and very well cooled.

The same must be finished in new float or liquors of Branne and water (such as ordinarilie are used for graine Colours) not above one day old, In manner following:

First, the Lead or Copper being prepared and filled with a quantitie of the said Float or liquor proportionable for such things as are purposed to be dyed, the Cloth or whatsoever else must be mathered, having an allowance of Mather, as followeth.

For	A Cloth for a French Russet of 60. l. waight, being a sad azure in woad, ——	Foure pound.
	A Violet of the like waight, being a sad watchet in woad, ——	Three pound.
	A Purple of the like waight, being a light watchet in woad, ——	Three pound.
	A Tawney of the like waight, being a light Azure in woad, ——	Eight pound.

And so proportionably for any colours to be dyed according to ÿ lightnes or sadnes of their Woads, Provided alwaies, that they be high ynough in woad for the colours purposed to be made.

Then being mathered as aforesaid, they are to be very cleane washed, and the Lead or Copper to be prepared with one third part of water according to the quantitie of Clothes or other things intended to be finished, And then a quantitie of our Compounded stuffe to be allowed, as followeth.

For	A French Russet of the waight of 60. l. ——	Five pound.
	A Violet of the like waight. ——	Twoo pound and a halfe.
	A Purple of the like waight. ——	Twoo pound.
	A Tawney of the like waight. ——	One pound.

Which Stuffe is to be put into the said third part of Water in the Lead or Copper, stirring the same continually till it boyle, which it must doe halfe an hower, and then put to the water, twice asmuch Float or liquor of a day old, or a day & a halfe at the most, this liquor must be sweet, and not made upon stale grounds.

Then your Lead or Copper being at a moderate or mathering heat, stirre about the liquor that the Stuffe may not lye at the bottome.

1 1 Then shoote in your Clothes putting in the saddest Colours first, for example, first a French Russet giving it 5 or 6 ends over the winche very quicke, Then a Violet giving it one or twoo ends in like sort, Then a Purple &c. when any of the aforesaid Clothes come to their height or sadnes, they must be landed, and very well cooled, continuing the fier till the Lead be ready to boyle:
2 2 Then stirre about the Liquor and shoote in the Clothes againe, and let them be handled 4 or 5 ends over the winche as shalbe thought fit, then let them boyle till they have taken the strength of the Stuffe, or till they come to the Colours desired, then land them and coole them very well.
3 3 And if they be once more shot into the Liquor and handled 4 or 5 ends over the winche, they will be the better and fairer.

Having thus finished any of the foresaid Colours, there may be dyed in the same Liquors (being refreshed with a little water or Float) Lavender colours, Silver colours, Ashe colours, or such like upon very light Woads, the same being boyled in Allom as aforesaid, and not washed out of their Boyling.

Figure 8-5. Instructions report for legal use of a dye.

The 10th *day of* Aprill 1655.

A true Report of the great number of poor Children, and other poor people maintained in the ſeverall Hoſpitalls by the pious care of the Lord Mayor, Commonalty and Citizens of the City of LONDON.

CHildren put forth Apprentices, and diſcharged out of *Chriſts Hoſpitall* the year laſt paſt — 68

Children buried the year laſt paſt — 16

Children now remaining under the care and Charge of the ſaid Hoſpitall (with them lately admitted) which are kept in the houſe, and divers places in *London*, and ſuburbs thereof, and at Nurſe in the Countrey, amounting in the whole to the number of — 948

The names of all which are Regiſtred in the Books kept in the ſaid Hoſpitall, and are there to be ſeen from what Pariſhes they have been from time to time admitted.

There being ſo great a number of Children as aforeſaid, under the care and Charge of the ſaid Hoſpitall; It is hoped ſeverall well affected perſons will freely Contribute towards the maintenance of them, the certain Revenew of the ſaid Hoſpitall being little more then the Moytie of the neceſſary Charges thereof.

THere hath been Cured this year laſt paſt, at the Charge of St *Bartholomews* Hoſpitall, of maimed Souldiers, Seamen, and other diſeaſed perſons, many whereof being Souldiers and Seamen have been relieved with money and other neceſſaries at their departure — 1154

Buried this year after much Charge in their ſickneſſe. — 211

Remaining under Cure, at this preſent, at the Charge of the ſaid Hoſpitall — 361

THere have been Cured at the Charge of St *Thomas* Hoſpitall, this year laſt paſt, of diſeaſed perſons, whereof many being Souldiers and Seamen, have been relieved with money, and other neceſſaries at their departure. — 1052

Buried this year after much Charge in the time of their ſickneſſe — 154

Remaining under Cure at preſent, upon the Charge of the ſaid Hoſpitall — 249

THere hath been brought to the Hoſpitall of *Bridewell*, within the ſpace of one whole year laſt paſt of wandring Souldiers, and other Vagrant people, to the number of — 688

Many whereof have been very chargeable to the ſaid Hoſpitall, for Apparell, ſick Dyet, and Surgery, beſides the ordinary dyet, and other proviſions and charges expended about them, which could not be avoided by reaſon of their neceſſities, and many of them have been paſſed into their Countries, with Clothing, Apparell, and other charges, which could not be avoided, conſidering their naked and miſerable condition, nor they thence paſſed without ſuch charge. And there are now kept and maintained in Arts and Occupations, and other uſefull and neceſſary works and labours, at the charge of the ſaid Hoſpitall, Apprentices, and other perſons, the number of — 130

THe Hoſpitall of *Bethlehem* is of great Antiquity, uſe, and neceſſity, for keeping and curing diſtracted perſons who are of all others moſt miſerable, by reaſon of their wants, both for ſoul and body, and have no ſenſe thereof.

The charge thereof is very great, there being kept and maintained with Phyſick, Dyet, and other relief, 51. diſtracted perſons conſtantly at leaſt, beſides the charge of ſervants to look to them, they being moſt unruly and not able to help themſelves; and the Rents and Revenues of the ſaid Hoſpitall being very ſmall, not amounting to two third parts of the yearly charge, and therefore this Hoſpitall is a fit object of charity, there having been of late, by the bleſſing of God, upon the charge of the ſaid Hoſpitall, and the care of thoſe that are intruſted with the ſaid diſtracted people, divers reduced to their former ſenſes.

Figure 8-6. Another hospital report—numbers of hospitalized poor, their numbers, and conditions.

the development of reports. Hartlib wanted to establish an Office of Address for Communications, which would become a center for proposals (a point discussed in the previous chapter); concept papers; reports on inventions; and information in general, but this effort did not come to fruition. Nevertheless, Hartlib attempted to become a conduit through which information reports on practical scientific matters, such as medicine, engineering, horticulture, and agronomy, were developed, printed, and disseminated. His report on *The Reformed Common Wealth of Bees* . . . illustrates his ability to synthesize the ideas of others with his own to produce what would be known as a research report. Hartlib's work on bees exemplifies one of many efforts to capture the best practices of a business for the purpose of improving commerce [12].

In addition, *The Reformed Common Wealth of Bees, Presented in Several Letters* . . . [12], also exemplifies an early research review. It combines information on how to raise bees and ensure that they produce honey. Hartlib opens with a summary composed of 28 main points taken from the work. The page number for each point appears in the summary. He then lists the current extant works on bees, a type of "literature review," and then presents his "research," a compilation of letters from people he knows who have knowledge and experience in working with bees. Intermingled with these letters, many excerpts are Hartlib's replies to those who wrote to him, along with his own observations on various procedures discussed in the letters. Of particular interest are the various designs for bee hives and the rationale for each design. Each appears as a woodcut with the parts labeled and discussed in the associated text (see Figure 8-7).

The importance of this report is its anticipation of the modern literature review in both its form (genre) and development. As literature reviews should do, Harlib's exemplifies effective extraction and then summation of information, a technique we require today.

ROYAL SOCIETY REPORTS—TECHNICAL DESCRIPTIONS AND PROCESS REPORTS

Other forms of research reports appear in the *Philosophical Transactions of the Royal Society*. These reports, particularly those printed before 1700, reflect examples of parts of the trades' histories advocated by Bacon. Others in the trades history category exemplify what we would call today process reports. From a historical perspective, the most important histories were Royal Society papers that discussed mining and metallurgy, a point discussed in Chapter 1. As historian Ochs states, the best study of mining was Samuel Colpresse's description of Devon and Cornish tin mines [13].

Daniel Colwall—History of Industrial Chemicals

Colwall's history of copperas, or green vitriol, manufacturing exemplifies the histories of chemical industries. Part 1 of his history describes how the

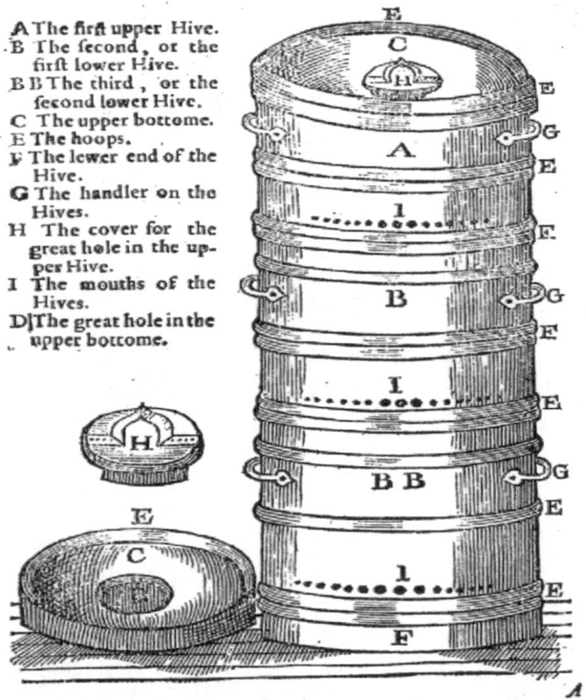

8 *The Common-wealth of Bees.*

A The firſt upper Hive.
B The ſecond, or the firſt lower Hive.
B B The third, or the ſecond lower Hive.
C The upper bottome.
E The hoops.
F The lower end of the Hive.
G The handler on the Hives.
H The cover for the great hole in the upper Hive.
I The mouths of the Hives.
D The great hole in the upper bottome.

The Common-wealth of Bees. 11

A Querie upon the deſcription of Dr. Brown's *new invented Bee-Hive.*

WHether the ſquare Figure may not prove the beſt, in that there may better be placed a bill or drawer in the bottome of the Hive, into which (being drawn forth) there may from time to time be food laid for any particular Hive, without any diſturbance to or from the reſt of the Hives, where every particular Hive may (if occaſion require) ſhut up and feed by it ſelf, which in the ordering of Bees may prove many times of good concernment.

A Letter diſcovering a new kind of excellent food for Bees.

SIR, Being much indebted to you for the gift of your Legacie, and other choice pieces, and underſtanding that you are about another of the like nature, which you intend to publiſh. I thought good in the mean time (till occaſion prompt ſome other meanes to ſerve you) to impart unto you this notable Secret, which I had from an old *Germane* Captain concerning Bees, that by long experience the Planting of Aniſe neer them proved the beſt meanes for multiplying and keeping of them, as alſo for their breeding of great ſtore of Honey. That the Hearb being taken, and the inſide of the Hives rubbed therewith, cauſeth great multitudes of Bees to enter, and become cloſe retainers to thoſe Hives, which ought to be placed directly againſt the Sun, ſo that the Sun-beames fall juſt into the Orifices of them. That the feeding upon this plant will cauſe each Stock to engender and thruſt out three young ones in one year, within which ſpace they doe elſe not uſe to doe ſo above once. That againſt the time of their thruſting forth, other Hives ought to be placed next to thoſe from whence they thruſt forth, which, as alſo the way leading to the Orifices, are to be rubbed with the Aniſe in ſuch ſort, that the juyce of the Hearb may come forth and ſtick thereunto, and the young Stocks that come forth from the old, will certainly enter into thoſe, and not repair any where elſe. But in caſe that upon the neglect any be ſwarved

C forth,

Figure 8-7. Example from Hartlib's research review report.

English produced alum from mining. Colwall uses a clear style with a step-by-step approach. Like modern technical process descriptions, it begins with a definition of Alum:

> *Alum* is made of a Stone digged out of a Mine, of a Seaweed, and Urine.
> . . .
> For the more convenient working of the Mine, which some times lies twenty yards under a furnace or Cap of Earth, (which must be taken off and borrowed away) they begin their work on the declining of a Hill, where they may also be well furnished with Water. They dig down the mine by stages, to save Carriage; and so throw it down the places where they Calcine it.
> The Mine, before it is Calcin's being exposed to the Air, will moulder in Pieces, and yield a Liquor wherof *Copperas* may be made: but being Calcin'd, is fit for *Alum.* As long as it continues in the Earth, or in Water, it remains a hard Stone.
> Sometimes a Liquor will issue out of the side of the Mine, which by the Heat of the Sun is turned into Natural *Alum.*
> The Mine is calcined with Cinders of *New Castle Coal*, Wood land Furzes. The Fire made about two feet and a half thick, two yards broad, and ten yards long. Betwixt every Fire, are stops made with wet Rubbish; so that any one or more of them may be kindled, without prejudice to the rest. [14, pp. 1052–1053]

His next part explains how to make copperas. Colwall uses a style that suggests he is talking with his readers. As in conversation, we can see informal phrases, as characterizes conversation:

> *Copperas* stones, which some call Gold-stones, are found on the Sea-shore in Essex, Hampshire, and to Westward. There are great quantities on the Cliffs; but not so good, as those on the Sore, where the Tides Ebb and Flow over them.
> The best of them are of a bright shining Silver Colour: The next, such as are of a rusty deep yellow. The worst, such as have Gravel and Dirt in them, of a saddor Umber Colour.
> In the midst of these Stones, are sometimes found the Shells of Cockles, and other small shell Fishes; small pieces of the Planks of Ships, and pieces of Seacoal.
> The brightest of these Stones they use for Wheel-lock Pistols and Fusies.
> In Order to the making of *Copperas*, they make Beds according as the ground will permit. Those at *Debtford*, are about an hundred feet long, fifteen feet broad at the top, and twelve feet deep, shelving all the way to the bottom.
> They ram the Bed very well, first with strong Clay, and then with the Rubbish of Chalk, whereby the Liquor, which drains out of the

> Dissulution of Stones, is conveighed into a Wooden shall Trough, laid in the middle of the Bed, and covered with a Board; being so boarded on all sides, land laid lower at one end than the other, whereby the Liquor is conveyed into a Cistern under the Boyling House.
>
> When the beds are indifferently well dryed, they lay on the Stone about two feet thick. [14, pp. 1056–1057]

Metals were of particular interest because of their importance to industry and commerce. Robert Moray introduced several innovative techniques for the blast furnace to help users conserve fuel, which also had to be mined.

According to Ochs, Colwall's account is informative and complete, as were many of the published articles related to the chemical industries [13, p. 138].

Robert Moray—Description of Mining Processes

Robert Moray, along with Colepresse, shared a trades' research interest in mining. In his 1665 *Philosophical Transactions* process description, "How Adits & Mines are Wrought at [The Town of] Liege without Air-Shafts," Moray describes how to remove water that often floods mines, how to remove unwanted (non-coal) minerals, and how to improve the resulting impure air generated during mining operations. Moray prepares this transactions paper via a report composed of an observation of various mining operations and an interview with Monsieur Du Son, who had developed a tool for cracking rock [15].

Moray begins by explaining the problem (paragraph 1) and then asserts that the methods used in the process he observed in the town of Liege offers the best solution to the problems faced by miners. In the next two paragraphs, he gives the process description. He presents his description in concise language, although he uses the mining nomenclature. Many of the devices described by writers in this time frame had no name, but the device and its purpose and technology was captured via the transaction reports.

> At the mouth or entry of the Adit there is a structure raised of Brick, like a Chimney, some 28 or 30 foot high in all: at the bottom, two opposite sides are (or may be) some 5 1/2 foot broad; and the other two, 5 foot: the wall 1 1/2 Brick thick. At the lower part of it, is a hole, some 9 or 10 inches square, for taking out of the Ashes, which when it is done, this Ashhold is immediately stopt so close, as Air cannot possibly get in at any part of it. Then, some 3 foot above ground or more, there is on that side, that is next to the Adit or Pit, a sqare hole of 8 or 9 inches every way, by which the Air enters to make the Fire burn: Into this hole there is fixed a square Tube or Pipe of Wood, whereof the Joints and Chinks are so stopt with Parchment pasted or glewed upon them, that the air can no here get into the pipe but at the end. [15, p. 80]

Following the description of the process and the brick device, Figure 1 in the document of the Chimney and "all the parts" (Figure 8-8), appears along with the description of the parts:

Moray's research includes his own observations and that of others, such as "A way to break easily and speedily the hardest Rocks," communicated by the same person from whom he received it, Monsieur Du Son, the inventor.

Figure 8-8. The chimney.

A. The Hole for taking out the Ashes.

B. The Square hole, into which the Tube or Pipe for conveying the Air is to be fixed.

C. The Border or ledge of Brick or Iron, upon which the Iron grate or Cradle, that holds the burning Cole, is to rest, the one being exactly fitted for the other.

D. The Hole here the Cradle is set.

E. The woodden Tube, through which the Air is conveyed towards the Cradle.

F. The Dore, by which the Grates and Cradlies let in, which is to be set 8, or 10 foot higher than the Hole D and the Shutter made of Iron, or Wood that will not hrink, that it may shut very clofe; this Dore being made large enough to receive the Cradle with ease.

G. The Grate or Cradle, which is narrower below than above, That the Ashes may more easily fall, and the Aire excite the Fire; the bottom being barred as the fides.

H. The Border or Ledge of the Cradle, that rests upon the Ledge C.

I. Four Chains of Iron fastened to the four corners of the Cradle, for taking of it up, and letting of it down.

J. The Chain of Iron, to which the other are fastened.

L. The Pulley of Iron or Brass, through which the Chain passeth.

M. A Hook, on which the end of the Chain is fastned by a Ring, the Hook fixed being placed in the fide of the Dore.

N. A Barr of Iron in the Walls, to which the Pulley is fastened. The higher the Shaft of the Chimney is, the Fire draws the Air the better. And this Invention may be made use of in the Pits or Shafts, that are Perpendicular, or any wise inclining towards it, when there is want of fresh Air at the bottom thereof, or any molestation by unwholsom

Figure 8-8. (Cont'd.) Description of the chimney parts.

Royal Society Instruction Reports

Many Royal Society Transactions illustrate research instructions, which are likely the first lab reports: for example, "Directions for Observations and Experiments to Be Made by Masters of Ships, Pilots, and Other Fit Persons" exemplify these Royal Society laboratory reports. The Contents operates as a descriptive abstract:

The Contents.

> Directions for Observations and Experiments to Be made by Masters of Ships, pilots, and other fit Persons in their Sea-Voyages; printed with Enlargements and Explications of what was formerly published of this kind; partly by Sir. R. Moray, partly by Mr. Hook; as, the several wayes of Observing, both at Sea and Land, the Declinations and Variations of the Needle: some ways of knowing the different Gravitie of Sea-water: A Form of a Scheme, representing at one view, to the eye, observations of the weather for a whole Month, &c. [16, p. 433]

The remainder of the report is partitioned into nine units. In these units, the author gives technical descriptions of several measuring devices, along with instructions for making them. As shown in six of the nine description titles, each one clearly states the content of the unit.

1. To observe the Declinations and Variations of the Compass or Needle from the Meridian exactly, in as many Places as they can, and in the same Places as they can, and in the same Places, every several Voyage,
2. To Carry Dipping-Needles with them.
3. To mark carefully the Flowing and Ebbings of the Sea, in as many places as may be.
4. To remark curiously the Situation, Figures, &. Of all dangerous Rocks, Sands, Channels, Entries, and Courses of Rivers, and all difficult Passages, and Courses in all places, to measure and describe the same Exactly, their distances, bearings, & As also the Prospects of remarkable Coasts, Promontories, Ports, Islands, & make Draughts, Plots, and Maps of them, with their Longitudes, Latitudes, Scales, & and all Beacons, Buoyes, Landmarks, Light-houses & which serve for directing and Course of Ships through narrow Channels, over Bars and Banks, into Rivers, Ports, Bayes, &. And to sound Depths near all Coasts, in all shallow Places, Roads, Etc.
5. To sound the deepest Seas without a Line, by the help of an Instrument, represented by Figure 2 in the document (see Figure 8-9).

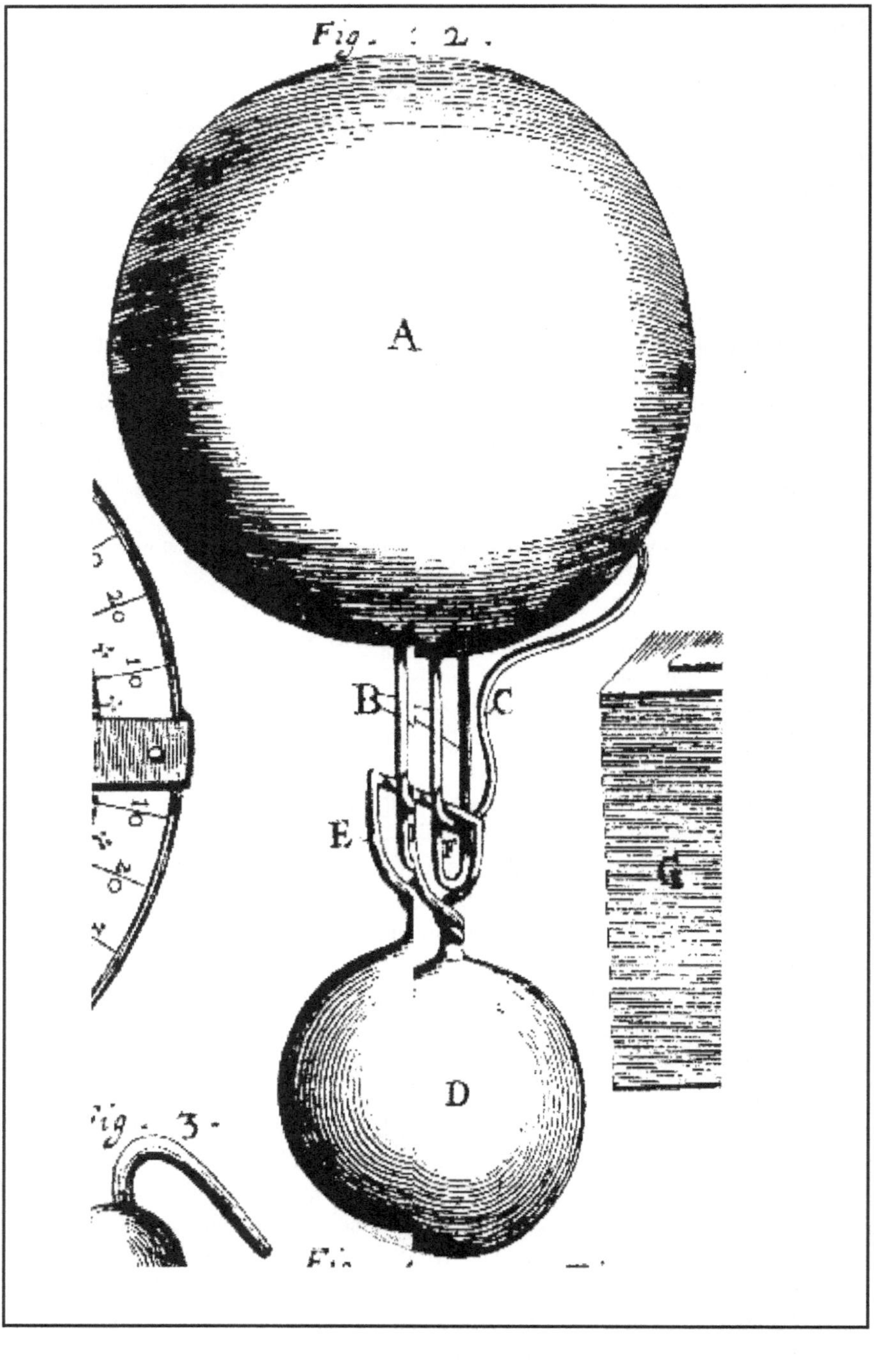

Figure 8-9. Experimental device to sound the deepest seas.

6. To keep a Register of all changes of Wind and Weather at all hoars by Night and by Day, Shewing the point the wind blows from, whether strong or weak: The Rains, Hail, Snow, and the like; the precise times of their beginnings and continuance; especially Hurricane and Spouts; but above all to take care to observe the Trade-Winds, about what degrees of Latitude and Longitude they first begin, where and when they cease or change, or grow stronger or weaker, and how much ; as near and exact as may be (see Figure 8-10).

The *Form* of a *Scheme*.

Which at one, view reprefents to the Eye Obfervations of the Weather, for a whole Month, may be fuch, as follows.

Days of the Moneth, and Place of the Sun	*Remarkable hours.*	*Age and Sign of the Moon at Noon.*	*The Quarters of the Wind, and its ftrength.*	*The Faces or vifible appearances of the Sky.*	*The Notableft Effects*	*General Deductions.* These are to be made after the fide is filled with Obfervations, as
June	4	27	W- - - -2	Clearblue, but yellowifh in the *N E.*	A great Dew	From the laft Quarter of the Moon to the Change, the weather was very temperate, but for the Seafon, cold ; the Wind pretty conftant between *N.* and *W.* *&c.*
14	8		- - - - - - - -3	Clouded toward the *South.*		
♊	12	♉ 9. 46	- - - - - - - -3½		Thunder far to the S.	
	4	*Perigeum*	- - - - - - - - - -	Checkered blue.		
12.46′	8		WSW 1		A very great Tyde.	
	12		- - - - - - - - - -			
15	8	28	N W 3	A clear sky all day, but a little checker'd about 4 *P. M.* At Sun-fet red and hazy.	Not by much fo big a Tyde as yefterday.	
♊	4		4		A great Thunder-Showre from the *N.*	
	6	♉ 24.51	N 2			
13.40′	12		1			
16.	10	*New Moon at* 7. 25. *A. M.*	S 1	Overcaft and very lowring, *&c.*	No dew upon the ground, but very much upon Marble-ftones, *&c.*	
♊		♊ 10.8	*&c.*			
14. 57		*&c.*				
&c.						

Figure 8-10. Example of a table to record changes in the wind and weather.

The example table, associated with Step 6, is shown in Figure 8-10. Like many of the tables discussed in Chapter 4, we see writers struggling with how to use tables to help readers relate information and synthesize it for ease of access. The problem seems to have been finding the best way to design a tabular format to best organize or "contain" the content.

The Figure 2 (see Figure 8-9), and the associated text, step 5, should remind us that all instruments, once designed and made by the inventor, had to be remade by others who wished to use them. Thus, instructions had to be clear for the benefit of artisans, whose skill and creativity were highly valued. Many worked at Gresham and then at Oxford to assist the researchers and then the members of the Royal Society who developed scientific instruments and measuring devices. The instructions for making and using in Figure 2 follow, after the picture of the device, which at this time has no name, as in Figure 8-9.

The writer, using the term *Note*, adds commentary about building this device. Figure 8-11 shows his effort to integrate text with visual.

Note, That care muſt be had of proportioning the weight and ſhape of the *Lead*, to the bulk, weight, and figure of the *Globe*, after ſuch a manner, as upon experience ſhall be found moſt convenient.

In ſome of the Trials already made with this Inſtrument, the Globe being of *Maple-wood*, well covered with Pitch, to hinder ſoaking in, was 5 11/16 inches in *Diameter*, and weighed 2 1/2 pounds; the Lead, of 4 1/2 pounds weight, was of a *Conical* (but is now uſed of a *Globous*) Figure 11 inches long, with the ſharper end downwards, 1 7/16 at the bottom in *Diameter*. And in thoſe Experiments made in the *Thames*, in the depth of 19 foot water, there paſſed between the Immerſion and Emerſion of the Globe, 6 Seconds of an hour; and in the depth of 10 foot water, there paſſed 3 1/2 Seconds, or thereabouts: From many of which kind of Experiments, it will likely not be hard to find out a method to calculate, what depth is to be concluded from any time of the like *Globes* ſtay under water: As for inſtance, if in the depth of 20 f. thom, meaſured by the Line, the Globe ſtay under water 15 Seconds.; then if the Ball ſtay 600 Seconds, the depth of the Sea is 933 fathom and 2 foot, if the Ball be found to move equal ſpaces in equal time.

T-

Figure 8-11. Excerpt from instructions for making the globe [16, pp. 439–440].

Royal Society Research Recommendation Reports

A number of transaction reports seem to have been written to provide research recommendations. This transaction report, likely prepared by John Evelyn or members of his group, because of its emphasis on agriculture and its classical allusions, compiles questions and topics relevant to the study of agriculture. It provides recommendations for the study of arable soil and meadows. Below is the recommended research dealing with meadows:

2. For Meadows.

1. How the above mentioned sorts of Soyl are prepared, when they are used for Pasture or Meadow?
2. The common Annoyances of these Pasture or Meadow Grounds being supposed to be, either Weeds, Moss, Sour-grass, Heath, Fern, Bushes, Bryars, Brambles, Broom, Rushes, Sedges, Gorse or Furzes; what are the Remedies thereof ?
3. What are the best waies of Drayning Marshes, Boggs, Fenns, &?
4. What are the several kinds of Grass, and which are counted the best?
5. What are the chief circumstances observable in the Cutting of Grass; and what in the making and preserving of Hay?
6. What kind of Grass is fittest to be preferved for winter-feeding? And what Grass is best for Sheep, for Cows, Oxen, Horses, Goats, &? [17, p. 43]

Royal Society Fellows were committed to planning and pursuing research related to topics under discussion at meetings. Many of these reports, disseminated to the Fellows, launched discussion and research during Society meetings.

Royal Society Technical Description Reports Introduced With Transmittal Letters

The June 30, 1670, *Philosophical Transactions* begins with a summary of contents. The opening document, "A Letter of John Evelvn Esq; to the Lord Vizcount Brouneker, Royal Society president, concerning the Spanish Sembrador or new Engin for Ploughing, Equal Sowing, and Harrowing, at once, together with a Description of the Contrivance and Use of this Instrument." Evelyn uses the transmittal letter, given below, to introduce the report that presents the planting tool, describes the Spanish inventor, and gives him full credit. The report then provides a technical description of the device and a drawing of it [18].

This transactions document has two segments: a transmittal letter from John Evelyn to Lord Brunecker, to introduce the "engine" designed to plow, equally sow all sorts of grains, and harrow—all at once. The transmittal letter from

John Evelyn to Lord Brunecker opens with a subject line, which precedes the salutation, a strategy used in modern technical writing.

> *A Letter of John Evelyns Esq;, to the Right Honourable the Lord Brouncker, Chancellor to her Majesty and President to the R. Society &. Concerning the Spanish Sembrador or New Engin For Plough, and Equal Sowing all sorts of Grain, and Harrowing, At once; by which a great quahtity of Seed-corn is saved, and a Rich increase yearly gained, together with a Description of the Contrivance and Uses of this Engin, English't out of Spanish, and Lately presented to the R. Society.*
>
> My LORD,
>
> . . . . .
>
> ......

The body of the transmittal letter explains that Evelyn found a paper written in Spanish that describes the Sembrador, this "engine." Evelyn translated the paper and now presents an abstract of it to the Royal Society for their consideration. Evelyn had the help of the Lord of Sandwich to bring the "engine" to England and place it with the RS curator, Robert Hooke. Evelyn's florid prose continues into the opening of the report, which explains (in seven points) the history of the Sembrador. Following the history, Evelyn provides a one-paragraph technical description of the Sembrador in which he alludes to the drawing of the engine located at the end of the transaction. This paragraph is followed by seven steps explaining how to use the engine. The next section—Larger Instructions for the Use of the New Sembrador—contains 13 steps. The Transaction closes with the following statement:

> *So for this Account,* If any desure to see this instrument, he may inquire for the Curator of the R. Society (Mr. Hooke,) or for their operator Mr. Shotgrave, in Gresham College, the latter of whom will also, in all likelyhood, be ready to provide the Engin itself for those who fill please to make use of it.

As I commented in Chapter 1, Evelyn remained a powerful influence on the Royal Society, from its intellectual inception to its solid position in England. However, his writing style in general lacked the power of Bacon's. Evelyn seemed to lack the ability to write directly with ease. However, this transaction shows Evelyn's ability to change styles, from the obsequious style of the transmittal letter to the precise technical description of the Sembrador and instructions for its use. His instructions for use stand in sharp contrast to the style of the transmittal letter. Here is a transmittal excerpt:

> My Lord of Sandwich is that Illustrious Person to whom the Society is obliged for this, and many other Favors, and Productions of his

> own more consummate Genius, which enrich their Registers. But, let me tell them, his Lordship hath made, and brought home with him, such other polite Notices and Particulars of Spain, and other Forrain Parts, as I know no Perfon of the moft refin'd and publick Spirit, who hath approached him, befides your Lordfhip: An Emulous and Worthy Example certainly to the rest of our Noblemen, and ministers Of State abroad. [18]

In contrast, examine these four steps from the first set of instructions for using the Sembrador:

> 4. In stiff ground the furrows ought to be 5 or 6 inches deep; In middle sort of ground, 6 or 7, and in light and sandy ground, 7 or e inches; according to this proportion the husbandman must govern himself deepening or showling the Plough, as the condition of thland shall require.
>
> 5. Especial care must be had, that the Wheels of the side of the Instrument do always turne round, a d never drag along without Turning, as also, that the Ears of the Plough be made somewhat Bigger then the ordinary ones.
>
> 6. "tis also convenient, that the Seed be well fitted and clean'd, That so the little Spoons may every time take up a grain, and the Seed be better distributed.
>
> 7. In Barley 'tis to be wel observed, that it be made clean in That manner that the straw and beards be broken off as near the Grain as may be, that so they hinder not the issuing of the grain Of the Instrument.
>
> 8. After See's-time done, furrows must be made to drain the Land of water, according to the use of each Country, without doing any More extraordinary, until the Harvest.

Evelyn knew Lord Brouncker well, and Evelyn's decision to prepare the letter in an ornate style seems odd, in that that it does not reflect the plain style required by the Society. The drawing of the Sembrador, labeled Table 1, appears at the end of the transaction (see Figure 8-12). This same device, integrated into a larger machine, would appear in John Woleridge's *Systema Agriculture* (see Figure 5-10 in Chapter 5).

Royal Society Laboratory Reports

Laboratory reports abound in the early *Philosophical Transactions* [16]. One, by Robert Boyle, discusses his research to determine how odors are produced [19]. Another, a collection of medicinal experiments, illustrates the practical goals of the Royal Society. Boyle used these experiments to create "A Collection

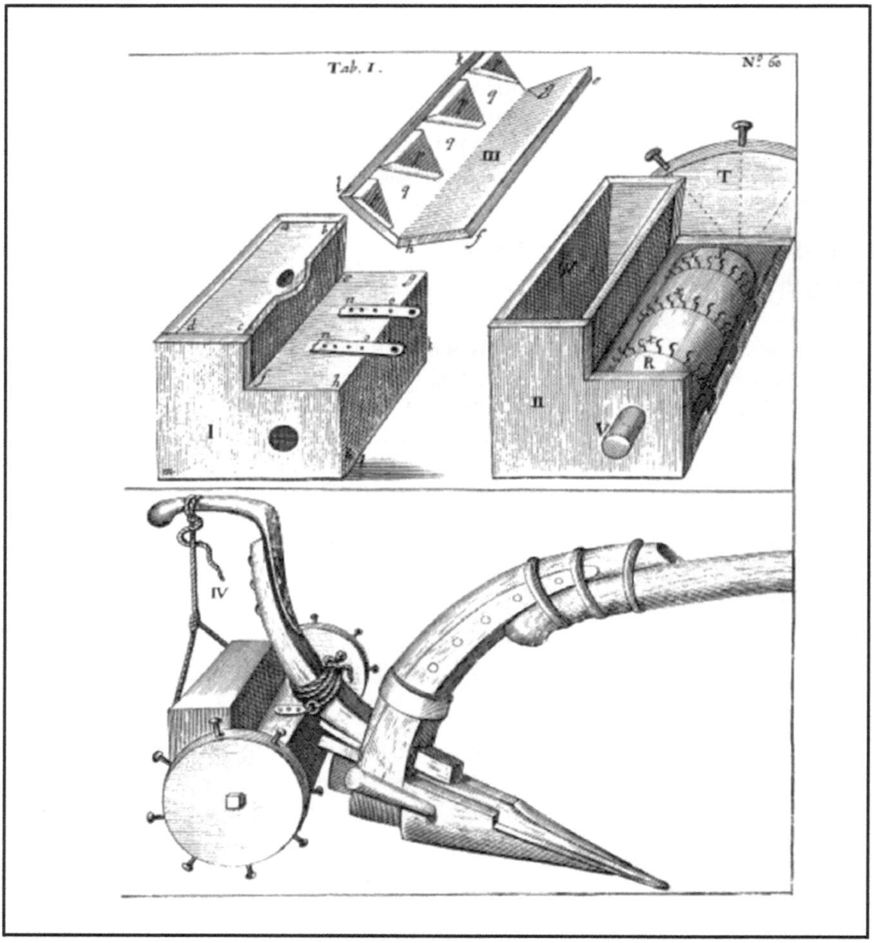

Figure 8-12. Parts of the Sembrador, an "Engin for Ploughing, Equal Sowing, and Harrowing" [18].

of Choice and Safe Remedies, for the Most Part Simple and Easily Prepared" for use by "Families, and Very Serviceable to Country People" [20]. Here is one recipe, shown in Figure 8-13.

The First Statistical Reports

The first occurrences of statistical reports can be found in reports prepared by William Petty and John Graunt, both Fellows of the Royal Society, who wrote extensively about the meaning of the Bills of Mortality and launched the fields

III. *To make Lime-water Uſe-ful in divers Diſtempers.*

C TAke one Pound of good *Quick-lime*, and ſlake it in a Gallon of warm Water, and let it ſtand 'till all that will ſubſide be ſettled at the bottom, and (Sepa-ration being made,)the Water ſwim clear at the top: (At which time it will often happen, that a kind of thin and brittle Subſtance,almoſt like Ice, will cover the Surface of the Liquor:) As ſoon as the Wa-ter is thus ſufficiently impregna-ted, delay not to pour it off wari-ly, and keep it very well ſtopp'd for Uſe.

Figure 8-13. One of Boyle's medicinal recipes. [20, p.11]

of economics and political economy. Both began their contributions with analysis of the Bills of Mortality. These reports also offer examples of 17th century English analytical reports. Their design anticipates modern analytical reports that interpret statistics.

Bills of Mortality

In English parishes, beginning in 1538, every burial required completion of a document now recognized as the precursor of the modern death certificate. These documents made the burial legal and allowed the deceased's estate to be legally settled. The number of deaths, compiled on both a weekly and an annual basis, became known as Bills of Mortality. In many parishes, the bills provide rough accounts of the causes of death, even though the reporting format and accuracy remained inconsistent. The procedure became more formal and systematic throughout England in 1603 and continued until it was superseded by the Births and Deaths Registration Act of 1836 [21].

In the years following the founding of the Royal Society in 1660, Graunt and Petty found these documents a fruitful source of information about the lives and deaths of the English people. John Graunt, a London haberdasher and amateur scientist, studied the impact of epidemic outbreaks of plague, the impact of death and its causes (men and women), and the relative merits of living either in a city such as London or in the country. Graunt published his analyses in Natural and Political Observations . . . on the Bills of Mortality (1662), a work that became the founding classic work of the modern sciences of vital statistics and epidemiology [21].

William Petty adopted an approach similar to Graunts' in his analyses, published in *Political Arithmetic* (1682) [21] and other works that documented the value of statistics in government planning. In addition to his pragmatism and his reputation as a Royal Society virtuoso (discussed in Chapter 1), William Petty is generally considered to be the founder of Economics [22]. In 1663, he became a charter member of the Royal Society in London. Petty's fame stems from his development of his "political arithmetic," by which he examined social problems mathematically.

Petty believed that population increases should be measured to determine the possibility of economic wealth as derived from sufficient labor. Petty urged the English and Irish governments to collect regular statistics on births, deaths, and total population. His analysis of the Bills of Mortality, which emerged in England in 1603, mark his early work in assessing these statistics [23]. In his *Treatise of Taxes* (1662) [24], he argued that the use of political arithmetic could rationalize tax collection and thus put the nation on more stable financial ground. Both Petty and Graunt, in pioneering the numerical study of society, investigated Bills of Mortality for a variety of purposes, ranging from determining the optimum number of physicians for England to demonstrating the superiority of England to France. In a series of pamphlets, Petty developed methods to estimate population from the number of houses and from the number of burials and christenings. He stated that the number of deaths due to contagious, acute, and chronic diseases would provide a measure of the health of a specific parish. He compared the mortality rates at London hospitals with those of Paris hospitals and concluded that London's were lower.

Petty, a prolific writer, published numerous books, pamphlets, and articles. Many of his other writings were published posthumously. He has been regarded by later writers, including Karl Marx, as the founder of English political economy [22]. More recently, historians have emphasized his contributions to the quantifying spirit of the 18th century and his advocacy of creating new methods of governance based on statistics that characterize modern societies [25]. As discussed in Chapter 1, Petty was a member of the precursor group of scientists and virtuosi who paid homage to Bacon's idea that knowledge should be useful. He was one of the utopian reformers who gathered around Samuel Hartlib and likely the most forceful of the group in stressing the critical importance of applied knowledge [26].

While Petty first emphasized and proposed improvements of education, in the second half of his life he investigated the structure of economies. He wanted to know how they functioned and believed that statistics provided a powerful tool to improve national efficiency and prosperity. He believed, as his mentor Thomas Hobbes did, that the government should secure both "peace and plenty" of the State. Statistical analysis of the economy could show how these goals might best be achieved. While a plethora of articles have been written on both Graunt and Petty to assess the contribution of each to the fields of political economy, economics, demography, and statistics, my goal here is to examine their methods of creating reports that contributed to their role in the establishment of each of these fields.

Petty's Analytical Report—*Observations Upon the Dublin Bills of Mortality*

Petty's report, *Observations upon the Dublin Bills of Mortality **1681*** [23], reflects his effort to design the best format for his statistical reports. After the title page, Petty has three tables, which he labels A, B, and C: "**A** Yearly Bills of Mortality, London and Dublin and Burials and Birth for London (1666–1680); **B** Dublin burials and births for 1671–1681; **C** The Parishes of Dublin with the numbers of families, hearths, births and Burials." Petty seems to be attempting to classify this information, to make sense out of it, based on different years. Trying to get comparable figures based on different ways of reporting made analysis of the numbers extremely difficult (see Figure 8-14).

Next, he has three separate pages of templates. The first page shows a template for recording 10 categories of information for 13 Dublin parishes. The second page provides a template for the same 13 parishes. He entitles this template **A** "Weekly Bill of Mortality for the City of Dublin." The third page, a foldout, he entitles "An Account of the People of Dublin for one year, Ending the 24th of March, 1681." This template he designed to track information for 10 categories. The bottom of the table includes a list of "Casualties and Diseases."

The report itself begins with a short overview, which explains from where he has derived the numbers. These three tables, "markt A, B, C; being indeed the A,B,C, of Publick Oeconomy, and even of that Policy which tends to Peace and Plenty." In his introduction to his report, Petty's presents an excellent example that anticipates modern introductions:

> The Observations upon the *London Bills of Mortality* have been a new Light to the World; and like Observation upon those of *Dublin*, may serve as Snuffers to make the same Candle burn clearer.
>
> The *London*-Observations flowed from *Bills* regularly kept for near One hundred years; but these are squeezed out of Six *stragling* Lonon-Bills, out of Fifteen *Dublin Bills*, and from a Note of the *Families* and *Hearths* in each *Parishes of Dublin*; which are all digested into the one *Table* or Sheet annexed, consisting of Three Parts, markt *A, B, C*; being indeed the *A, B, C,* of *Public Oeconomy*, and even of that *Policy* which tends to Peace and Plenty.

Yearly Bills of Mortality for LONDON and DUBLIN. — LONDON.

Anno.	London and Dublin: London Burials.	London and Dublin: London Births.	London and Dublin: Dublin Burials.	London and Dublin: Dublin Births.	LONDON Burials: Males.	LONDON Burials: Females.	LONDON Births: Males.	LONDON Births: Females.
1680	21053	12747	1826	1096	11039	10044	6543	6199
1679	21730	12288	1397	1061	11154	10576	6247	6041
1678	20678	12601	1401	1045	10681	9977	6568	6033
1674	21201	11851	2106	942	11000	10196	6113	5738
1672	18230	12563	1436	987	9560	8070	6443	6120
1668	17278	11633	1699	1026	9111	8167	6073	5566
	120170	73683	9865	6157	62545	57030	37992	35697
[illegible]	20028	12280	1644	1026	10424	9505	6332	5940

A

DUBLIN.

Anno	Burials.	Births.	In Ternaries of years.	
1666	1480	952		
1667	1642	1001	4821	2979
1668	1699	1026		
1669	1666	1000		
1670	1713	1067	5353	3070
1671	1974	1203		
1672	1436	967		
1673	1531	933	5073	2842
1674	2106	942		
1675	1578	823		
1676	1391	952	4328	2672
1677	1359	897		
1678	1401	1045		
1679	1397	1061	4624	3202
1680	1826	1096		
	24199	14765	24199	14765
[illegible]	1613	984	1613	984

B

	The Parishes of *DUBLIN.*	Anno 1671. Families.	Anno 1671. Hearths.	An. 1670, 71, & 72. at a Medium. Births.	An. 1670, 71, & 72. at a Medium. Burials.
1	St. *Katherins* and St. *James,*	661	2399	161	290
2	St. *Nicholas* without,	490	2348	207	262
3	St. *Michans,*	656	2301	127	221
4	St. *Andrews* with *Donabrook,*	483	2123	108	178
5	St. *Bridgets,*	416	1989	70	100
6	St. *Johns.*	244	1337	70	138
7	St. *Warbrough,*	267	1650	54	108
8	St. *Audaens,*	216	1081	53	121
9	St. *Michael,*	140	793	41	59
10	St. *Keavens,*	106	433	61	133
11	St. *Nicholas* within,	93	614	28	34
12	St. *Patrick's* Liberties,	52	255	21	44
13	Christ-Church and *Trinity*-Colledge per estimate.	26	197		1
		3850	17500	1013	1696
	Houses built between 1671, and 1681, per estimate,	150	550		
		4000	18150		

C

Figure 8-14. Petty's worksheets for his analytical report on the Bills of Mortality.

The report moves from the overview above to the discussion, beginning on page 2. The report discussion has three segments: Observations upon the Table A and then similar observations on both Table B and Table C. As shown in Figure 8-15, an excerpt of his Observation upon the Table A, he lists and numbers his observations, a common device. He follows this method for Table B and then for Table C. His conclusion, "A Postscript to the Stationer," discusses the problems he encountered in trying to compile and then analyze these numbers. This report, which has a title page, an overview, a discussion, and a conclusion (instructions to the printer), provides a recognizable example of a modern analytical report [27].

John Graunt—Statistical Analysis of the Bills of Mortality

John Graunt developed his statistical reports in a fashion similar to Petty, but his reports provide an edifying example of how the plain style required of the Royal Society existed side by side with the high style of the Commonwealth. In his report, *Natural and Political Observations Mentioned in a following Index, and made upon the Bills of Mortality*, Graunt begins with two dedicatory prefaces: one to John Lord Roberts of the Privy Council, and one to Robert Moray, president of the Royal Society [26]. In a sense, we can see that Graunt uses the dedicatory prefaces as transmittal letters in which he asks both to consider the legitimacy of his analysis. However, while the content remains the same, his style shifts substantially.

The first letter to Roberts announces that Graunt has information that Roberts could use in governing London. His letter to Moray really served as a request to the Royal Society members to examine his work as a contribution to Bacon's experimental learning because it added to the store of natural facts [25, p. 343]. Graunt develops each letter with careful attention to the specific reader and the style appropriate to each based on their positions. Both letters illustrate Richardson's point that the letter format served for many other purposes [27]. Even in early printed works, transmittal letters often served to introduce the work.

Since John Roberts was a baron, Lord Privy Seal, and a member of the King's Privy Council, a large power distance separated Graunt, the Crown, and Roberts. Thus, the letter to Roberts exemplifies the florid, ostentatious style first seen in petitions prepared by commoners to the King in the early 14th century.

> My Lord, As the favours I have received from your Lordship oblige me to present you with some token of my gratitude: so the especial Honour I have for your Lordship . . . [I] have reduced several great confused Volumes into a few perspicuous Tables, and abridged such Observations as naturally come from them, into a few succinct Paragraphs, without any long Series of multiloquious Deducations. [27]

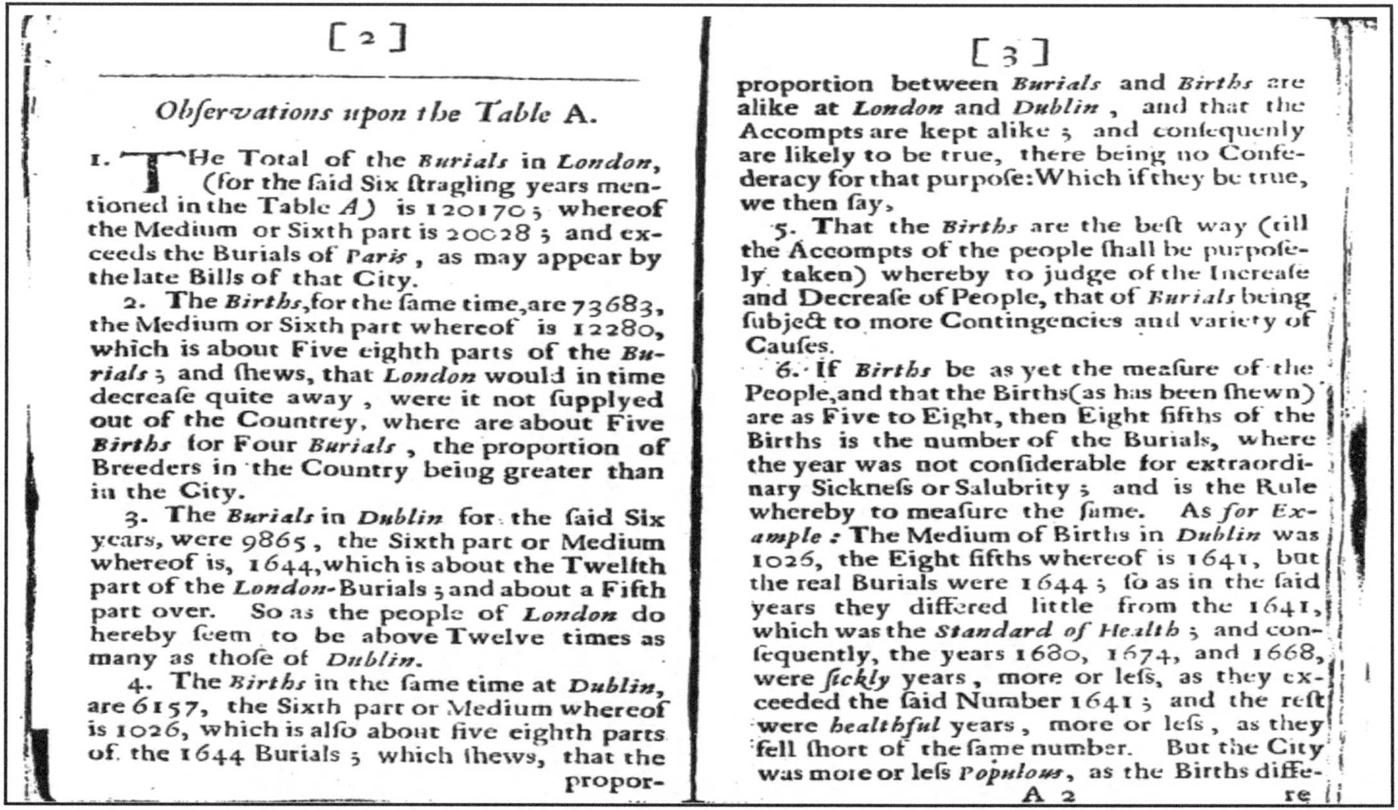

[2]

Obſervations upon the Table A.

1. THe Total of the *Burials* in *London*, (for the ſaid Six ſtragling years mentioned in the Table *A*) is 120170; whereof the Medium or Sixth part is 20028; and exceeds the Burials of *Paris*, as may appear by the late Bills of that City.

2. The *Births*, for the ſame time, are 73683, the Medium or Sixth part whereof is 12280, which is about Five eighth parts of the *Burials*; and ſhews, that *London* would in time decreaſe quite away, were it not ſupplyed out of the Countrey, where are about Five *Births* for Four *Burials*, the proportion of Breeders in the Country being greater than in the City.

3. The *Burials* in *Dublin* for the ſaid Six years, were 9865, the Sixth part or Medium whereof is, 1644, which is about the Twelfth part of the *London*-Burials; and about a Fifth part over. So as the people of *London* do hereby ſeem to be above Twelve times as many as thoſe of *Dublin*.

4. The *Births* in the ſame time at *Dublin*, are 6157, the Sixth part or Medium whereof is 1026, which is alſo about five eighth parts of the 1644 Burials; which ſhews, that the propor-

[3]

proportion between *Burials* and *Births* are alike at *London* and *Dublin*, and that the Accompts are kept alike; and conſequently are likely to be true, there being no Confederacy for that purpoſe: Which if they be true, we then ſay,

5. That the *Births* are the beſt way (till the Accompts of the people ſhall be purpoſely taken) whereby to judge of the Increaſe and Decreaſe of People, that of *Burials* being ſubject to more Contingencies and variety of Cauſes.

6. If *Births* be as yet the meaſure of the People, and that the Births (as has been ſhewn) are as Five to Eight, then Eight fifths of the Births is the number of the Burials, where the year was not conſiderable for extraordinary Sickneſs or Salubrity; and is the Rule whereby to meaſure the ſame. As *for Example*: The Medium of Births in *Dublin* was 1026, the Eight fifths whereof is 1641, but the real Burials were 1644; ſo as in the ſaid years they differed little from the 1641, which was the *Standard of Health*; and conſequently, the years 1680, 1674, and 1668, were *ſickly* years, more or leſs, as they exceeded the ſaid Number 1641; and the reſt were *healthful* years, more or leſs, as they fell ſhort of the ſame number. But the City was more or leſs *Populous*, as the Births diffe-

A 2 re

Figure 8-15. Petty's analysis of data set A.

Graunt then says that that he does not wish to bore Roberts with information Roberts already knows from his immense learning, but he wishes to share with him what he has learned from the Bills of Mortality. Graunt uses a late-medieval letter strategy, from what Richardson calls the Bulletin Letter [27, p. 130], a listing of 17 points from his research of the Bills of Mortality that will be useful to Roberts. Each point begins with "That":

> That . . . how few starve of the many that beg; That the irreligious Proposals of some, to multiply people by Polygamy, is withal irrational, and fruitless: That the troublesome seclusions in the Plague-time are not a remedy to be purchased at vast inconveniences . . . That the greatest are equally, and quickly repaired from the Country: That the wasting of Males by Wars and Colonies do not prejudice them and Females.

He concludes his list by stating,

> That your Lordship is no stranger to these Positions; yet because I knew not, that your Lordship had ever deduced them from the Bills of Mortality, I hoped it might not be ungrateful to your Lordship, to see unto how much profit that one Talent might be improved, besides the many curiosities concerning the waxing and waning of Diseases, and the relation between healthful and fruitful Seasons, the difference between the City and Country Air, &c. All which being new, to the best of my knowledge.

Graunt then adds that his report can be quickly read and will show the value of his analysis of these mortality accounts.

His letter to Robert Moray uses a style that stands in stark contrast to the style Graunt uses in his letter to Roberts. In the second letter, Graunt "gets to the point" immediately and uses an unadorned style expected by members of the Royal Society. He makes clear that his analysis of the Bills fits with Bacon's goal [26].

The greeting states that Graunt targets Moray and the "Royal Society of Philosophers meeting at Gresham College, and to the rest of that honourable Society."

> The Obsevations which I happened to make (for I designed them not) upon the Bills of Mortality, have faln out to be both Political and Nautral, some concerning Trade and Government, others concerning the Air, Countries, Seasons, Fruitfulness, Health, Diseases, Longevity, and the portions between the Sex and Ages of Mankind. All which (because Sir Francis Bacon reckons his Discourses of Life and Death to be Natural History; and because I understand your selves are also appointing means, how to measure the Degrees of Heat, Wetness, and

> Windiness in the several Parts of His Majestie's Dominions) I am humblyi bold to think Natural History also, and consequently that I am obliged to cast in this small Mite into your great Treasury of that kind. . . .
>
> I do desire your leave to present the same unto You also, as it relates to Natural History, and as it depends upon the Mathematicks of my Shop-Arithmetic. For you are not only His Majestie's Privy Council for Philosophy, but also His Great Council. You are the three Estates, viz. the Mathematical, Mechanical, and Physical. You are his Parliament of Nature. [26]

Graunt cautiously praises the wisdom of the Royal Society and states in the concluding line, "I also desire to be owned as one of you, and that no longer than I continue a faithful Friend and Servant of your Designs and Persons." In short, he wants the Royal Society to value his work as a contribution to Bacon's vision.

His "Index of the Positions, Observations, and Questions contained in this discourse," which follows the letters, provides 106 brief descriptions or summaries of his findings and the associated page in the report for each point. Graunt moves to a brief introduction (Preface) composed of four numbered paragraphs in which he explains his compilation method. Chapter 1 explains the history of the Bills of Mortality and their development. The remaining 11 chapters each begin with a descriptive assessment of the tables that follow in the chapter. Each paragraph is numbered in each chapter, just as Petty did (see Figure 8-15). Graunt first lists the chapters and their content. And in Figure 8-16, we see an example of his analysis.

Chapter 2	General Observations Upon the Casualties
Chapter 3	Of Particular Casualties.
Chapter 4	Of the Plague. [Figure 8-16] illustrates how Graunt presents his analysis
Chapter 5	Other Observations on the Plague and Casualties
Chapter 6	Of the Sickness, Healthfulness, and Fruitfulness of Seasons
Chapter 7	Of the Difference between Burials and Christenings
Chapter 8	Of the Difference between the numbers of males and females
Chapter 9	Of the Growth of the City
Chapter 10	Of the Inequality of Parishes
Chapter 11	Of the Number of Inhabitants
Chapter 12	Of the Country Bills

The report ends with a conclusion, in which Graunt mulls the value of his analysis. He argues that statistics help government plan to provide for its people. The 50-page appendix includes the tables, two foldout tables he has used—he extracts data from these tables for his chapters—and additional commentary.

From a historical perspective, Graunt's report [26] exemplifies a fine example of an analytical report designed to interpret material from the mortality bills and to present it in an interesting, useful way (see Figure 8-17). His prose

(46)

CHAP. IV.

Of the Plague.

1. BEfore we leave to diſcourſe of the *Caſualties*, we ſhall add ſomething concerning that greateſt *Diſeaſe* or *Caſualty* of all, *The Plague*.

There have been in *London*, within this Age, four times of great *Mortality*, that is to ſay, the years 1592 and 1593, 1603, 1625 and 1636.

There died *Anno* 1592, from *March* to *December*,	25886
Whereof of the *Plague*	11503
Anno 1593,	17844
Whereof of the *Plague*	10662
Chriſtned in the ſaid year	4021
Anno 1603, within the ſame ſpace of time, were Buried	37294
Whereof of the Plague	30561
Anno 1625, within the ſame ſpace	51758
Whereof of the *Plague*	35417
Anno 1636, from *April* to *Decemb.*	23359
Whereof of the *Plague*	10460

2. Now

(47)

2. Now it is manifeſt of it ſelf, in which of theſe years moſt died; but in which of them was the greateſt *Mortality* of all Diſeaſes in general, or of the *Plague* in particular, we diſcover thus. In the Years 1592, and 1636, we find the proportion of thoſe dying of the *Plague* in the whole to be near alike, that is, about 10 to 23, or 11 to 25, or as about 2 to 5.

3. In the Year 1625, we find the *Plague* to bear unto the whole in proportion as 35 to 51, or 7 to 10, that is almoſt the triplicate of the former proportion; for the *Cube* of 7 being 343, and the *Cube* of 10 being 1000, the ſaid 343 is not ⅖ of 1000.

4. In *Anno* 1603, the proportion of the *Plague* to the whole was as 30 to 37, *viz.* as 4 to 5, which is yet greater than the laſt of 7 to 10: For if the year 1625 had been as great a *Plague* year as 1603, there muſt have died not only 7 to 10, but 8 to 10, which in thoſe great numbers makes a vaſt difference.

5. We muſt therefore conclude the year 1603 to have been the greateſt *Plague* year of this Age.

6. Now to know in which of theſe four was the greateſt Mortality at large, we reaſon thus:

Anno

Figure 8-16. Example analysis from Graunt, Opening of Chapter 4, Of the Plague.

style represents the modern, concise, direct prose requested by the Royal Society and developed by the middle class of medieval England [28]. Note that he numbers each paragraph.

Reflections—The Origin of Reports

Examining these examples of Renaissance and 17th century English reports suggests the source of many of our modern report practices and developmental components of technical composition. We see definitions serving as a foundation for many reports, letters of transmittal used to introduce reports, clearly stated titles for reports, use of overviews and introductions to open reports, excellent examples of technical description carefully linked to drawings, examples of process analysis and instructions, examples of literature review reports, letter reports, and the beginnings of recommendation and analytical reports. In short, the foundation of what we now call technical writing, much of it developed or exemplified by the Royal Society, was firmly in place by 1700.

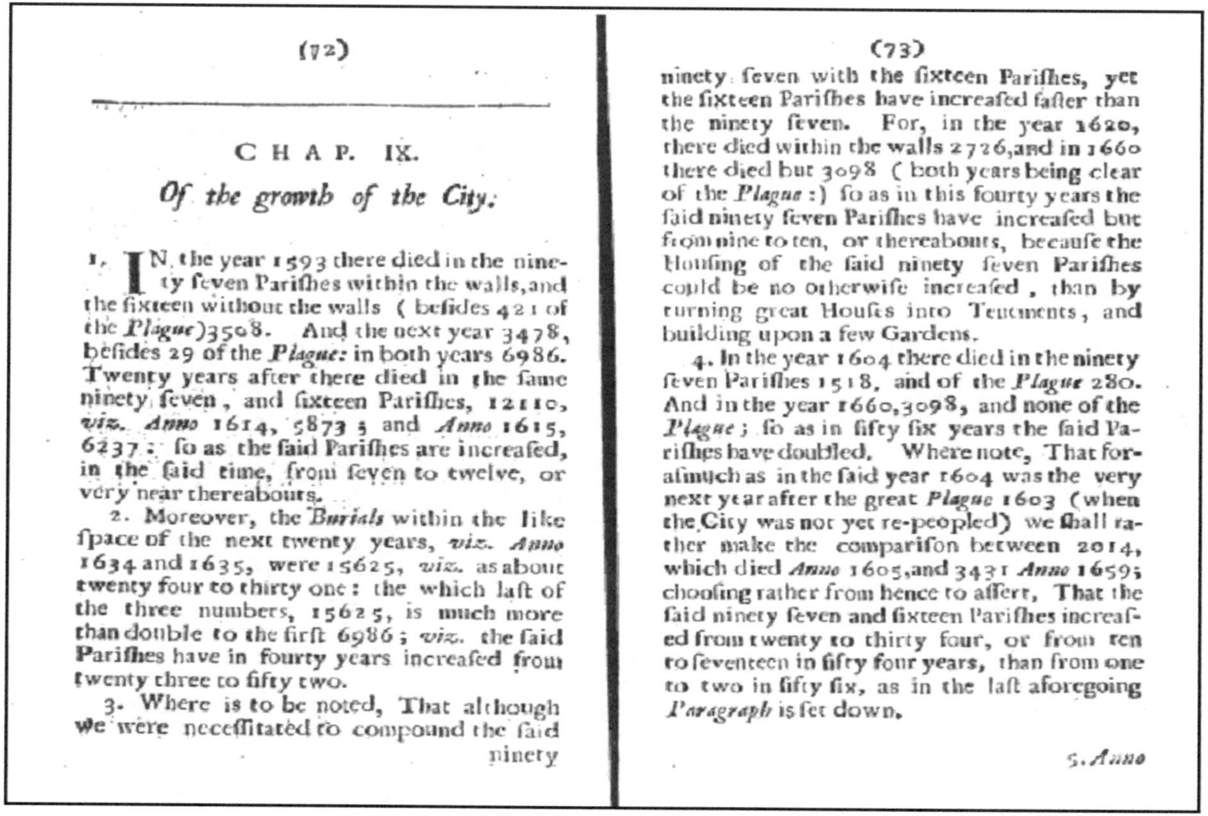

(72)

CHAP. IX.

Of the growth of the City.

1. IN the year 1593 there died in the ninety ſeven Pariſhes within the walls, and the ſixteen without the walls (beſides 421 of the *Plague*) 3508. And the next year 3478, beſides 29 of the *Plague*: in both years 6986. Twenty years after there died in the ſame ninety ſeven, and ſixteen Pariſhes, 12110, *viz.* *Anno* 1614, 5873; and *Anno* 1615, 6237: ſo as the ſaid Pariſhes are increaſed, in the ſaid time, from ſeven to twelve, or very near thereabouts.

2. Moreover, the *Burials* within the like ſpace of the next twenty years, *viz.* *Anno* 1634 and 1635, were 15625, *viz.* as about twenty four to thirty one: the which laſt of the three numbers, 15625, is much more than double to the firſt 6986; *viz.* the ſaid Pariſhes have in fourty years increaſed from twenty three to fifty two.

3. Where is to be noted, That although we were neceſſitated to compound the ſaid ninety

(73)

ninety ſeven with the ſixteen Pariſhes, yet the ſixteen Pariſhes have increaſed faſter than the ninety ſeven. For, in the year 1620, there died within the walls 2726, and in 1660 there died but 3098 (both years being clear of the *Plague*:) ſo as in this fourty years the ſaid ninety ſeven Pariſhes have increaſed but from nine to ten, or thereabouts, becauſe the Houſing of the ſaid ninety ſeven Pariſhes could be no otherwiſe increaſed, than by turning great Houſes into Tenements, and building upon a few Gardens.

4. In the year 1604 there died in the ninety ſeven Pariſhes 1518, and of the *Plague* 280. And in the year 1660, 3098, and none of the *Plague*; ſo as in fifty ſix years the ſaid Pariſhes have doubled. Where note, That forasmuch as in the ſaid year 1604 was the very next year after the great *Plague* 1603 (when the City was not yet re-peopled) we ſhall rather make the compariſon between 2014, which died *Anno* 1605, and 3431 *Anno* 1659; chooſing rather from hence to aſſert, That the ſaid ninety ſeven and ſixteen Pariſhes increaſed from twenty to thirty four, or from ten to ſeventeen in fifty four years, than from one to two in fifty ſix, as in the laſt aforegoing *Paragraph* is ſet down.

5. *Anno*

Figure 8-17. Example of analysis of mortality bills by Graunt.

REFERENCES

1. G. Dodd, *Justice and Grace, Private Petitioning and the English Parliament in the Late Middle Ages*, Oxford University Press, United Kingdom, 2007.
2. W. M. Ormrod, Agenda for Legislation, 1322-1340, *The English Historical Review, 105*, p. 414, 1990.
3. J. H. Fisher, M. Richardson, and J. L. Fisher, *An Anthology of Chancery English*, University of Tennessee Press, Knoxville, Tennessee, 1984.
4. Queen Elizabeth I, London, 1561.
5. *A True Report of the Great Costs and Charges of the Five Hospitalls in the City of London under the Care of the Lord Mayor*, April 1, 1653.
6. R. Mansell, *The True State of the Business of Glasse of All Kinds, as it now standeth both in the price of Glasse and Materialls*, how sold these fifteen yeers last past, and how formerly, the price of materialls as they are now bought, and what hath been formerly paid, with a report of the condition of all kindes of glasses, London, 1641.
7. J. Houghton, *An Account of the Acres & Houses, with the Proportional Tax, &c of each County in England and Wales, Humbly Presented to the Lords and Commons,* London, 1693.
8. *The Particulars of all the Late Bloody Fight at Sea On Thursday and Friday Last,* London, June 6, 1653.
9. N. Grew, *New Experiments, and Useful Observations Concerning Sea—Water Made Fresh, According to the Patentees Invention,* London, 1684.
10. *Directions Given by the Patentees, how their Compounded Stuffe Is to be Used in Dyeing,* London, 1604
11. The brief Contents of the Bill exhibited against Logwood, and abuses in dyeing.
12. S. Hartlib, *The Reformed Common Wealth of Bees. Presented in Severall Letters and Observations to Samuel Hartlib with The Reformed Virginian Silk-Worm. Containing Many Excellent and Choice Secrets, Experiments, and Discoveries for Attaining of National and Private Profits and Riches*, London, 1655
13. K. H. Ochs, The Royal Society of London's History of Trades Programme: An Early Episode in Applied Science, *Notes and Records of the Royal Society of London, 39*:2, pp. 129-158, 1985.
14. D. Colwall, An Account of the Way of Making English Green-Copperas, *Philosophical Transactions (1665-1678)* (Vol. 12), Royal Society, London, pp. 1056-1059, 1678.
15. R. Moray, An Account, How Adits & Mines are Wrought at Liege without Air-Shafts, Communicated by Sir Robert , *Philosophical Transactions (1665-1678)* (Vol. 1), Royal Society, London, pp. 79-82, 1665-1666).
16. Directions for Observations and Experiments to Be Made by Masters of Ships, Pilots, and Other Fit Persons in Their Sea-Voyages, *Philosophical Transactions (1665-1678)* (Vol. 2), pp. 433-448, 1666-1667.
17. Enquiries Concerning Agriculture, *Philosophical Transactions (1665-1678)* (Vol. 1), pp. 91-94, 1665-1666).
18. A Letter of John Evelyn Esq;, to the Right Honourable the Lord Viscount Brouncker, Chancellor to Her Majesty and President to the R. Society & c. Concerning the Spanish Sembrador or New Engin for Ploughing, and Equal Sowing all Sorts of Grain, and Harrowing, at Once; by Which a Great Quantity of Seed-Corn is Saved, and a

Rich Increase Yearly Gained, together with a Description of the Contrivance and Uses of This Engin, . . . Author(s): John Evelyn Reviewed work(s): Source: *Philosophical Transactions (1665-1678)* (Vol. 5), pp. 1055-1065, 1670.

19. R. Boyle, *Experiments and Observations about the Mechanical Production of Odors*, London, 1675.
20. R. Boyle, *Medicinal Experiments, of A Collection of Choice and Safe Remedies, for the Most Part Simple and Easily Prepared: Useful in Families, and Very Serviceable to Country People,* London, 1693.
21. J. M. Last, Bills of Mortality, *Encyclopedia of Public Health, 2002,* Encyclopedia. com. July 12, 2011. http://www.encyclopedia.com.
22. T. Aspromourgos, The Life of William Petty in Relation to His Economics: A Tercentenary Interpretation, *History of Political Economy, 20*:3, pp. 337-356. 1988.
23. W. Petty, *Observations upon the Dublin Bills of Mortality, 1681*, London, 1683.
24. W. Petty, *A treatise of taxes and contributions shewing the nature and measures of [brace] crown-lands, assessments, customs, poll-moneys, lotteries, benevolence, penalties, monopolies, offices, tythes, raising of coins, harth-money, excize, &c. : with several intersperst discourses and digressions concerning [brace] warres, the church, universities, rents and purchases, usury and exchange, banks and lombards, registries for conveyances, beggars, ensurance, exportation of money/wool, free-ports, coins, housing, liberty of conscience, &c. : the same being frequently applied to the present state and affairs of Ireland*, printed for N. Brooke, London, 1662.
25. T. Aspromourgos, The Invention of the Concept of Social Surplus: Petty in the Hartlib Circle, *European Journal History of Economic Thought, 12*:1, pp. 1-24, 2005.
26. R. Kargon, John Graunt, Francis Bacon, and the Royal Society, The Reception of Statistics, *The Journal of the History of Medicine and Allied Sciences, 18*:4, pp. 337-348, 1963.
27. J. Graunt, *Natural and Political Observations Mentioned in a following Index, and made upon the Bills of Mortality*, London, 1676.
28. M. Richardson, *Middle-Class Writing in Late Medieval London*, Pickering & Chatto, London, 2011.

CHAPTER 9

Reflections, Perspectives, and Research Recommendations

If this monograph is successful, technical communication faculty will have material to share with students about the history of this genre of writing. And I believe technical communication is a genre to people who use it and a field of study for students and faculty who have developed undergraduate and graduate degrees in technical communication. From having taught technical writing for over three decades, I find that students have no idea that technical, business, or practical writing is the oldest form of writing, or among the oldest. Literature faculty, to paraphrase Malcolm Richardson and M. T. Clanchy, don't like to admit that people needed writing to keep track of their possessions and to ensure that work occurred long before they needed literature. As Goody remarked, the Ugarit tablets contained more administrative writing than literature. The preponderance of practical writing has been noted by major scholars, such as Bush, Adolph, Clanchy, Miller, Prestwich, Richardson, and Wright, to name only a few historians. Without practical writing, the world of work on which even literary scholars depend could not survive. Literature existed in oral form longer than business, technical, or practical writing. Literature, with its rhyme, could be remembered. Chapter 5, on the history of instructions, shows that technical writing, at least through the 16th century, incorporated rhyme and poetry. Tusser's *Five Hundred Points of Good Husbandry*, discussed in Chapter 5, contained abstracts of his advice, also in poetry, in the modern sense of the word. From my research, I find that Tusser's use of *abstract* is the first example of its use in English.

My research into the history of technical communication and my four decades in the classroom suggest to me that the status of technical communication, as a discipline or even a field of study, is questionable at this time. Unlike medicine, law, agriculture, and the major sciences, we have no history to show our sustained existence in the world—just a collage of articles and a few monographs. My goal in writing *Emergence of a Tradition* and now *Flowering of a Tradition* has been to provide more foundation for the presence and the importance of what now exists as modern technical writing. I am indebted to Malcolm Richardson, whose recent monograph establishes the primacy of practical writing among the English middle class in late medieval England. Studies by many other medieval scholars support this premise.

My hope is that historical work will continue on technical writing in both England and the United States during the 18th and 19th centuries. Overarching books on practical writing in England, 1700–1750; a similar book on the 1750–1800 period and also 1800–1900; and similar books on practical writing in the United States would also add to our claim for disciplinary status. In addition, books on history should provide the beginnings of a consensus reading list similar to Albert C. Baugh's *Literary History of England* [1], once required reading for English graduate students in many universities, along with many multivolume histories, such as Spiller's *Literary History of the United States* [2].

Without more survey works on the history of technical communication to establish its history, we will remain a collection of courses that too often lack continuity. We have no clear philosophy, no common body of knowledge, no theory (except what we borrow from other disciplines). However, we have made solid connection with Aristotelian rhetoric that has worked and continues to work well in the world of work. The importance is determining audience(s), purpose(s), and context(s) for writing at work, which helps ensure the effectiveness of what we write. And Aristotelian rhetoric has saved us from the destructive viciousness of postmodern theories that aver that the world of work, major corporations, and business of all kinds is evil and destructive to society. For all documents we either develop or analyze, we can focus on readers, purposes, contexts, and design principles without moving into the weeds of politics that come and go like the winds. In unearthing—pillaging the tombs of noncanonical texts—to find and describe technical/practical writing from the 13th through the 17th centuries in England, I have found that Aristotelian rhetoric enables me to make sense of what I have found and continue to find. And if technical communication faculty disparage the world of work, then we must show how our departments and our universities are not, in essence, large businesses, even multi-campus corporations. What I cannot recommend is presenting articles featuring new historical texts, hidden for centuries, and discussing them from a postmodern perspective or a feminist or an environmental rhetoric standpoint. I have seen too many historical texts distorted in terms of their originating context, intended readers, and their rhetorical purpose to argue for some current trend. Writing history requires ethics and as accurate an analysis as possible.

While some of my work in the history of practical writing has involved gender, I have avoided gender in this monograph. Technical, business, and practical writing in general are global, and the focus on gender has been adequately discussed over the past two decades. Again, what technical communication scholars need to do is concentrate on establishing our history. Therefore,

I urge scholars and graduate students to go beyond what I have begun: correct errors you find in my seminal efforts; explore archives of the documents of business, technical, and research organizations that lie quietly in public and university archives across the United States and the United Kingdom. Explore the development of practical genres through these archives. For example, the history of business letters from the Renaissance onward has yet to be written, although many excellent articles exist. Intense analysis of procedures and instructions needs to continue. Not all procedures developed in the same way. Agricultural instructions provided a good beginning, as agriculture was fundamental to all early cultures and to England through the 18th century. However, military science, navigation, medicine, and science procedures all developed with unique strategies. I believe that analysis of visuals (Chapter 4) offers some initial illumination on how early technical writers, themselves scientists and practitioners in a variety of fields, struggled to explain how to do their work and to help readers trying to apply new knowledge.

While *Emergence of a Tradition* shows the sustained existence of technical writing in the English Renaissance, *Flowering of a Tradition* shows how the initial goals of the Royal Society contributed to the development of technical writing in terms of clear style, reports, and proposals. While plain style had existed for centuries, the efforts of the Royal Society, with the admonition of Wilkins, Glanvill, and the Royal Society Statutes, provided the focus needed to move plain style into the mainstream. As examples show, writers like John Graunt and John Evelyn could move from the ornate "high style," to the plain or "low style." What the Royal Society did was rescue English style from the scholastic sophistry of Oxford and Cambridge. The objections of the English Church (Archbishop Laud) and the universities actually nurtured the development of the new science at Gresham College. The support for the practical, clearly exemplified by Bacon and William Petty, encouraged the Greshamites to move forward. The disdain of the university intelligencia steeped in scholastic tradition and resistant to change reminds me of the disdain that many technical communication faculty still receive from the literati within their universities who believe that theory should take precedence over what is useful and needed by students. We are often held in veiled contempt because we are tarnished with their postmodern contempt for the workplace.

In monographs, authors often have chapters that they believe provide substantial contributions to knowledge. Chapter 3 clearly shows that the modern paragraph existed in England in the 15th century, despite claims of compositionists who limit paragraphs to the 19th century, when speech and writing split into separate fields. This error by the first compositionists again exemplifies what happens when the study of English style is limited to literary and humanistic works, a position that implies that the language of literature has more value than the language of the workplace, despite the quantity of workplace writing.

In Chapter 4, I wanted to show the critical role that page design and format have played in the development of technical writing, along with advances in typography. Writers like Vesalius knew that accessibility of material was critical to those whom he was trying to reach. Vesalius, with his anatomical revelations, used page design, typeface, and exquisite drawings to bypass Galen's cherished position and to establish the study of anatomy as an art form and a discipline. While the assertion of influence remains dangerous turf, Vesalius' appearance in an English version likely helped spread the significance of his methods for conveying anatomy.

Yet I can hear comments from many voices across academe: Why is she so concerned about format? It's the content that's important. With satisfaction, I can show, from a historical perspective, what the Document Design Institute discovered in the early 1980s [3] and what has been heavily documented by Karen Schriver [4]: accessibility of text and readability on all levels are significant in establishing the worth of the text, in addition to its ease of being remembered by its users. In short, Vesalius, with his foldout drawings, artistic use of woodcuts, and integration of text and visuals, may have continued to influence writers such as Hooke, military science writers such as Robert Barrett's *The Theorick and Practike of Moderne Warres* (1598) and John Bingham's *The Art of Embattailing an Army* (1629). Both, discussed in Chapter 4, are available in facsimile and worthy of analysis. This consideration poses an interesting research topic: a detailed study of the evolution of page design from the beginning of print to the end of the 18th century. The fact that even late Renaissance technical writers understood the value of page design shows that these writers understood, as I have stated, that readers who need text to learn how to perform tasks need format and page design to help them access and interiorize the content.

Chapter 5, on this history of instructions in agricultural and estate management, opened for me the world of the formulary and its role in the development of writing. From Goody's work on the Ugarit tablets, I have theorized that instructions emerged from lists; lists were reshaped into formularies; and fully developed sentences and paragraphs then developed to capture what formularies could not. Yet the perseverance of the oral tradition continued until the quantity and complexity of information suppressed the use of verse. While this proposition may not be correct, the shift to the textual tradition needs more scrutiny. However, technical communication faculty owe much to Ong, Clanchy, Miller, Prestwich, Wright, and Richardson (all cited throughout this monograph) for their monumental efforts to establish the shift from orality to the primacy of text.

The continuity of orality is nowhere better shown than in the ship industry, called shipwrightery, in the Renaissance. Much of the naval and shipwright nomenclature is still used and has the richest lexical history of all disciplines. The technical writing of shipbuilding struggled to develop amid the closed culture of the shipyards, designed to protect the secrecy of ship design. Much research into the development of the technical writing of naval architecture and navigation remains to be pursued, along with research on military science and military architecture. My decision to study the technical writing of shipbuilding emerged from the lack of texts about this important topic, a problem I had not encountered in agriculture, medicine, accounting, and a number of other fields. After researching the history of shipbuilding, I discovered the reason for the primacy of oral instructions: fear that instructions would fall into the hands of competitors. I examined most of Baker's notebook and saw the cryptic quality of each drawing. As I stated in Chapter 6, definitive books on shipbuilding methods did not emerge until the 18th century. Those works, as they evolved from the late 17th century materials, also remain an area for research. Those interested in the lexicology of technical writing will find shipbuilding an interesting field. Many 17th century writers felt they had to focus on basic arithmetic and geometry before approaching the applications to ship design.

Many nautical terms, used in navigation, also have a rich heritage. And the technical writing of navigation also offers areas of study. Navigation history has been developed much more than shipbuilding history, and Renaissance navigation books, for example, offer some of the first printed maps and allow historians to see the development of navigation maps, tables, and charts. For both shipbuilding and navigation, Chapter 4, which deals with typography and visual communication, offers a point of departure.

I began my study on the history of proposals with great trepidation. I had found many examples of proposals, as I mentioned in *Emergence of a Tradition*, and my 1999 *Journal of Technical Writing and Communication* article, "The Flowering of a Tradition," which launched my research into technical writing that appeared during the 1640–1700 period. I owe much to Malcolm Richardson, who suggested that I study medieval petitions. I took that suggestion and found, from 1360 onward, an increasing number of not just recommended actions to resolve the petitioners' problems but detailed proposals. The tone of many of these "proposals" was "humble," but these were more than petitions: they contained specific recommendations. Thus, we see the blend of the proposal/recommendation report. For example, "An Humble Proposal to the Parliament and Citie in this Time of Danger," 1643 [5], recommends 16 actions to protect London from war. The title may be "humble," but the recommendations lack humility in any form. A proposal for regulating the brewing of beer both proposes and recommends. It seems more like a recommendation report than a proposal [6]. The visual of this document, Figure 7-4, shows the close relationship between recommendation reports and proposals. Assigning genre to many early works is difficult.

Proposals, as we now know them, emerged in the second half of the 17th century. Many proposals, as I discussed in Chapter 1 and then in Chapter 7, were prepared by Royal Society members who understood the importance of improved education for children. I am sure I have omitted a number of 15th and 16th century proposals that in structure, content, and purpose played a significant role in other fields. I did find that, by the 17th century, the word *petition* had been replaced with *proposal* and that the two words seem to have been used interchangeably.

My work on the evolution of the report, which links the proposal to the report, needs much more research and the study of additional examples from other fields. However, I believe that Petty's reports on the Bills of Mortality provide excellent examples of technical reports that have a decidedly modern form, as do Graunt's reports. However, Petty's reports exemplify the modern routine report, while Graunt's exemplify the modern formal report. I strongly encourage technical writing historians to pursue an intensive study of the English report, particularly those in the 18th century, as they may have evolved from reports provided by Royal Society members and Parliament. What I offer in this chapter provides only a beginning history of the major genre of technical writing—the report. While I have not discussed correspondence, the development of the business letter that links extant research published in the 1970s and 1980s to the large trove of printed letters available in online collections would yield an important dissertation and a resulting monograph.

In short, *Flowering of a Tradition* provides more historical background to students of technical communication as well as research opportunities. A wealth of information remains, and the challenge is and will be selection. The history of technical, business, and practical communication has many unstudied gaps in the milieu of major eras in English and American history.

REFERENCES

1. A. C. Baugh, *Literary History of England* (2nd Edition), Appleton-Century Crofts, New York, 1967.
2. R. E. Spiller, *Literary History of the United States* (4th Edition; 2 vols.), Macmillan, New York, 1974.
3. D. B. Felker, *Document Design, A Review of the Relevant Research,* Document Design Center, American Institutes for Research, Washington, DC, 1980.
4. K. A. Schriver, *Dynamics in Document Design: Creating Text for Readers*, Wiley, New York, 1997.
5. An *Humble Proposall to the Parliament and Citie, in this time of present danger*, London, 1643.
6. L. Rawlins, *To the Right Honorable the Knights, Citizens, and Burgesses in Parliament Assembled . . . Whether the Restraining of Retailers of Beer and Ale within the Kingdome of England and Dominion of Walres from Brewing, may not only prove profitable to the Frown, but be expedient; upon due considerations of the subsequent reasons*, London, 1659.

Foundation Research for the *Flowering of a Tradition: Technical Writing in England, 1641–1700*

Books of Secrets—Authors and Their Perception of Audience in Procedure Writing of the English Renaissance, *Issues in Writing, 3*, pp. 41-67, Fall/Winter 1990.

Visual Language: Format and Page Design in English Renaissance Technical Writing, *Journal of Business and Technical Communication, 5,* pp. 246-274, July 1991.

The Evolution of Technical Description in English Renaissance Technical Writing: From Orality to Textuality, *Issues in Writing, 4,* pp. 59-109, Fall/Winter, 1991.

Ramus, Visual Rhetoric, and the Emergence of Page Design in Medical Writing of the English Renaissance: Tracking the Evolution of Readable Documents, *Written Communication, 8*:4, pp. 411-443, 1991.

Renaissance Epistolography and the Origins of Business Communication, 1568-1640: Implications for Modern Pedagogy, *Journal of Business and Technical Communication, 6*, pp. 75-98, Jan. 1992.

with Jimmie Killingsworth, Expanding and Redirecting Historical Research in Technical Writing: In Search of Our Past, *Technical Communication Quarterly, 1*:2, pp. 5-32, Spring 1992.

with Mary M. Lay, Images of Women in Technical Books from the English Renaissance, *IEEE Transactions on Professional Communication, 35*:4, pp. 196-207, December 1992.

Technical Writing for Women of the English Renaissance: Technology, Literacy, and the Emergence of a Genre, *Written Communication, 10*:2, pp. 164-199, April 1993.

From Orality to Textuality in Accounting Books: 1556-1680, *Journal of Business and Technical Communication, 7*:3, pp. 322-359, July 1993.

with Mary M. Lay, The Emergence of the Feminine Voice, *Journal of Advanced Composition, 15*:1, pp. 53-82, 1995.

Elizabeth Tebeaux et al., Studying the History of Business Communication, *Business Communication Quarterly, 59*:2, pp. 109-127, 1995.

Women and Technical Writing, 1475-1700: Technology, Literacy, and Development of a Genre, *Women, Science and Medicine, 1500-1700 and Science. In Honor of the 500th Anniversary of Gresham College*, Lynette Hunter, Margaret Pelling, and Sarah Hutton (eds.), University of Leeds, Sutton, United Kingdom, pp. 29-62, 1997.

The Voices of Renaissance Women and the History of Technical Communication, 1641-1700: Literacy, Roles, and Values, in *Three Keys to the Past: The History of Technical Communication, Volume 7: ATTW Contemporary Studies in Technical Communication,* Teresa Kynell and Michael G. Moran (eds.), Ablex, pp. 105-122, 1998.

Technical Writing in Seventeenth-Century England: The Flowering of a Tradition, *Journal of Technical Writing and Communication, 29*:3, pp. 219-262, 1999.

Visual Texts: Format and the Evolution of English Accounting Texts, 1100-1700, *Journal of Technical Writing and Communication, 30*:4, pp. 307-341, 2000.

Pillaging the Tombs of Noncanonical Texts: Technical Writing and the Evolution of English Style, *Journal of Business and Technical Communication, 18*:2, pp. 165-197, 2004.

Technical Writing in English Renaissance Shipwrightery: Breaching the Shoals of Orality, *Journal of Technical Writing and Communication, 38*:1, pp. 3-25, 2008.

Instructions: The Oldest Form of Technical Writing: From Oral to Textual Instructions in English Estate Management, 1200-1700, *Technical Communication Quarterly, 19*:4, pp. 352-378, 2010.

with Michael Moran, A Bibliography of Works Published in the History of Professional Communication from 1994-2009: Part 1, *Journal of Technical Writing and Communication, 41*:2, pp. 193-214, 2011.

Technical Writing and the Development of the English Paragraph 1473-1700, *Journal of Technical Writing and Communication, 41*:3, pp. 3-25, 2011.

with Michael Moran, A Bibliography of Works Published in the History of Professional Communication, from 1994-2009, Part 2, *Journal of Technical Writing and Communication, 42*:1, pp. 457-496, 2012.

Index